SUMMA TECHNOLOGIAE

ELECTRONIC MEDIATIONS

Katherine Hayles, Mark Poster, and Samuel Weber, Series Editors

(continued on page 411)

SUMMA TECHNOLOGIAE

Stanisław Lem

TRANSLATED BY

Joanna Zylinska

Electronic Mediations
Volume 40

University of Minnesota Press
Minneapolis
London

Originally published in Polish in 1964 as *Summa
technologiae* by Wydawnictwo Literackie. Copyright
2000 Wydawnictwo Literackie, Kraków. Copyright 2010
Barbara Lem and Tomasz Lem. http://www.lem.pl.

English translation and Translator's Introduction copyright
2013 by the Regents of the University of Minnesota.

Published by the University of Minnesota Press
111 Third Avenue South, Suite 290
Minneapolis, MN 55401-2520
http://www.upress.umn.edu

Library of Congress Cataloging-in-Publication Data
Lem, Stanisław.
[Summa technologiae. English]
Summa technologiae / Stanisław Lem; translated
and with an introduction by Joanna Zylinska.
(Electronic mediations; 40)
Includes bibliographical references and index.
ISBN 978-0-8166-7576-0 (hc : alk. paper)
ISBN 978-0-8166-7577-7 (pb : alk. paper)
1. Technology and civilization. I. Zylinska, Joanna. II. Title.
CB478.L4313 2013
306—dc23
2012043827

Printed in the United States of America on acid-free paper

The University of Minnesota is an equal-
opportunity educator and employer.

20 19 18 17 16 15 14 13   10 9 8 7 6 5 4 3 2 1

CONTENTS

TRANSLATOR'S INTRODUCTION

EVOLUTION MAY BE GREATER THAN THE SUM OF ITS PARTS, BUT IT'S NOT ALL THAT GREAT: ON LEM'S *SUMMA TECHNOLOGIAE*

JOANNA ZYLINSKA

Is the human a typical phenomenon in the Universe or an exceptional one? Is there a limit to the expansion of a civilization? Would plagiarizing Nature count as fraud? Is consciousness a necessary component of human agency? Should we rather trust our thoughts or our perceptions? Do we control the development of technology, or is technology controlling us? Should we make machines moral? What do human societies and colonies of bacteria have in common? What can we learn from insects? For answers to all these questions and more, Stanisław Lem's *Summa Technologiae* is undoubtedly the place to go.

Lem (1921–2006) is best known to English-speaking readers as the author of the novel *Solaris* (1961), the film versions of which were directed by Andrei Tarkovsky (Grand Prix at the 1972 Cannes Film Festival) and Steven Soderbergh (2002). However, science fiction aficionados all over the world have been reading Lem's original and often surprising novels—translated into over forty languages—for years. Be that as it may, the Polish writer's attitude to science fiction was not unproblematic. Witness his spat with the Science Fiction and Fantasy Writers of America association, which was incensed by Lem's unabashed critique of the majority of the works within the genre as unimaginative, predictable, and focused on a rather narrow idea of the future. Lem's own novels take a rather different approach. Drawing on scientific research, they are deeply philosophical speculations about technology, time, evolution, and the nature (and culture) of humankind. What makes Lem's writings particularly distinct is his ironic writing style, which is full of puns, jokes, and clever asides. Yet, on another level, his gripping stories about space travel, alien life, and human enhancement are also complex philosophical parables about human and nonhuman life in its past, present, and future forms.

The philosophical ambition of Lem's fiction is carried through to what is probably his most accomplished and mature work: a treatise on futurology, technology, and science called *Summa Technologiae* (1964). With a title that is a pastiche of Thomas Aquinas's *Summa Theologiae*, Lem erects a secular edifice of knowledge aimed at rivaling that of his scholastic predecessor. His *Summa* sets out to investigate the premises and assumptions behind the scientific concepts of the day and, in particular, the idea of technology that underpins them. As Lem writes in the book's opening pages: "I shall focus here on various aspects of our civilization that can be guessed and deduced from the premises known to us today, no matter how improbable their actualization. What lies at the foundation of our hypothetical constructions are *technologies, i.e., means of bringing about certain collectively determined goals that have been conditioned by the state of our knowledge and our social aptitude*—and also those goals that no one has identified at the outset."

Despite having been written nearly fifty years ago, *Summa* has lost none of its intellectual vigor or critical significance. Some specific scientific debates may have advanced or been corrected since Lem published *Summa* in 1964, yet it is actually surprising to see how many things he did get right, or even managed to predict—from the limitations of the Search for Extraterrestrial Intelligence (SETI) program through to artificial intelligence, bionics, the theory of search engines (Lem's "ariadnology"), virtual reality (which he terms "phantomatics"), and nanotechnology. However, it is in the multiple layers of its philosophical argument that the ongoing importance of his book lies. Biophysicist Peter Butko, who published an explicatory essay on *Summa* in 2006, describes the book as "an all-encompassing philosophical discourse on evolution: not only evolution of science and technology . . . but also evolution of life, humanity, consciousness, culture, and civilization."[1]

Lem's investigation into the parallel processes involved in biological and technical evolution, and his exploration of the consequences of such parallelism, provides an important philosophical and empirical foundation for concepts that many media theorists use somewhat loosely today, such as "life," "entanglement," and "relationality," while also stripping

On November 23, 2011, Google marked the sixtieth anniversary of the publication of Lem's first book, *The Cyberiad,* by including an animation created by Daniel Mróz on its main page.

these concepts of any vitalist hubris. For Lem, evolution "just happened," we might say. Given the current renewed interest in the works of Henri Bergson and his idea of creative evolution, the rereading of Bergson provided by Gilles Deleuze (of whose philosophy Lem probably would not have been a great fan, as we shall see later), as well as the multiple engagements with, and reconceptualizations of, Darwin's work, Lem's critical investigation into the different strands of, and stories about, evolution and the emergence of life on Earth has not lost any of its significance or timeliness. His postulate that we should examine the two types of evolution—biological and technical—together is more than just an argument by analogy; there is also a clear pragmatic dimension to such parallelism. In his reflections on *Summa* titled "Thirty Years Later," Lem explains that the key idea behind the book was "a conviction that life and the processes examined by the biological sciences will become an inspirational gold mine for future constructions in all phenomena amenable to the engineering approach."[2] It is interesting, then, that we can find an (unwitting) echo of such an entangled evolutionary trajectory in Bernard Stiegler's important work on originary technicity developed in his *Technics and Time, 1* and inspired by the paleontological research of André Leroi-Gourhan—work that has become one of the cornerstones of contemporary philosophy of technology and media theory.

This way of thinking about the emergence of life on Earth is no doubt a blow to anthropocentrism, which positions the human as the pinnacle of all creation. For Lem, not only did evolution not have any plan or overarching idea behind its actions, it also seems to have moved in a series of jumps that were full of mistakes, false starts, repetitions, and blind alleys. He argues that "it would be futile to search for a straight genealogical line of man, since attempts to descend to earth and walk on two feet were made over and over again a countless number of times." As Polish critic and author of many publications on Lem, Jerzy Jarzębski, points out, Lem also draws an important distinction between biological evolution and the evolution of reason, rejecting the assumption that an increase in the latter automatically means improved design capacity. Predating Richard Dawkins's idea of evolution as a blind watchmaker by over two decades, Lem's view of evolution is not just nonromantic; it is also rather ironic—as manifested in the closing chapter of *Summa*, "A Lampoon of Evolution." Evolution is described there as opportunistic, shortsighted, miserly, extravagant, chaotic, and illogical in its design solutions. The product of evolution that is of most interest to us—that is, the human himself—is seen by Lem as "the last relic of Nature," which is itself in the process of being transformed beyond recognition by the invasion of technology the human has introduced into his body and environment. There is no mourning of this impending change on the part of Lem, though, no attempt to defend "Nature's ways" and preserve the essential organic unity of the human, since the latter is seen to be both transient and, to some extent, fictitious. As Butko puts it, "philosophically Lem is a pragmatist who knows that for most humans the measure of all things is humanity.... [Yet t]here is no pedestal for humanity in *Summa*: we are not the crowning achievement of evolution, and it would indeed be strange if evolution stopped now."[3]

To develop this point, we could combine Lem's thinking about evolutionary design with Stiegler's idea of *originary technicity*, whereby the human is seen as always already technical, having emerged in relation with technology—from flint tools and fire through to steam engines and the Internet. Stiegler explains this exteriorizing movement on the part of the human toward the world by what he terms a *technical tendency*, which supposedly already exists in the older, zoological dynamic. It is this very tendency that makes the (not-yet) human stand up and reach for things in the world and to start making things. "For to make use of his hands, no longer to have paws, is to manipulate—and what hands manipulate are tools and instruments. The hand is the hand only insofar as it allows access to art, to artifice, and to *tekhnē*," writes Stiegler.[4] In the Stieglerian framework, the traditional Aristotelian model of technology as a mere tool is expanded to include the whole environment—a theoretical maneuver that allows the French philosopher to posit our being in the world as inherently technical. It also invalidates any attempt to condemn technology tout court and to return to an imaginary place of nature as supposedly primary and hence more authentic, truthful, and pure, coming, as it supposedly does, *before* technology. In a similar way, Lem does not permit us to retain any such illusions about nature's workings. For him, this process of human emergence is still ongoing, although it has arguably undergone an acceleration since the age of the Industrial Revolution. The information deluge that Lem talks about at length in *Summa* is one of the consequences of this acceleration.

The framework for Lem's argument in the book is provided by the (then nascent) discipline of cybernetics, as evidenced by the following passage:

> Every technology is actually an artificial extension of the innate tendency possessed by all living beings to gain mastery over their environment, or at least not to surrender to it in their struggle for survival. Homeostasis—a sophisticated name for aiming toward a state of equilibrium, or for continued existence despite the ongoing changes—has produced gravity-resistant calcareous and chitinous skeletons; mobility-enabling legs, wings, and fins; fangs, horns, jaws, and digestive systems that enable eating; and carapaces and masking shapes that serve as a defense against being eaten. Finally, in its effort to make organisms independent of their surroundings, homeostasis implemented the regulation of body temperature. In this way, islets of decreasing entropy emerged in the world of general entropic increase.

The very history of our civilization, with what Lem terms "its anthropoid prologue and its possible extensions," is thus seen as a cybernetic process of expanding the range of homeostasis—which is another way of defining humanity's transformation of its environment—over several thousand years. Now that cybernetic thinking has made serious inroads into media studies, science and technology studies, and digital humanities, thanks to the pioneering work of N. Katherine Hayles, Cary Wolfe, and Bruce Clarke,[5] among others, it is fascinating to be able to see Lem as a willing yet already critical adopter of this particular framework for thinking about the world and about the natural and technical processes taking place within it. Interestingly, Lem is also able to situate the study of cybernetics in the political context of the Cold War period, with its impending nuclear threat, and to tell the story of the relationship between science and politics from the perspective of the "Eastern bloc," which he sees as undergoing a series of oscillations between conflict and détente with its Western counterpart. In this way, *Summa* serves as an important companion to Hayles's *How We Became Posthuman,* which traces the origins of cybernetics via the Macy conferences in the context of the research funding and the bipolar systemic thinking enabled by the Cold War.

Moving beyond the anthropocentric framework in which the human is seen as occupying the very top of the chain of beings, Lem nevertheless spends a good deal of time considering humans' singularity in the cosmic universe as well as their moral and political responsibility. As pointed out by Peter Swirski, in *Summa,* Lem distances himself from "Enlightenment ideals, according to which humanity could transcend tribalism and build a better future on a planetary scale."[6] Instead, he focuses on "our ubiquitous and seemingly unstoppable drive for conflict and aggression."[7] The Polish author remains skeptical as to "the rationality of Homo sapiens, whom—like Jonathan Swift before him—he sees as *Homo rationis capax.*"[8] As Lem himself puts it rather ominously in the first chapter of *Summa,* "man knows more about his dangerous tendencies than he did a hundred years ago, and in the next hundred years, his knowledge will be even more advanced. Then he will make use of it." It becomes quite clear from both *Summa* and Lem's other writings (in particular, the short pieces on science and philosophy he wrote in the later part of his life)[9] that he is not very optimistic about the human as the product of evolution—not just in terms of our future developmental prospects, as mentioned previously, but also in terms of our current ethicopolitical situation. This is perhaps unsurprising since, as explained earlier, evolution for Lem cannot be trusted with "knowing what it is doing." Neither, seemingly, can we—at least not always or consistently. This limited knowledge results from an underlying conflict Lem identifies "between a conscious mind that can think and an underlying program that determines action," that is, genes, as explained by Hayles. The latter goes on to suggest that the crisis of agency in Lem's writings is therefore "bodied forth as an inescapable and tragic condition of thinking mind(s)."[10]

Indeed, it is the ethical aspect of his discourse on nature, science, and technology that arguably raises the most thought-provoking and timely questions. Outlining his ethical standpoint in a long interview with Swirski conducted in 1994, Lem starts from the premise that "the traditionally inherited types of ethics are all rapidly becoming impotent."[11] Living through the collapse of various forms of authority, secularization, the emergence of both extreme nationalisms and extreme regionalisms, and the pathologies of escapism, the human being in the second half of the twentieth century finds himself for Lem in a kind of *horror vacui,* "giving us as a result a new type of 'man without conscience.'"[12] Such pessimism and sorrow about the human condition is obviously a familiar trope in both philosophy and literature. Yet we have to distinguish here between the pessimistic view of the human as encapsulated by many metaphysical narratives, including those of the dominant religions, whereby man is suffering from some kind of original sin or some other innate fault that predisposes him to doing evil, and the more skeptical–realist argument, which evaluates human faults empirically, so to speak, on the basis of historical experience. Furthermore, this positing of the potential to do evil is an argument through subtraction: the human will eventually make use of his knowledge, as Lem claims, and put it to various uses, including harmful ones, because there is nothing inherent either in him or in the world to stop this course of action. Having acquired technical knowledge through the parallel processes of biological and technical evolution, the human nevertheless lacks any inherent political wisdom, or *sophia,* as Stiegler puts it,[13] which is why there is nothing to stop him producing weapons rather than utensils, to prevent him from making war rather than love. Political systems, state and organizational policies, moral codes, and cultural values may serve as barriers to such negative and damaging turns of events. However, in most cases, politics and ethics find it difficult to catch up with the development of science. As a consequence, they arrive too late to prevent various events from happening. This restricted freedom with regard to his own agency, combined with the lack of knowledge about being with others in the world, contribute to the human's tragic condition.

Morality for Lem is a "genuinely human contribution to human history" that endows "the immorality of the Darwinian model with a human touch, a structural sense," while also allowing the author to close the literary work "with a clear-cut point," as argued by Jarzębski.[14] At the same time, Jarzębski indicates some logical weaknesses in Lem's narrative, especially with regard to the latter's belief in the unconstrained development of reason—which, Lem claims, will at some point overcome human intelligence and move in yet unspecified but possibly dangerous directions: toward a cosmic death, say. While pointing out that traditional theology deals better with issues of impending apocalypse and the end of the world, Jarzębski highlights the eschatological aspects of Lem's own thinking around such issues. This perhaps explains Lem's tendency "to construct worlds equipped with a kind of umbilical cord, or a gateway, to transcendence understood in physical and definitely lay terms. It makes it possible to remove eschatological questions into the other world, disburdening us from the duty of answering them within the bounds of the known cosmos."[15] I would argue that in such cases, Lem resorts, unwittingly perhaps, to smuggling some shards of humanism through the back door of his argument. His at times Swiftian misanthropy covers up a sorrow over the human condition and a desire for it to be better—something that can only ever be enacted in metaphysics or in fiction. This is where Lem's sharp critique of the sociopolitical quietism of Buddhism as outlined in *Summa* becomes to some extent applicable to his own technoscientific speculations.

Yet we should not underestimate the power of Lem's multifaceted critique, which, as with the technologies he writes about, could in the right hands become a force of a true intellectual and cultural transformation in a world in which science needs to be treated seriously but where any authority is ultimately constrained and fallible. Indeed, throughout his argument, Lem mercilessly applies Occam's razor to many of the assumptions and premises held by science, while also remaining fascinated by scientific debates and discoveries. Although he himself remains rather skeptical with regard to any kind of -isms in literary theory and seems rather short-tempered about many thinkers and schools of thought—he is very critical of "this lunatic Derrida," considers Hegel an "idiot," and sneers at patients of Freudian psychoanalysts who dream in sex symbols, while those of Jungian ones supposedly dream in archetypes—his skepticism (or rather "fallibilism," as Paisley Livingston suggests[16]) with regard to the nature of cognition is combined with an exploration of some nonhuman forces at work in the world that escapes human control. These are predominantly natural forces, as evidenced in the turns and twists of evolution, but Lem also shows interest in individual human agency being overcome by, competing with, but sometimes also working in collaboration with an agency of "the system"—be it biological, social, or political. This perhaps explains his repeated engagement with questions of accident, chance, and luck, with game theory—so popular in the 1960s—providing him with a useful framework to analyze the goings-on of the modern world. Yet, as mentioned earlier, we should not go back to Lem for historical reasons alone. What distinguishes Lem from the vast majority of philosophers of technology is his wit—understood as both sharp intellect and sense of humor—as well as his awareness of the narrative, storytelling character of both his philosophy and his science. (This is not to say that science is "made up," only that it relies not just on mathematical language but also on culturally specific semiotic descriptors and that its conventions and assumptions change over time.)

Summa is therefore an example of a different kind of philosophy of technology, one that combines rigorous intellectual analysis with a linguistic playfulness more readily associated with literature. Even though science, with its methodology rooted in objectivity and the rational method,[17] provides an unabashed foundation for the standpoint Lem adopts throughout *Summa,* he is arguably more interested in signaling certain problems and posing questions about them[18] than in offering any determined visions of either the present or the future. Combining scientific rigor with philosophical speculation, Lem humbly declares, "I do not see myself as subjectively infallible."[19] He is thus a skeptic, as much concerned with looking at developments in science and technology as he is with exploring indeterminacy in science and the limits of human cognition. This may also explain his lack of generosity toward certain schools of thought that do not comply with his adopted framework (or that are simply not to his liking). For example, Lem has no time for structuralism in literary theory, although he does not bat an eye about game theory's structuralist foundations; admits to not reading philosophy too intensely as it is, for him, a mere derivative of science (which obviously limits philosophy to its analytical incarnation); and considers women an "unnecessary complication"[20] in both literary and scholarly endeavors.

Arguably, literature offers a unique space in which such deliberations can occur most productively and most freely—with his "fictions about science" strongly rooted in science (as opposed to science fiction, which most often maintains only a cursory relation to science) providing a testing ground for many of his thought experiments. Although in its philosophical style and the nature of its scholarly argument, *Summa* belongs to a different genre than Lem's novels, they are all bound by a unique literariness, which manifests itself in their author's engagement with language as a plastic material of a creative cultural process that yields itself to experimentation, while also showing some "resistance of material" at times. Jarzębski goes so far as to suggest that for Lem, evolution itself is also a "narrative."[21] Again, this does not mean that it did not really happen, only that it requires various narrative renditions to be conveyed as a concept—or transmitted, to borrow from the communications discourse favored in *Summa*—to human receivers in their particular sociocultural and philosophical milieux. A certain paradox emerges here because evolution, as Jarzębski has it, cannot be comprehended by reason—"hence the only way to cope with it is to impose a human, quasi-sensible narrative onto it," which is why it must be "absorbed into human history, compared with something we know."[22] Arguably, *Summa* is an attempt to deal with this very paradox.

Speaking of Lem's aforementioned literariness, many critics have commented on what Andy Sawyer calls the writer's "overwritten style" as well as his "love of the grotesque, of imagistic and linguistic excess"—which, for Sawyer, makes Lem a "master of the baroque."[23] This baroque and frequently playful style can be a source of joy to Lem's readers but also a source of frustration to his translators. The long line of excellent decoders of Lem's linguistic and conceptual experiments—Michael Kandel, Antonia Lloyd-Jones, Peter Swirski—have demonstrated true linguistic mastery in being able to translate many of his neologisms and complex turns of phrase. Lem himself seemed very much aware that the translator's task is to some extent open-ended when he said, "Optimal strategies of literary interpretation never exist uniquely, but come in interrelated and correlated sets. It is pretty much the same with translations. There are several different translations of Shakespeare, most of them very good, yet not only are they not identical, but in fact quite divergent. And there is no way to avoid this. Some readers will always like this translation of *Hamlet*, while others will find a different one much more congenial."[24] We could easily replace *Hamlet* with *Summa* here, of course—a reassuring note from an author who was supposedly famous for writing "divorce letters" to his many translators!

In the light of the preceding, translating *Summa* has been an interesting intellectual and linguistic adventure for me. On being approached by Professor Mark Poster and the University of Minnesota Press back in 2009 to consider translating the book, I found it an extremely tempting proposition, which is not to say I jumped at the opportunity or that my subsequent decision to take it on was not accompanied by any trepidation on my part. (That said, Lem's anecdote about a visit he had from a "young lady" who wanted to translate Summa but, "for all the effort, ...was forced to give up"[25] only served to spur me on!) As a regular Lem reader, a translator of science and humanities academic texts prior to my academic career, and now a scholar working in the very areas of philosophy of technology and ethics with which Summa concerns itself, I was aware I would be dealing with an exciting yet difficult text that would no doubt permeate my own philosophical spectrum and linguistic repertoire. This is indeed what has happened: I have been thoroughly "Lemmed." For example, Lem's thinking haunts the book Life after New Media: Mediation as a Vital Process,[26] which I cowrote with Sarah Kember while translating Summa. It helped me outline what we could describe as critical vitalism, whereby evolutionary processes are treated seriously but not idolatrously and where life in its biological and philosophical enactments requires a human intervention to make sense of it and to control its random unfolding.

First published in Polish in 1964, Summa was reissued again 1964, with subsequent editions (to which Lem made some alterations) coming out in 1967 and 1974. This translation is based on the fourth edition of the book from 1974, as this is arguably the most mature and up-to-date edition. Butko explains that "Summa is essentially a work in progress: Lem corrected mistakes and updated his thinking, based in part on feedback from readers who are often scientists and experts in their respective fields."[27] Summa was reissued in Poland in 2000, and then once again, after Lem's death, as part Lem's collected works published in 2009–11 by Gazeta Wyborcza's publisher, Agora SA—a phenomenon that testifies to the ongoing popularity and significance of Lem in his native country. Lem himself seemed to remain convinced about the long-lasting timeliness of Summa's argument, a timeliness that went well beyond the fulfillment of any particular scientific prophecies of his, when he confessed to the Polish literary critic Stanisław Bereś that "from my discursive books Summa Technologiae is the only one I am happy with. This does not mean that one should not change it; but, if one does not have to, one should not do it. It has survived and is still very much alive."[28] Yet Lem also admitted in 1991 that he would gladly publish "a new critical edition of Summa, much enlarged to include—in the margins, footnotes, or in some other way—my commentary on the things I wrote in the 1960s."[29] His two essays, "Twenty Years Later" (written in 1982 and appended to Summa's fourth edition, in which Lem offers his reflections on futurology) and "Thirty Years Later" (in which he engages with a critical review of his book by Polish philosopher Leszek Kolakowski), reveal a certain sadness on the author's part that the book "sank without a trace,"[30] while also updating the examples from *Summa* with their later scientific equivalents (e.g., synthetic biology and virtual reality devices such as Eye Phone and Data Glove). Those latter efforts were aimed at demonstrating how many of the inventions and discoveries Lem did manage to predict correctly in *Summa*—notwithstanding his suspicions toward humanity's futurological aspirations. "Nothing ages as fast as the future,"[31] he quipped.

The aim of this translation, prepared specially for the University of Minnesota Press's Electronic Mediations series—a series that has been instrumental in setting new pathways in studies of new technologies and new media over more than a decade—is to offer an accurate rendition of Lem's text in its mature fourth edition from 1974, while also attending to the quirks of his language. While some of the scientific material or even terminology is now out of date, there was no attempt here to "update" Lem for the twenty-first century because the book has indeed "survived and is still very much alive." As every translator inevitably does, I had to make a number of decisions with regard to style, grammar, and a linguistic rendition of particular concepts. I was mindful of Lem's rather strong position on adjusting the third person singular *he*, which he expressed in the following terms: "I am terribly irritated by the contemporary injunction in North America that when one writes about someone, say a physicist, it has to be in the form 'he' or 'she.' I am thoroughly opposed to this, and when they requested my permission to use this in the American translations, I categorically refused. I told them that they could print my texts, but only as they were. This is the same kind of absurdity as referring to God as a 'she'—a peculiar concept, given that he is male in all monotheistic religions. I don't see a reason for changing this; I did not start the convention."[32] The Polish term *rozum*, which is to some extent foundational to the argument of the book, can be translated either as "reason" or "intelligence" (also in phrases such as *istota rozumna*, "rational" or "intelligent being"). I have opted for the latter translation because of its ubiquity in astrophysics and artificial intelligence research. Lem's *Konstruktor* has been rendered here as a "Designer," but I also want readers to be aware of the engineering connotations of this term. Lem's use of capitalization throughout the volume seems to have been a conscious stylistic and visual feature, which deserves special attention. Words such as *evolution, designer, history,* or *nature* normally appear in lowercase throughout the text, yet suddenly one of them will be capitalized—presumably to draw the reader's attention to the importance of this particular concept at a given moment. I have retained the use of capitals as per the original.

I received advice on scientific concepts and ideas, and on their historical alterations, from many various sources: from Witold Maciejewski at the Astrophysics Research Institute at Liverpool John Moores University through to arXiv.org. Thanks are also due to Gary Hall and Sarah Kember, who were careful and patient readers of different sections of the translation, and to my home institution, Goldsmiths, University of London, for allowing me time to work on this project. Last, but not least, I am grateful to McGill University (where the Lem spirit is very much alive) for inviting me as their Beaverbrook Visiting Scholar in 2011.

SUMMA TECHNOLOGIAE

I

DILEMMAS

We are going to speak of the future. Yet isn't discoursing about future events a rather inappropriate occupation for those who are lost in the transience of the here and now? Indeed, to seek out our great-great-grandsons' problems when we cannot really cope with the overload generated by our own looks like a scholasticism of the most ridiculous kind. If we could at least use the excuse that we are trying to find some optimism-enhancing strategies, or acting out of love for the truth, which is to manifest itself clearly in the future. (In our vision, such a future would be free from all kinds of storms, both metaphorical and literal ones, after our climate has been brought under control.) But the justification for my argument does not lie in scholarly passion or in an unshakeable optimism that would guarantee a favorable turn of events, no matter what happens. My justification is even simpler, more sober and probably also more modest, because, in setting off to write about tomorrow, I am only doing what I am capable of doing—no matter how well, since this is my only ability. And if this is the case, then my labor will not be any less or any more unnecessary than any other kind of work, as they are all based on the fact that the world exists and that it will continue to exist.

Having thus demonstrated that my intention is free from indecency, let us look into the subject matter and method of this book. I shall focus here on various aspects of our civilization that can be guessed and deduced from the premises known to us today, no matter how improbable their actualization. What lies at the foundation of our hypothetical constructions are *technologies, i.e., means of bringing about certain collectively determined goals that have been conditioned by the state of*

3

our knowledge and our social aptitude—and also those goals that no one has identified at the outset.

The mechanism of individual technologies, both actual and possible ones, does not interest me much. I would not have to look into it if man's creative activity were free, in a godlike manner, from being polluted by unknowledge—if, now or in the future, we could fulfill our goal in the purest way possible by being able to match the methodological precision of Genesis; if, in saying "let there be light," we could obtain as a final product light itself, without any unwanted additives. However, the previously mentioned splitting of goals, or even the replacement of one goal with another, often an undesirable one, is a classic phenomenon. Malcontents are able to see a similar kind of disturbance even in God's work—especially ever since the launch of a prototype of the intelligent being and the subsequent passing of the *Homo sapiens* model to the production stage. But we shall leave this aspect of our deliberations to theotechnologists. It is enough to say that man hardly ever knows what he is actually doing—at least he does not know for sure. Let me illustrate this point with a rather extreme example: the destruction of Life on Earth, which is entirely possible today, was not actually the goal of any of the discoverers of atomic energy.

It is therefore somewhat out of necessity that technologies are of interest to me, since a given civilization embraces everything society has desired, but also everything else that has not been part of anyone's plan. At times, frequently even, a technology is born from an accident—like when one was looking for the philosopher's stone but invented porcelain instead—but the role of intention, or conscious purpose, in all the causative efforts oriented toward technology increases with the growth of knowledge itself. In becoming more infrequent, surprises can actually reach almost apocalyptic dimensions. This is what has been stated previously.

There are few technologies that could not be classified as double-edged, as illustrated by the example of the scythes that were attached to the wheels of the Hittite battle carts or the proverbial plowshares that had been beaten into swords.[1] Every technology is actually an artificial extension of the innate tendency possessed by all living beings to gain mastery over their environment, or at least not to surrender to it in their struggle for survival. Homeostasis—a sophisticated name for aiming toward a state of equilibrium, or for continued existence despite the on-going changes—has produced gravity-resistant calcareous and chitinous skeletons; mobility-enabling legs, wings, and fins; fangs, horns, jaws, and digestive systems that enable eating; and carapaces and masking shapes that serve as a defense against being eaten. Finally, in its effort to make organisms independent of their surroundings, homeostasis implemented the regulation of body temperature. In this way, islets of decreasing entropy emerged in the world of general entropic increase. Biological evolution is not limited to this process since it builds higher entities—not islets anymore, but whole islands of homeostasis—from organisms, from phyla, classes, and plant and animal species. In this way, it shapes the surface and atmosphere of the planet. Animate nature, or the biosphere, involves both a collaboration and a mutual voraciousness; it is an alliance that is inextricably linked with struggle—as indicated by all the hierarchies examined by the ecologists. We can find among those hierarchies, especially among animal forms, pyramids at the top of which stand great predators. The latter feed on smaller animals, which in turn feed on even smaller ones. Only at the very bottom, at the base of the state known as "life," a green transformer that is omnipresent on land and in oceans converts solar into biochemical energy. Through a trillion ephemeral reeds, it maintains within itself a bulk of life that changes in its form but that never completely disappears.

Using technologies as its organs, man's homeostatic activity has turned him into the master of the Earth; yet he is only powerful in the eyes of an apologist such as himself. Faced with climatic disturbances, earthquakes, and the rare but real danger of the fall of meteorites, man is in fact as helpless as he was during the last ice age. Of course, he has come up with ways of helping those who have been affected by various cataclysms. He can even predict some of them—although not very accurately. It is still a long way to homeostasis on a planetary scale, not to mention stellar homeostasis. Unlike most animals, man does not adjust to his surroundings but rather transforms those surroundings according to his needs. Will he ever be able to do this with the stars? Will a technique of remotely directing inner transformations of the Sun emerge at some point, perhaps in a far-away future, so that creatures that are transient when compared with the duration of the solar mass can freely direct its billion-year fire? It seems possible to me, and I am saying this not to worship the rather excessively venerated human genius but, on the contrary, to open up the possibility of a contrast. So far, man has not enlarged himself in any way. But what has increased significantly is his ability to do good and evil to others. Someone who is capable of switching stars on and off will also be capable of annihilating whole inhabited globes, transforming himself in this way from an astrotechnician to a starbuster—and thus a criminal on a large cosmic scale. If the former is possible (no matter how improbable it seems and how little chance it has of coming about), so is the latter.

Such improbability, I should explain, does not derive from my belief in the necessary triumph of Ormuzd over Ahriman.[2] I do not trust any promises, nor do I believe in any assurances built on so-called humanism. The only way to deal with technology is with another technology. Man knows more about his dangerous tendencies than he did a hundred years ago, and in the next hundred years, his knowledge will be even more advanced. Then he will make use of it.

The acceleration of scientific and technological development has become so obvious that one does not have to be an expert to notice it. I think the mutability of the living conditions caused by it is one of the factors that negatively affect the formation of homeostatic systems that regulate our customs and norms. When the entire life of a future generation ceases to be a repetition of their parents' lives, what kinds of lessons and instructions can the old, experienced as they are, offer to the young? This disruption to the models of activity and their ideals by the very element of constant mutation is actually masked by another process, one that is more distinctive and that certainly has more serious immediate consequences. This process involves accelerating the oscillations of the self-awakening system of positive feedback with a very small negative component, that is, the East–West system—which, in recent years, has been oscillating between world crisis and détente.

It goes without saying that it is precisely thanks to this acceleration in the accumulation of knowledge and in the emergence of new technologies that we have an opportunity to take a serious look at our current topic. No one is questioning the fact that the changes that are occurring are rapid and sudden. Anyone who would say that the year 2000 is going to be very much like our present times would immediately become a laughingstock. Yet such attempts to project the (idealized) present condition into the future have not always been considered absurd, as evidenced by the example of Bellamy's utopia (1960), which described the 2000s from the perspective of the second half of the nineteenth century. Bellamy consciously ignored all the possible inventions that were still not known to his contemporaries. As a righteous humanist, he thought that changes caused by technoevolution were not significant either for the functioning of societies or for an individual psyche. Today we do not have to wait for the arrival of our grandchildren to make someone laugh at the naïveté of such prophesying. Anyone can have some fun by just putting in a drawer for a few years what is currently being described as a believable image of tomorrow.

And thus the rapid speed of change, which becomes an impetus for deliberations such as ours, also reduces the viability of any prophecies. I am not even thinking here of innocent popularizers, as the blame lies with their learned masters. P. M. S. Blackett, a well-known English physicist and one of the founders of operational research (the early work on mathematical strategy)—and thus a kind of professional foreteller—in his book of 1948 predicted the future development of atomic weapons and their military consequences up to the year 1960 as inadequately as one could only imagine. Even I was familiar with a 1946 book by the Austrian physicist Thirring, who was the first to have publicly described the theory of the hydrogen bomb. And yet it seemed to Blackett that nuclear weapons would not exceed a kiloton range because megatons (a term which, incidentally, did not exist at the time) would not have any targets that would be worth destroying. Today there is increasing talk of begatons (a billion tons of TNT). The prophets of astronautics did not fare any better. There were, of course, also reverse errors: around 1955, it was believed that the fusion of hydrogen into helium observed in the stars would generate industrial energy in the near future. Current estimates situate the production of the microfusion cell[3] in the 1990s, or even later. But it is not so much the development of a particular technology that is at issue but rather the unknown consequences of such a development.

We have so far discredited any predictions regarding progress, thus in a way clipping the branch on which we have been trying to undertake a series of daring exercises—mainly the casting of an eye into the future. Having shown how hopeless such an undertaking can be, we would be better off occupying ourselves with something else. Yet we are not going to give up that easily. The risk exposed can actually stimulate further debate. Besides, even if we are to make a series of gigantic mistakes, we will find ourselves in exquisite company. For an infinite variety of reasons that turn making predictions into a thankless task, I shall list several such mistakes that are particularly displeasing to the artist.

First, transformations that lead to a sudden turn in existing technologies sometimes burst forth like Athena from Zeus's head—to the surprise of everyone, including the experts. The twentieth century has already been taken aback several times by the newly emergent hegemonies, such as, for example, cybernetics. Enamored of the scarcity of means and considering—not incorrectly—that similar maneuvers are one of the cardinal sins against the art of composition, the artist hates such deus ex machina devices. But what shall we do when History turns out to be so easy to please?

Furthermore, we are always inclined to extend the course of new technologies into the future by means of straight lines, hence "the world full of balloons" or "the total steam world" of the nineteenth-century utopians and draftsmen—both of which seem most amusing today. Hence also the contemporary attempt to populate outer space with space "ships," including a brave "crew," "watch officers," "helmsmen," and so on, on board. It is not that one should not write like this, but this kind of writing belongs to a genre of fantasy; it is a form of "reverse" nineteenth-century historical novel. In the way one used to ascribe the motivations and psychological traits of contemporary monarchs to the pharaohs, one represents today the corsairs and pirates of the thirtieth century. We can surely amuse ourselves like this, provided we remember we are only playing. However, History has nothing to do with such simplifications. It does not show us any straight paths of development but rather uses the curvilinear zigzags of nonlinear evolution. This means that the conventions of elegant design have to be abandoned, unfortunately.

Third, a literary work has a beginning, a middle, and an end. The entanglement of plots, the interweaving of temporalities, and the use of other devices aimed at modernizing fiction have not so far liquidated this structure. Generally speaking, we have a tendency to place any phenomenon within a closed schema. Let us imagine a 1930s thinker who has been presented with the following imaginary situation: in 1960, the world will be divided into two antagonistic parts, each one of which will be in the possession of some terrible weapons, capable of annihilating the other half. What will the outcome be? Our thinker would no doubt answer: total annihilation or total disarmament (but he would also most certainly add that our idea is weak because it is so melodramatic and unbelievable). Yet so far this kind of prediction has not delivered much. Please note that it has been over fifteen years since the emergence of the "balance of fear":[4] three times as long as it had taken to manufacture the first atomic bombs. To a certain degree, the world is like a sick man who believes that either he is going to get better soon or die very quickly and to whom it does not even occur that he can, moaning as he does, and going through some short-term ups and downs, go on living until old age. Yet this comparison is rather shortsighted...unless we invent some medication that will manage to cure this man's sickness completely but that will pass on to him some new problems resulting from the fact that even if he is given an artificial heart, it will be placed on a little trolley connected to it with a bendy pipe. This is, of course, nonsensical, but we are talking about the price of such a total cure. Liberation from oppression (humanity's atomic independence from the limited oil and coal resources, say) has its price, while the amount of expected repayment and its period, as well as the method of its delivery, usually come as a surprise. The mass application of atomic energy for peaceful aims carries with it a gigantic problem of radioactive waste, which we still do not know what we should do about. The development of nuclear weapons can quickly lead to a situation in which today's proposals for disarmament, together with our "annihilation proposals," will turn out to be anachronistic. It is hard to determine whether it is going to be a change for the better or for the worse. The overall threat may increase (e.g., inner striking power will get bigger and will thus require us to build shelters from reinforced concrete), but the possibility of its actualization will decrease—or the other way round. Other combinations are also possible. In any case, the global system remains unbalanced, not only in the sense that it can tip toward war, as there is nothing new about that, but primarily because it is evolving as a whole. At the moment, it seems somewhat "scarier" than it was in the era of kilotons, since we have megatons today, but it is only a transitional phase. Contrary to popular belief, one should not think that an increase in charge power and in the velocity of their carriers, or the development of "antimissile missiles," represents the only possible gradient of this evolution. We are entering higher and higher levels of military technology, as a result of which it is not only conventional battleships and bombers, strategies and staff, that are becoming obsolete: so is the very idea of global antagonism. I have no idea how it is going to evolve. Instead, I am going to introduce briefly a novel by Olaf Stapledon,[5] the "plot" of which spans over two billion years of human civilization.

Martians, a species of viruses capable of aggregating into jellylike "intelligent clouds," attack the Earth. People fight the invasion for a long time, without knowing that they are dealing with an intelligent life-form and not with a cosmic cataclysm. The "victory or defeat" alternative does not ensue. After centuries of struggle, the viruses undergo a transmutation so deep that they enter the human genotype, which leads to the emergence of a new kind of *Homo sapiens*.

I think this is a beautiful model of a historical phenomenon on a yet unknown scale. The probability of this phenomenon taking place is not that relevant; I am more interested in its structure. History does not deal with tripartite closed schemas entailing "a beginning, a middle, and an end." It is only in a novel that a character's life gets immobilized into a certain image, before the words "The End" appear, thus filling the author with aesthetic delight. It is only in a novel that we must have an end, a happy or an unhappy one, but certainly one that closes things off on the level of composition. Yet the history of humankind does not know, and I hope *will* not know, such definitive closures or "final ends."

2

TWO EVOLUTIONS

It is difficult for us to understand the process whereby ancient technologies emerged. Their utilitarian character and their teleological structure remain undisputed, yet they did not have any individual designers or inventors. Trying to get to the origins of early technologies is a dangerous task. Successful technologies used to have myth or superstition as their "theoretical foundation": their application was either preceded by a magic ritual (medicinal herbs supposedly owed their properties to a formula that was being recited while collecting or applying those herbs), or they themselves became a form of ritual, in which a pragmatic element was irrevocably linked with a mystical one (the ritual of shipbuilding, in which the production process was a form of liturgy). When it comes to becoming aware of the ultimate goal, the structure of a collective task can today approach the realization of a task achieved by an individual. But it was different in the old days, when one could only speak of technical goals of ancient societies metaphorically.

The shift from the Paleolithic to the Neolithic period, the Neolithic revolution—which rivals the atomic one in terms of its cultural significance—was not the consequence of an idea of farming "popping into the head" of some kind of Einstein of the Stone Age, who then "convinced" his contemporaries about the advantages of this new technology. It was an extremely slow process, a creeping transformation, exceeding the life span of many generations—from using the plants one found as nutrition through to nomadic hunting and gathering and then sedentarization. The changes occurring within the lifetime of a single generation were hardly noticeable. In other words, each generation would encounter an apparently unchanged technology, as "natural" as sunrises and sunsets.

This mode of emergence of technological practice has not disappeared altogether since the cultural significance of every great technology reaches much further than just the lifetime of each individual generation—which is why its future-oriented consequences of a systemic, habitual, and ethical kind, as well as the very direction in which it is pushing humanity, not only are not a subject of anyone's conscious intention but also effectively defy the recognition of the existence of such significance or the definition of its nature. With this terrifying sentence (terrifying in style, not content), we are opening a section on the metatheory of the gradients of man's technical evolution. We say "meta" because it is not the delineation of its direction or the determination of its consequences that preoccupies us for the time being but rather a more general and overarching phenomenon. Who causes whom? Does technology cause us, or do we cause it?[1] Does it lead us wherever it wishes, even to perdition, or can we make it bend before our pursuit? But what drives this pursuit if not technical thought? Is it always the same, or is the "humanity–technology" relation itself historically variable? If the latter is the case, then where is this unknown quantity heading? Who will gain the upper hand, a strategic space for civilization's maneuvers: humanity, which is freely choosing from the widely available arsenal of technological means, or maybe technology, which, through automation, will successfully conclude the process of removing humans from its territory? Are there any thinkable technologies that are impossible to actualize, now and in the future? What would determine such a possibility—the structure of the world or our own limitations? Is there another potential direction in which our civilization could develop, other than a technical one? Is our trajectory in the Universe typical? Is it the norm—or an aberration?

We shall try to find answers to these questions—although our search will not always yield unequivocal results. Our starting point will be provided by a graphic chart illustrating the classification of effectors, that is, systems capable of acting, which Pierre de Latil included in his book, *Thinking by Machine* (1956). He distinguishes three main classes of effectors. The first class of determined effectors contains simple tools (such as a hammer), complex tools (an adding machine, a classic machine), and tools linked to the environment (with no feedback) such as, for example, an automatic fire detector. The second class—that of organized effectors—contains feedback systems: machines with built-in determinism of action (self-regulators, e.g., those of a steam engine), machines with a variable goal of action (programmable from outside, e.g., an electric brain),

and self-programming machines (systems capable of self-organization).
Animals and man belong to the latter class. There are also systems that
possess one more degree of freedom—systems that, to achieve their goals,
are capable of changing themselves. (De Latil calls this a "who" freedom,
in the sense that whereas man has been "given" his organization and body
material, systems of this higher class, which do not just possess freedom
on the level of their building material, can radically transform their own
systemic organization. A living species that is undergoing biological
evolution would be an example of such systems.) De Latil's hypothetical
effector of an even higher class also has freedom on the level of the choice
of material from which it "builds itself." As an example of such an effector with the highest degree of freedom, de Latil proposes the mechanism
of the spontaneous generation of cosmic matter as outlined in Hoyle's
theory.[2] It is easy to see that technical evolution is a far less hypothetical
system of this kind, one that is also much easier to verify. It manifests
all the characteristics of a feedback system that is programmable "from
within," that is, self-organizing, and that also possesses freedom with
regard to its total transformation (the way a living and evolving species
does), as well as freedom of choice with regard to its building material
(since technology has at its disposal everything the Universe contains).

I have summarized the classification of systems with increasing degrees of freedom as proposed by de Latil, removing from it some highly
contentious details regarding their categorization. Before we move on
with our discussion, we should perhaps add that this classification, in
the form presented here, is not complete. One can imagine systems possessing yet another degree of freedom, since selection from among the
materials contained within the Universe is necessarily limited to the "parts
catalog" the Universe possesses. Yet we can also think of a system that,
not being content with what is provided, creates materials "off-catalog,"
that is, materials that do not exist in the Universe. A theosophist would
perhaps be inclined to designate such a "self-organizing system with
a maximum degree of freedom" as God. However, such a hypothesis
is not necessary for us since we can conclude, even on the basis of the
limited knowledge we have today, that creating "off-catalog" parts (e.g.,
subatomic particles that the Universe does not contain "normally") is
possible. Why? Because the Universe does not actualize all of its possible
material structures. As we know, it does not create, say, in the stars—or
anywhere else—typewriters; yet the "potentiality" of such typewriters is
contained within it. The same applies, we can assume, to the phenomena

that contain the (at least yet) unrealized states of matter and energy in the Universe, in the space and time that carry them.

SIMILARITIES

We know nothing certain about the origins of evolution. What we do know rather well is the dynamics of the emergence of a new species—from its birth through its developmental peak to its decline. There are almost as many evolutionary paths as there are species, and they all have numerous characteristics in common. A new species appears in the world unnoticed. Its appearance seems to come from what already exists, and this borrowing seems to testify to the inventive inertia of the Designer. At the beginning, there is not much indication that this upheaval in its inner organization, to which a species will owe its later development, has in fact already taken place. The first specimens are usually small; they also sport a number of primitive features, as if their birth had been hurried and fraught with uncertainty. For a period of time, they vegetate in a semisecretive state, barely managing to compete with the established species—which are already optimally adapted to the tasks of the world. Then, eventually, prompted by the change to the general equilibrium resulting from the seemingly insignificant transformations in the environment (whereby a species's environment includes not only the geological world but also all the other species vegetating in it), a new kind of expansion takes off. Entering the already occupied territories, a species openly shows its lead over its competitors in the struggle for life. When it enters an empty unconquered space, it bursts into evolutionary radiation, which in one go initiates the emergence of a whole range of variations. In these variations, the disappearance of the remnants of primitivism in a species is accompanied by the emergence of new systemic solutions that are ever more bravely dominating its outer appearance and its new functions. This is the route a species takes to reach its developmental peak. Through the process, it gives a name to the whole epoch. The period of rule on land, in the sea, or in the air lasts a long time. Then a homeostatic equilibrium is eventually disturbed once again—yet this still does not signal defeat. The evolutionary dynamics of a species gains some hitherto unobserved new traits. In its core branch, the specimens are getting bigger, as if gigantism was to provide protection against the threat. Evolutionary radiations start to take place again, this time often marked by hyperspecialization.

The lateral branches attempt to penetrate into environments in which competition is comparatively weaker. From time to time, that latter maneuver culminates in success. Then, when all the traces of the giants—whose emergence was a defense strategy on the part of the core species against its extinction—have disappeared, when all the simultaneous efforts to the contrary have also failed (as some evolutionary lines promptly head toward dwarfism), the descendants of the lateral branch, having happily encountered propitious conditions inside the peripheral area of their competition, continue their existence almost without change. In this way, they serve as the last proof of the primeval abundance and power of a species.

Please forgive my somewhat pompous style, a rhetoric that has not been supported with any examples. Any vagueness here stems from the fact that I have been talking about two kinds of evolution at the same time: biological and technical.

As a matter of fact, their dominant characteristics show a great number of surprising analogies. It is not only that the first amphibians were similar to fish, while the mammals resembled small lizards. The first airplane, the first automobile, or the first radio owed its appearance to the replication of the forms that preceded it. The first birds were feathered flying lizards; the first automobile was a spitting image of the coach with a guillotined shaft; the airplane had been "copied" from the kite (or even the bird), the radio from the already existing telephone. Those prototypes tended to be rather undersized, while their primitive design left a lot to be desired. The first bird, the ancestor of the horse or the elephant, was quite tiny; the first steam locomotive was not much bigger than a regular cart, while the first electric locomotive was even smaller. A new principle of biological or technical design initially deserves pity more than enthusiasm. Ancient mechanical vehicles moved more slowly than horse-driven ones; the first airplane could barely manage to lift itself off the ground; and listening to radio broadcasts was no fun even when compared with the tinny sound of the gramophone. Similarly, the first land animals were no longer good swimmers, yet they still had not mastered the art of walking. The feathered lizard—the archaeopteryx—did not as much fly as flutter. Only during the process of perfecting those traits did the previously mentioned "radiations" take place. Just as the birds conquered the sky and the herbivorous mammals the steppe, the combustion engine vehicle took mastery over the roads, thus giving rise to ever more specialized varieties. In the "struggle for life," the automobile not only pushed out the stagecoach but also "gave birth" to the bus, the truck, the bulldozer, the fire engine, the tank, the off-road vehicle, and dozens of other means of transport. The airplane, taking control over the "ecological niche" of the airspace, was probably developing even faster, changing several times the already fixed shapes and types of drive (the piston engine was replaced by the turbo engine, then by the turbo-piston engine, and eventually by the jet engine; when it came to shorter distances, the winged airplane found a serious rival in the helicopter; and so on). It is worth noticing that just as the predator's strategy affects the strategy of its prey, so the classical airplane defends itself against the helicopter's invasion. It does this by creating a prototype of the winged plane, which—owing to the change of direction of the thrust—is capable of taking off and landing vertically. It is a struggle for the maximum universality of function, one that is very well known to any evolutionist.

The two means of transport discussed earlier have not yet reached their developmental peak, which is why we cannot talk about their late forms. It was a different story with the piloted hot air balloon, which, when threatened by machines whose weight exceeded that of air, developed symptoms of elephantiasis, so typical of the predecline blossom of dying evolutionary branches. The last zeppelins of the 1930s can be easily compared with the atlantosauruses and brontosauruses of the Cretaceous period. Gigantic size was also achieved by the last exemplars of the steam-driven freight train, before it was made obsolete by diesel and electric locomotives. When looking for signs of descending evolution, which is attempting to overcome the danger it faces with secondary radiations, we can turn to radio and cinema. The competition from television led to a sudden "radiation of variations" among radio sets and to their appearance in new "ecological niches." In this way, miniaturized, pocket, and other kinds of radio sets—including those affected by hyperspecialization, such as high-fidelity radios with stereophonic sound, integrated hi-fi recorders, and so on—came into being. Cinema itself, in its battle against television, has considerably increased the size of the screen and is even demonstrating a tendency to "surround" the viewer with it (videorama, Circarama).[3] One can also imagine further development of mechanical vehicles, which will make wheel drive obsolete. When the current automobile is eventually replaced by an "air-cushion vehicle," it is quite likely that, say, a small combustion-driven lawnmower will be the last descendant of the classical car—still vegetating in a "lateral branch." Its design will be a remote reflection of the automobile period, just as certain specimens of lizards from the archipelagos of the Indian Ocean are the last remaining descendants of the large Mesozoic reptiles.

The morphological analogies between the dynamics of bio- and technoevolution, which can be represented with a slowly ascending curve that is then to descend from its peak toward decline on a graph chart, do not cover all the convergences between these two large areas. One can identify some other, even more surprising convergences. For example, in certain living organisms, there exist a high number of exceptionally unique traits whose origin and survival cannot be explained by their adaptation values. We can mention here, alongside the well-known cockscomb, the fantastic plumage of some male birds, such as the peacock or the pheasant, and even the saillike spinal outgrowths of fossil reptiles.[4] Similarly, the majority of products of a given technology possess seemingly superfluous or nonfunctional traits that cannot be explained by the way these products work or by what they are supposed to do. We can posit here an extremely interesting but also somewhat amusing similarity with an invasion, which is taking place inside biological and technological construction processes: in the former case, it is an invasion by the criteria of sexual selection, in the latter by fashion. If, for the sake of clarity, we limit our analysis to the example of the contemporary motor vehicle, we shall see that the car's main features are forced on the designer by the current state of technology. So, for instance, when using rear wheel drive together with a front-mounted engine, the engineer has to place the transmission tunnel of the propeller shaft within the car's interior. However, between the requirement not to change the design of the vehicle's "systemic" organization, on one hand, and the requirements and tastes of the vehicle's recipient, on the other, lies a vast space of "inventive freedom" because the said recipient can be offered various shapes and colors for his vehicle, various shapes and sizes of the windows, additional embellishments, chrome plating, and so on. In bioevolution, the counterpart of product differentiation, which results from fashion pressure, is provided by the multifarious shapes of secondary sexual characteristics. Those characteristics were initially the result of accidental changes, that is, mutations. They became fixed in the generations that followed due to the privileging of their owners as sexual partners. And thus the counterpart of the car's tail, chrome-plated embellishments, fantastically shaped cooling pipes, and head and rear lights can be found in mating colors, crests, unusual outgrowths, and last but not least, in the unique distribution of fat tissue, combined with the facial features that are sexually appealing.

Of course, the sluggishness of "sexual fashion" is much greater in bioevolution than in technology since Nature the Designer is incapable of altering the models it produces each year. However, the essence of this phenomenon, that is, the unique influence of "impractical," "insignificant," and "nonteleological" factors on the shape and development of living creatures and technological products, can be detected and verified in a large number of randomly selected examples.

We could find some other, even more inconspicuous similarities between these two large evolutionary trees. For example, in bioevolution, there is the phenomenon of mimicry, that is, the adaptation of an appearance of one species to that of another, which proves beneficial for the "imitators." Nonvenomous insects can be a spitting image of some remote yet dangerous species; they can even "pretend" to be a single body part of a creature that has nothing in common with insects—I have in mind here the amazing "cat's eyes" on the wings of certain butterflies. Analogies to mimicry can also be found in technoevolution. The lion's share of nineteenth-century metalwork and smithery emerged as a result of imitating botanical forms (the ironwork in bridge structures, handrails, street lanterns, and fences, or even the funnel "crowns" of old locomotives, all "pretended" to be botanical designs). Everyday objects such as fountain pens, cigarette lighters, lamps, and typewriters often have aerodynamic shapes these days, masquerading as forms designed in the aerospace industry—which is an industry of high-speed technologies. Yet this kind of mimicry lacks the profound justification that is evident in its biological counterpart. What we have here is the influence of dominant technologies over lower-level, secondary ones. Fashion also has much to say in this respect. In any case, it is usually impossible to detect to what extent a given shape has been determined by the designer's supply and to what extent by the buyer's demand. We are faced here with circular processes in which causes become effects and effects causes and in which numerous instances of positive and negative feedback are at work: living organisms in biology or subsequent industrial products in the technical civilization are only tiny elements of these higher processes.

The preceding statement also reveals the genesis of the similarity between the two evolutions. Both are material processes with almost the same number of degrees of freedom and with similar dynamic laws. These processes take place in a self-organizing system—a term that applies to the Earth's biosphere as a whole as much as it does to the totality of man's technical activities. A system of this kind is characterized by the phenomenon of "progress," that is, an increase in homeostatic capability, which has an ultrastable equilibrium as its direct goal.[5]

Drawing on examples from biology will also turn out to be useful and productive in our further deliberations. Alongside the similarities, these two types of evolution demonstrate some far-reaching differences, a thorough examination of which may reveal both the limitations and the deficiencies of that perfect Designer that Nature is supposed to be and the unexpected opportunities (as well as dangers) that the rapid development of technology poses in human hands. I say "in human hands" because technical evolution is not, for the time being at least, people-free; it only achieves its full being having been "complemented by mankind." It is here perhaps that the biggest difference lies: bioevolution is beyond all doubt an amoral process, which is something we cannot say about technical evolution.

DIFFERENCES

The first difference between our two evolutions is genetic and centers on the question of their driving forces. Nature is the "cause" of bioevolution, Man of technical evolution. An attempt to explain the "starting point" of bioevolution is still causing great difficulty. The problem of the origin of life occupies a significant place in our discussion, as solving it will need to involve more than just determining the causes of a given historical fact related to the Earth's remote past. We are not so much concerned with the fact itself as with the consequences it still bears on any further development of technology. Its development has resulted in a situation in which any further progress will not be possible unless we gain accurate knowledge about extremely complex phenomena—phenomena as complex as life itself. This is not to say that we want to "imitate" a living cell. We do not imitate the mechanics of bird flight even if we do fly ourselves. It is not imitation that is at stake here but understanding. And it is this attempt to understand biogenesis "from the designer's point of view" that is causing such immense difficulties.

Traditional biology appeals to thermodynamics as a reliable arbiter. The latter declares the shift from greater to lesser complexity to be the norm. Yet the emergence of life was a reverse process. Even if we do accept as a general law the hypothesis regarding "a minimal complication threshold"—the crossing of which results in a material system not only maintaining its current organization despite external disturbances but even passing it on, unchanged, to the organisms of its descendants—this hypothesis does not explain anything on a genetic level. At some point an organism must have crossed this threshold for the first time. What is most significant here is whether this was as a result of a so-called accident or of causal necessity. In other words, was the "beginning" of life an exceptional event (like winning the lottery) or a typical one (like losing it)?

When talking about the spontaneous generation of life, biologists say that it must have been a gradual process, consisting of a number of stages, with each subsequent stage on the way to the emergence of the first cell having its own determined probability. The formation of amino acids in the primeval ocean as a result of electrical discharges was, for instance, quite probable, while the formation of peptides from them was a little less probable, yet it still had quite a good chance of taking place. The spontaneous synthesis of enzymes—those catalysts of life, the steersmen of its biochemical reactions—was, in turn, a rather unusual occurrence (even if it was necessary for the emergence of life). In an area ruled by probability, we are faced with statistical laws. Thermodynamics is actually one such law. From its point of view, the water in a pot placed over fire will boil, but this is not absolutely certain. There is a possibility of this water freezing over fire, even if the likelihood of this happening is astronomically low. The argument that even phenomena that seem most improbable from a thermodynamic point of view always eventually take place, as long as one waits for them patiently—and the evolution of life was "patient" enough, as it took billions of years—sounds convincing, provided we do not expose it to mathematical reasoning. Indeed, thermodynamics can swallow even the spontaneous formation of proteins in amino acid solutions, but it does not allow for the biogenesis of enzymes. If the whole of the Earth was an ocean of protein solution, if its radius was five times bigger than it actually is, it still would not have had enough mass for an accidental emergence of the kinds of highly specialized enzymes that are needed to engineer life. The number of all possible enzymes is higher than the number of stars in the whole Universe. If the proteins in the primeval ocean had had to wait for their spontaneous generation, this could have taken all eternity. And thus, to explain the emergence of a certain stage of biogenesis, we have to resort to postulating a highly improbable phenomenon: "winning the jackpot" in the previously mentioned cosmic lottery.

Let us be honest: if all of us, including the scientists, were intelligent robots and not flesh-and-bone creatures, then the number of scientists who would be willing to accept such a probabilistic version of the hypothesis about the origin of life could be counted on the fingers of one hand. The fact that the number of such scientists is higher does not so much result from the general conviction about the truthfulness of this hypothesis as from the simple fact that we are alive, which means that we ourselves represent a persuasive, albeit indirect, argument to support biogenesis. It is because two or even four billion years is enough to form a species and its evolution but not to form a living cell by means of repeated, blind "draws" from the statistical bag of all possibilities.

The issue presented in this way is not only improbable from the point of view of scientific methodology (which deals with typical phenomena, not accidental ones that border on unpredictability) but also announces a rather unequivocal verdict. It declares that any attempt to "engineer life," or even to "engineer very complex systems," is futile, given that the latter's emergence is determined by an exceptionally rare accident.

Luckily, this is a false approach. It is based on the fact that we only know two types of systems: very simple ones, such as the machines we have built so far, and extremely complex ones, such as all living beings. The lack of any intermediate links has led us to hold on too tightly to a thermodynamic explanation of phenomena, without taking into account the gradual emergence of systemic laws in systems aiming to achieve equilibrium. If the situation is as constricted as it is in the example of the clock—where equilibrium amounts to the stopping of its pendulum—we do not have enough material to extrapolate to systems with multiple dynamic possibilities, such as a planet on which biogenesis is starting to take place or a laboratory in which scientists are constructing self-organizing systems.

Such systems, still relatively simple today, are the indirect links for which we have been looking. Their emergence, for example, in the form of living organisms, is no "jackpot in the lottery of accidents." It is rather a manifestation of the necessary states of dynamic equilibrium within systems featuring many different elements and tendencies. And thus the process of self-organization is not unique but rather typical, while the emergence of life is only one possible enactment of the process of homeostatic organization, which is widespread in the Universe. This process does not disturb in any way the thermodynamic equilibrium in the Universe, as this equilibrium is global, allowing as it does for the occurrence of many phenomena—such as, for example, the formation of heavier (and thus more complex) elements from lighter (and thus simpler) ones.

And thus a Monte Carlo–type hypothesis of the cosmic roulette—which is a naïve methodological extension of thinking based on the knowledge of extremely simple mechanisms—is replaced by a theory of "cosmic panevolutionism." The latter transforms us from beings condemned to wait passively for an arrival of some extremely rare circumstances into designers capable of making choices from among the staggering overabundance of possibilities. Those possibilities are contained in the so far rather general instructions for building self-organizing systems of ever increasing degrees of complexity.

What the frequency of the cosmic occurrence of these "parabiological evolutions" postulated previously is, and whether they actually culminate in the emergence of what our human understanding calls "psyche," is a different matter. But this is a subject for a separate discussion, one that would require us to draw on an extensive assembly of facts from the field of astrophysical observation.

Nature, the Great Designer, has been conducting its experiments for billions of years, developing from the once obtained material (although this point is still debatable) everything that is possible. Spying on its tireless activity, man as the son of Mother Nature and Father Chance has been wondering for centuries about the meaning of this deadly serious game, in all its finality. This is certainly a pointless activity if he is to continue asking this question forever. It will be a different story, though, as soon as he starts answering the question himself, taking over Nature's convoluted secrets and initiating Technical Evolution in his own image.

The second difference between the two evolutions under discussion concerns methodology, that is, the "how" question. Biological evolution can be divided into two stages. The first one covers the period from the "start" from inanimate matter to the emergence of living cells—which became clearly demarcated from their surroundings. While we are quite familiar with the laws and general behavior of evolution during its second stage—that of the emergence of the species—as well as with the many specific paths it took, we cannot really say anything certain about its initial phase. This initial stage remained underappreciated for a long time, both for its scope and for the range of phenomena that had taken place during it. Today we can estimate that it lasted for at least half of the total evolutionary time, that is, around two billion years, yet some experts still complain about its short duration. Importantly, it is then that the cell as an elementary building block of biological material emerged, manifesting the same core structure both in the trilobite from a billion years ago and in chamomile, the hydra, the crocodile, and the human being of today. The universality of this building material is most astonishing and in fact rather hard to comprehend. The cells of the paramecium, of the mammalian muscle, of the plant leaf, of the snail's lymph glands, and of the abdominal gland of an insect all have the same basic components that a nucleus has—with its mechanism that allows for the transfer of hereditary information, which has been perfected on a molecular level—and also that the enzymatic system of mitochondria and the Golgi apparatus have. Every one of such cells contains the potential for dynamic homeostasis, for selective specialization, and also for the hierarchical construction of multicellular organisms. One of the fundamental laws of bioevolution is the short-term planning of its activities, since every change directly services current adaptive needs. Evolution cannot enact the kinds of changes that would only work as a preparatory introduction for the changes that are to take place in millions of years. It does not "know" anything about what is going to happen in millions of years since it is a blind designer, working by "trial and error." Unlike an engineer, it cannot "stop" the faulty machine of life to embark on its radical redesign, having reconsidered its design principles.

This is why we are so astonished and shocked by its "original foresight," which it showed during the prologue to the multiact drama of the species, when it created the building material so versatile and malleable that it is unlike any other. Since, as we said earlier, evolution is incapable of carrying out any sudden radical reconstructions, all of its hereditary mechanisms—its ultrastability, together with the accidental element of mutation (without which there would be no change, and hence no development) that interferes with this ultrastability, the division of the sexes, reproductive potential, and even those characteristics of the living tissue that manifest themselves most clearly in the central nervous system—all of these had already been inserted, we may say, into the Archeozoic cell billions of years ago. This kind of farsightedness was demonstrated by a nonhuman and mindless Designer, who seemingly just cared about the most transient state of affairs, that is, the survival of a generation of proto-organisms that existed at the time, slimy microscopic droplets of protein that were capable of only one thing: continuing to exist in the fluid equilibrium of physicochemical processes and passing on the dynamic structure of their existence to others!

We do not know anything about the early dramas of that phase, which served as preparation for the actual evolution of the species, as it left absolutely no traces. It is quite likely that during those millions of years, various forms of proto-life, which would have been very different both from the present life-forms and from the most ancient fossils, came into being and then subsequently disappeared. We can speculate about the repeated emergence of larger, "almost-living" conglomerates, which could have taken some time to develop (a period once again measured in millions of years, no doubt), and only at a later stage in the struggle for survival would those creatures have succumbed to being inevitably driven out of their ecological niches by more universal and hence fitter beings. This would represent a theoretical possibility, or even probability, of the initial variation and diversification among the paths followed by self-organizing matter, whereby ongoing extinction would be equivalent to the idea of designing ultimate universality. The number of designs that perished is no doubt thousands of times bigger than the handful of those that survived the ordeal.

The construction method of technical evolution is completely different. Figuratively speaking, Nature had to presuppose in its biological building material all the potentialities that were to be actualized much later. Man, in turn, tended to initiate his technologies and then abandon them to move on to some new ones. Being relatively free with regard to his choice of building material, having at his disposal high and low temperatures; metals and minerals; gases, solids, and liquids, he was seemingly capable of achieving more than Evolution did. Evolution was always limited to what had been given to it—tepid water solutions, gluey multiparticle substances, a relatively limited number of elements that appeared in the Archeozoic seas and oceans—but it squeezed out absolutely everything that was possible from such a limited starter kit. Ultimately, for the time being, the "technology" of living matter is head and shoulders above our human engineering—which is being supported by all the resources of socially acquired theoretical knowledge.

In other words, the universality of our technologies is minimal. Technical evolution has so far been moving in a kind of reverse direction to biological evolution in that it has only been creating narrowly specialized devices. The human hand was a model for the majority of tools, yet each time it was only one of its movements or gestures—finger clenching, one straight finger revolving around a longitudinal axis thanks to the movements in the wrist and elbow joints, a fist—that was respectively imitated by the pliers, drill, and hammer. So-called universal machine tools are in fact rather highly specialized. Even automated factories—the construction of which is only just starting to take off—lack the plasticity of behavior that characterizes simple living organisms. The potential for universality seems to lie in a further development of the theory of self-organizing systems—which would be capable of adaptive self-programming. Their functional similarity to man is, of course, not accidental.

The end of this road does not lie, as some claim, in the "duplication" of the human design or the design of some other living organisms, inside the electrical circuits of digital machines. For now, life's technology is far ahead of us. We have to catch up with it—not to ape its results but to exceed its seemingly unmatched perfection.

A separate chapter of evolutionary methodology covers the relationship between theory and practice, between abstract knowledge and actualized technologies. This relationship does not, of course, exist in bioevolution, since Nature evidently "doesn't know what it's doing." It just simply actualizes that which is possible, that which spontaneously emerges from given material conditions. It has not been easy for man to accept this state of things, not least because he is also one of Mother Nature's "unwanted" and "accidental" children.

Yet this is not actually a chapter but rather an enormous library. Any attempt to summarize its contents seems rather hopeless. Threatened with such an explicatory abyss, we must be particularly laconic. Prototechnologists did not have any theoretical knowledge because, among other things, they did not know that such a thing was even possible. For thousands of years, theoretical knowledge developed without experimentation, through magic thinking—which is a form of induction, but one that is applied falsely. A conditioned response—that is, an "if A, then B" type of reaction—is its animal antecedent. Of course, both a response of this kind and magic need to be preceded by observation. It would often happen that a working technology was contradicted by the false theoretical knowledge of the time. A chain of pseudoexplanations would then follow, aimed at reconciling the two (the fact that water could not be raised by means of pumps above the height of ten meters was "explained" by Nature's fear of a vacuum). In its contemporary understanding, knowledge is an examination of the laws of the world, whereas technology is a way of using these laws to satisfy human needs—which are actually still the same as they were in the Egypt of the pharaohs. Its tasks involve cladding and feeding us, providing a roof over our heads, transporting us from one place to another, protecting us against illnesses. Knowledge cares about facts—atomic, particle, or stellar ones—and not about us, or at least not in a way that would be driven by the direct usefulness of results. We should point out that the disinterestedness of theoretical investigations used to be purer than it is today. Thanks to our experience, we now know that there is no such thing as useless knowledge in the most pragmatic sense because one never knows when a certain piece of information about the world is going to come in handy or even when it will turn out to be particularly needed and valuable. One of the most "useless" branches of botany, lichenology—which deals with fungi—turned out to be vital, literally, after the discovery of penicillin. Researcher–ideographers, tireless collectors of facts, describers and classifiers, did not hope for such successes in the old days. Yet man—a creature whose impracticality only every so often matches his curiosity—was more interested in counting the stars and studying the structure of the Universe than in the theory of agriculture or in the workings of his body. From the painstaking, at times even manic, effort of gatherers and collectors of observations, the great edifice of nomothetic sciences, which generalized facts into systemic laws of phenomena and things, slowly emerged. As long as theoretical knowledge lags far behind technological practice, human engineering efforts are in many ways similar to the "trial and error" method used by Evolution. Just as evolution "tries out" adaptive capabilities of animal and plant mutant "prototypes," an engineer tests the real capabilities of new inventions, flying devices, vehicles, and machines—often resorting to constructing reductionist models. This method of empirically eliminating false solutions and then restarting the efforts underpins the emergence of nineteenth-century inventions: the bulb with a carbon fiber filament, the phonograph, Edison's dynamo, and, prior to that, the locomotive and the steamship.

All this led to the popularization of the idea of an inventor as someone who did not need anything apart from talent, common sense, persistence, pliers, and a hammer to achieve his goal. Yet this is an extravagant way of inventing things—almost as extravagant as the working of bioevolution, whose billion-year-long empirical practices, that is, its "false solutions" to the problem of preserving life posed by the new conditions, claimed hecatombs of victims. The essence of technology's "empirical era" lay not so much in its lack of theoretical solutions as in the derivative character of those solutions. First we had the steam engine, then thermodynamics, just as first we had the airplane and then flight theory, or just as we first built bridges and then learned how to make calculations for them. We could risk saying that technological empiricism develops as far as it can. Edison tried to invent something like an "atomic engine," but this did not—and could not—come to much because, whereas a dynamo can be built through trial and error, an atomic reactor cannot.

Technological empiricism does not naturally mean blindly tossing from one badly thought-out experiment to another. The empiricist inventor usually has an idea, or rather—thanks to what he has already achieved (or what others had achieved before him)—he can see a short stretch of the road ahead of him. The sequence of his actions is regulated by negative feedback (the failure of an experiment explains every time that this is the wrong way). As a result, the road he takes is like a zigzag, but it is certainly going somewhere and has a definite direction. Gaining theoretical knowledge allows for a sudden leap ahead. During the Second World War, the Germans did not have a theory of supersonic rockets' ballistic flight. They derived the shape of their V2 from multiple empirical trials (conducted on reductionist models in an aerodynamic tunnel). Knowing the correct formula would have, of course, made building all those models redundant.

Evolution does not possess any other "knowledge" apart from "empirical," which is contained in the genetic inscription of information. It is a dual type of "knowledge." First, it is a knowledge that defines and determines in advance all the possibilities of a future organism (i.e., the "innate knowledge" of tissues telling them how to function to ensure the smooth running of the living processes, how certain types of tissues and organs need to behave in relation to others, but also how the organism as a whole is to behave in relation to its environment. This latter piece of information stands for "instincts," defense mechanisms, tropisms, etc.). Second, there is a "potential" type of knowledge that is not species based but rather unique to an individual. It is not determined in advance but can be acquired in the course of one's lifetime, thanks to the possession of the nervous system (the brain) by the organism. Evolution is capable of accumulating the first type of knowledge to a certain extent—but only to a certain extent, since the structure of contemporary mammals reflects a million-year-long "experience" of constructing sea and land vertebrates, which preceded it. It is true at the same time that evolution at times "loses" along the way some rather excellent solutions to biological problems. This is why the design plan for a given animal (or human being) is not just a sum of all the optimal solutions identified so far. Witness the fact that we lack not only the muscle strength of a gorilla but also the regenerative potential of amphibians or of so-called lower fish, or the mechanism of the constant renewal of dentition possessed by rodents, or, last but not least, the universal adaptability to the water environment that characterizes mammals living both on land and in water. This is why we should not overestimate the "wisdom" of biological evolution, since it often led whole species down a developmental blind alley. It tended to repeat not just beneficial solutions but, equally often, erroneous ones that would lead to a decline. Evolution's knowledge is empirical and short term. Its apparent perfection is a consequence of the long stretches of space and time it has traversed. On balance, it has had more failures than successes. Human knowledge is only just emerging—and still not in all disciplines—from the empirical period, with the process being the slowest in biology and probably also in medicine. However, we are already noticing today that if the only thing something needed to be actualized was patience and persistence, interspersed with flickers of intuition, then it has already been accomplished in principle. Everything else—that is, everything that requires the highest clarity of theoretical thinking—is still ahead of us.[6]

The last problem we have to deal with concerns the moral aspects of technoevolution. Its productivity has already attracted severe criticism since it is widening the gap between the two main spheres of our activity: the regulation of Nature and the regulation of Humanity. From this point of view, atomic energy found itself in human hands prematurely. Man's first step into space was also premature, especially since already in the early days of astronautics, it demanded great resources, thus depleting the already unfair distribution of the Earth's global income. As a result of the drop in mortality figures, medical progress has led to a sudden increase in population numbers, which, owing to the lack of birth control, cannot be halted. The technologies that facilitate living are becoming a tool for life's impoverishment because the mass media are turning from their role of a compliant duplicator of spiritual goods to that of a producer of cultural junk. We are told that, culturally, technology is at best barren. I say "at best" because the unification of humanity it promotes takes place at the expense of the spiritual heritage of the past centuries and also at the expense of the ongoing creative efforts. Subjugated by technology, art begins to be dominated by economic laws, showing signs of inflation and devaluation. Above the technical pool of mass entertainment—which has to be easily accessible because general accessibility is the mantra of Technologists—only a handful of creative types survive. Their efforts are focused on ignoring or deriding the stereotypes of mechanized life. Briefly put, technoevolution brings more evil than good, with man turning out to be a prisoner of what he himself has created. The growth of his knowledge is accompanied by the narrowing down of possibilities when it comes to deciding about his own fate.

I believe I have been honest here in reporting, carefully albeit briefly, the position that condemns technological progress.

Can we, or should we, debate this view at all? Should we explain that technology can be put to both good and bad uses? Should we point out that we cannot demand contradictory things of anyone or anything and that this also applies to technology (things such as the protection of life and thus its growth and its decline; an elite culture that would simultaneously be a mass culture; the kind of energy that could move mountains but that would not harm a fly)?

This would probably not be very sensible. Let us say right from the start that technology can be approached in different ways. At first glance, technology is a result of man's and Nature's activity because it enacts that to which the material world gives its silent consent. We will thus perceive it as a tool for achieving various goals, the choice of which depends on the stage of development of a given civilization and on its social system and is therefore subject to a moral judgment. But it is only this choice that is subject to judgment, not the technology itself. So it is not a question of condemning or praising technology but rather of examining to what extent we can trust its development and to what extent we can influence its direction.

Any other type of argument would be based on a (quietly accepted) false premise, one that sees technoevolution as an aberration of development, its false and deadly direction.

Yet this is not true. Indeed, the direction of its development had not been programmed by anyone—either before the Industrial Revolution or after it. This direction ran from Mechanics, that is, the "classic machines," with astronomy conceived in mechanical terms serving as an example for the copyist–designer; through to Heat, with its chemical fuel engines; Thermodynamics; and Electricity. On a cognitive level, it involved a passage from singular to statistical laws, from rigid causality to probability, and, as we only understand it now, from simplicity (which was "artificial" in the sense that nothing in Nature is simple) to complexity (the increase in which has made it clear to us that Regulation will be the next key task).

As we can see, this was a transition from simple to more elaborate, that is, complex, solutions. It is only when we analyze individual steps on this journey—discoveries, inventions—separately and in isolation that we can see them as having resulted from happy coincidences, accidents, and strokes of luck. Taken in its entirety, the journey would surely look extremely probable and typical if we were to compare the Earth's civilization with some hypothetical outer space ones.

We have to see it as inevitable that such generative capacity should reveal, in its cumulative effect after centuries of activity, some undeniably harmful consequences alongside desirable ones.

This is why the condemnation of technology as a source of evil should give way not to an apology but to our understanding of the fact that the preregulation era is coming to an end. Moral rules should accompany our further actions, offering advice on how to make choices from among the alternatives that are presented to us by their producer—amoral technology. Technology provides us with means and tools, but credit or blame for putting them to good or bad use lies with us.

This is quite a common view. It only works as a brief introduction but not as much more than that. The preceding distinction cannot be maintained, especially long term. It is not because we construct technology; it is because technology constructs us and our views, including our views on morality. Of course, it does this through social systems, which serve as its production base, but my current concerns lie elsewhere. Technology can also act—and is indeed acting—directly. We are not used to seeing direct relations between physics and morality—and yet this is the case. Or, at least, it can be the case. To avoid sounding hollow, the moral judgments of acts depend first of all on their irreversibility. If we could resurrect the dead, the killing—while continuing to be seen as a negative act—would stop being a crime, the same way slapping somebody in anger is not a crime. Technology is more aggressive than we normally think. Its influence over our emotional life, issues concerning the synthesis and metamorphosis of our personality—which we shall cover later on—currently constitute only an empty class of phenomena.

This empty class will be filled by future progress. Many moral imperatives that are considered unshakeable today will perish then, while new problems and new ethical dilemmas will emerge.

This would suggest that there is no such thing as transhistorical morality. Phenomena differ when it comes to the scale of their duration, but eventually even mountain ranges collapse and turn into sand because this is the way the world is. Man, a transient creature, keenly uses the concept of eternity. Certain spiritual goods, great artworks, or moral systems are depicted as eternal. Yet let us not delude ourselves: they are all mortal. We are not talking about replacing order with chaos or inner necessity with randomness. Morality changes slowly, but it does change, which is why the greater the temporal separation between two moral codes is, the more difficult it is to compare them. We are close to the Sumerians, yet the morality of the people from the Levallois culture would terrify us.

We shall attempt to demonstrate that there is no system of transhistorical judgment, just as there is no absolute Newtonian frame of reference or no absolute simultaneity of phenomena. This does not mean imposing a ban on making such evaluations with regard to past or future phenomena: man has always made value judgments that transcend his position and actual capabilities. It only means that every period has its own position, which can be agreed or disagreed with but which should be first of all understood.

THE FIRST CAUSE

We are living in the times of accelerated technoevolution. Does this mean that man's whole past, from the last ice age through the Paleolithic and the Neolithic, the ancient times and the Middle Ages, was actually a preparation, a gathering of strength, for the jump that is carrying us today to an unknown future?

The model of a dynamic civilization emerged in the West. It is surprising when, in studying history, we learn how various nations got close to the edge of the "technological start" and how they then stopped at this edge. Contemporary steelmakers could take lessons from the patient Indian craftsmen, who created the famous corrosion-resistant iron pillar in the Qutb complex by using powder metallurgy—which was discovered for the second time only in our times.[7] Everyone knows about the invention of gunpowder and paper by the Chinese. The scientifically indispensable thinking tool, mathematics, owes its development to Arab scholars. Yet all these inventions, revolutionary as they were, did not lead to any kind of civilization push or initiate any rapid progress. Today the whole world is adopting the developmental model of the West. Technology is being imported by nations that can be proud of having more ancient and more complex cultures than those that gave rise to this technology. This raises a fascinating question: what would have happened if the West had not undertaken a technological revolution, if it had not mobilized its Galileos, Newtons, and Stephensons toward the Industrial Revolution?

This is a question about "the first cause." Do the origins of technical evolution not lie in military conflict? The influence of wars as vehicles of technoevolution is well known and widely recognized. Over time, military technology lost the specialist character that used to separate it from the other domains of knowledge and became universal. Whereas ballistae and rams were still just military tools, gunpowder could already be put to industrial uses (e.g., in mining). This is even more the case with transportation technology because there are no means of transport—be it a wheel cart or a rocket—that, on their modification, could not serve some peaceful goals. Atomic, cybernetic, and astronautic technologies, in turn, demonstrate an almost complete integration between their military and peace potentials.

Yet man's belligerent tendencies cannot be considered a driving force behind technical evolution. They usually accelerated it and served as a great application of the resources made available by the theoretical knowledge of the time, but we need to distinguish here between accelerating and initiating factors. All the military tools owe their emergence to the physics of Galileo and Einstein, to eighteenth- and nineteenth-century chemistry, and to thermodynamics, optics, and atomistics, but it would be absurd to look for a military genesis of those theoretical fields of enquiry. The course of technoevolution, once it has started, can without a doubt be accelerated or slowed down. The Americans decided to invest twenty billion dollars in the landing of their first people on the Moon around the year 1969. If they had been prepared to move the date forward by about twenty years, the carrying through of the Apollo program would have surely cost far less because a primitive technology in its early phase claims disproportionate resources—compared with the resources needed to achieve the same goal when the technology has reached its maturity.

However, if the Americans had decided to invest two hundred billion dollars rather than twenty, they certainly would not have landed on the

Moon within six months, just as no matter how many billions of dollars we spend on interstellar travel, it will not be possible to undertake it for some time yet. By investing large sums of money and concentrating our efforts, it is therefore possible to reach the limit of technoevolution's speed, after which any further spending will not have any impact. This statement, which borders on the obvious, corresponds to analogous rules that apply to bioevolution. The latter also knows the maximum speed at which it can evolve, which cannot be exceeded under any circumstances.

But we have inquired about "the first cause" and not about the maximum speed of a process that is already active. Trying to go back to the origins of technology for this purpose is a rather hopeless task; it is a journey deep inside history that only acknowledges facts but that does not explain their causes. Why is it that this massive tree of technical evolution—a tree whose roots may reach as far as the last ice age, whose crown stretches to future millennia, and whose growth during the early stages of our civilization, in the Paleolithic and the Neolithic, was more or less the same all over the world—experienced its proper full bloom in the West?

Lévi-Strauss (1952) attempted to answer this question, but only qualitatively, without any mathematical analysis—the latter approach not being possible owing to the complexity of the phenomenon. He considered the emergence of technoevolution statistically, using probability theory to provide a genetic explanation of it.

The technologies of steam and electricity, then of chemical syntheses and the atom, were initiated by numerous inquiries. Initially independent from one another, those inquiries had traveled a long and occasionally convoluted way, sometimes via routes leading from Asia, to inspire the minds of the Mediterranean area. During the period of several hundred years, a "hidden" accumulation of knowledge took place, which culminated in events such as the abolishing of Aristotelianism as a dogma and the adopting of empiricism as a directive for all cognitive activity; turning a technical experiment into a socially significant phenomenon; or popularizing mechanistic physics. Those processes were accompanied by the emergence of inventions for which social demand already existed. This last phenomenon is extremely important because each nation had some potential Einsteins or Newtons, but they lacked the appropriate soil, the right conditions, or the group resonance that would have amplified the outcomes of their individual activities.

For Lévi-Strauss, a group enters a path of accelerated progress thanks to a particular sequence of successive events. There exists a kind of critical quantity, a coefficient of the "multiplication" of ideas and of their social actualizations (the construction of the first steam machines, the emergence of the coal power industry and of thermodynamics, etc.), which eventually leads to a rapid increase in discoveries conditioned by that multiplication. In the same way, there exists a critical quantity of the coefficient of the "multiplication" of neutrons, which leads to a chain reaction in the mass of a heavy metal after a certain threshold has been crossed. We are now living through a civilization's equivalent of such a reaction, or even perhaps through a "technological explosion," one that is undergoing full expansion. Whether a given society enters such a path or starts off a chain reaction is, according to the French ethnographer, purely accidental. From a probabilistic point of view, just as a player rolling dice can expect to have a sequence of all sixes provided he plays long enough, every social group has, at least in principle, an equal chance of entering the path of rapid material progress.

We must point out that Lévi-Strauss's concerns were different from ours. He wanted to demonstrate that even the most dissimilar civilizations, including nontechnical ones, were equally legitimate and that we should not apply value judgments to them, that is, consider some of them to be "higher" than others only because they had been lucky enough in the previously named "game," as a result of which they arrived at the starting point of the chain reaction. This model is beautiful thanks to its methodological simplicity. It explains why singular discoveries, even if they are significant, can become suspended in a vacuum when it comes to their technogenerative social effects—as was the case with the powder metallurgy of the Indians or the gunpowder of the Chinese. Subsequent links that were needed to start a chain reaction were not there. This hypothesis clearly demonstrates that the East was simply a "less lucky" player than the West, at least when it came to technological leadership. It thus follows logically that if the West had been absent from the historical scene, then the East would have taken its path sooner or later. Let us argue about the correctness of this proposition later on. For now, we are going to focus on the probabilistic model of the emergence of technical civilizations.

Going back to our major analogy, biological evolution, we have to recognize that in the course of evolution, varieties, species, and families emerged on separate continents, sometimes in a parallel way. We can establish a correspondence between the herbivores or predators of the

Old World and some forms in the New World that are not related to the former but that had been shaped by evolution in a similar way, because evolution had acted on their ancestors through similar environmental and climactic conditions. The evolution of phyla, in turn, tended to be monophyletic, at least according to a great majority of experts. Vertebrates emerged only once within the scope of the Earth, fish too, and so did amphibians, reptiles, and mammals. This raises several questions. As we can see, a substantial transformation of the corporeal organization, this kind of "designer's feat," only happened once on the scale of the planet.

We can see this phenomenon too as being subject to statistical laws: the emergence of mammals or fish was of such low probability that a similar kind of "jackpot," one that required much "luck" and the convergence of many causes and circumstances, was an extremely rare phenomenon. And the rarer the phenomenon, the less of a chance there is of it being repeated. There is also another common characteristic that can be observed in both types of evolution. Both have produced higher and lower forms, more complex and less complex ones, which *have survived up to this day.* On one hand, fish certainly appeared before amphibians, which in turn preceded reptiles, but today we have representatives of all these classes. On the other hand, the family and the tribal system preceded slavery and feudalism, which then preceded capitalism. Even if they are not all simultaneously present anymore, until very recently, all those different systems, including the most primitive ones, certainly coexisted in the world. The remnants of the latter can still be discovered in some archipelagoes of the southern seas.

When it comes to bioevolution, this phenomenon can be easily explained: a change is always caused by a need. If there is no environmental demand, if the environment allows for the existence of single-cell organisms, they will then keep producing further generations of protozoa for the next hundred or five hundred million years.

But what causes the transformation of social systems? We know that a change to the means of production, that is, to technologies, is its driving force. We are thus back to square one as it is clear that systems do not change if they keep using traditional technologies, even if those technologies date back all the way to the Neolithic.

We shall not solve this problem once and for all. Yet we may conclude that the probabilistic hypothesis of the "chain reaction" does not take into account the particularity of the social structure in which such a reaction is to occur. Systems with very similar production bases can show considerable differences in their cultural superstructures. The abundance of sophisticated social rituals, of sometimes painfully complicated behavioral norms of family life, tribal life, and so on—norms that have become accepted and that are being rigorously imposed—is accidental. The anthropologist, fascinated as he is with myriads of those inner dependencies among various civilizations, needs to be replaced by the sociologist–cyberneticist. Consciously ignoring the inner cultural and semantic meanings of all such practices, the latter will examine their structures as if they constituted a feedback system that aims at an ultrastable equilibrium, and whose dynamic task lies in regulation aimed at ensuring the perpetuation of this equilibrium.

It is highly probable that some of those structures, of those mutually dependent human relations, can successfully impede any scientific and technological progress through the restrictions they impose on the freedom of thought and action. It is also probable that there exist some other structures that—even if not supporting such inventiveness—can at least open a space, however limited, for it. It goes without saying that the basic characteristics of European feudalism were surprisingly similar to the feudalism of nineteenth-century Japan. And yet both models of this same system—the Asian and the European—also showed some specific differences, which played a secondary or even tertiary role in the actual social dynamics. Those differences nevertheless resulted in the fact that it was Europeans rather than the Japanese who destroyed feudalism with their new technology and who then constructed the foundations of industrial capitalism on its ruins.[8]

From this perspective, a technological chain reaction is seen as having been initiated not by a sequence of *uniform* occurrences (e.g., by subsequent discoveries of a certain kind) but rather by an overlap between two sequences of events, the first one of which (i.e., a cybernetically understood superstructure) has a mass statistical character of a higher kind than the second one (i.e., the emergence of empirical and technical interests in individuals). These two sequences must cross for the possibility of technoevolution to arise. If this encounter does not take place, then the Neolithic civilization may turn out to be the ultimate threshold that cannot be crossed.

This brief outline is certainly full of simplifications, but it is only through further research that the issue will become clearer.[9]

SEVERAL NAÏVE QUESTIONS

Every sensible person makes plans for the future. That person has freedom of choice—within certain limits—with regard to his education, career, and way of life. He can change his job and even, to some extent, his behavior, if he decides to do so. We cannot say the same about a civilization. At least until the end of the nineteenth century, civilization was not part of anyone's plan. It emerged spontaneously, speeding up through the technological jumps of the Neolithic era and the Industrial Revolution and coming to a standstill for thousands of years. Some cultures grew and waned, others formed over their ruins. A civilization "does not itself know" at which particular moment its development enters a path of increased acceleration as a result of a series of scientific discoveries and their social exploitation. This development is manifested in an increase in homeostasis, in the growth of energy used, and in the ever more efficient protection of the individual and the collectivity against all sorts of threats (diseases, natural disasters, etc.). It allows for the subsequent taming of the natural and social forces through regulatory activity, but at the same time, it takes control over, and gives shape to, the human fate. A civilization does not act the way it wants but rather the way it has to act. Why should we be developing cybernetics? Because, among other things, we are probably going to hit against "the information barrier" soon. This will hold back the development of science, unless we bring about a revolution of minds on a scale similar to that which has taken place in the domain of physical labor over the last two hundred years. Oh, so this is the case! We will therefore not be doing what we want but rather what the current stage of civilization's dynamics requires of us. A scientist will say that this is where the objective activity of a developmental gradient manifests itself. Yet cannot a civilization gain the freedom of choice with regard to its future path the way an individual can? But what conditions would need to be fulfilled for this freedom to take place? Society must become independent from the technologies of elementary problems. Basic issues that each civilization has to deal with—food, clothing, transportation, but also the initiation of life, the distribution of goods, the protection of health and property—have to disappear. They have to become invisible, like air—the abundance of which has so far been the only excess in human history. No doubt this can be achieved. But this is only a preliminary condition because it is only then that the question "What now?" will present itself with full intensity. Society gives the meaning of life to an individual. But who or what gives this meaning, or specific life content, to a civilization? Who decided about its hierarchy of values? It does this by itself. This meaning, or content, comes from it as soon as it enters the realm of freedom. How can we imagine this freedom? This is, of course, freedom from disasters, from poverty, from misfortunes—but does this lack, this absence of familiar inequalities, of unsatisfied yearnings and desires, signify happiness? If that was to be the case, such a worthy ideal would be reached by a civilization that consumes the maximum amount of goods it can produce. Yet there is widespread doubt about the possibility of such a consumer paradise on Earth making us happy. It is not that we should be consciously aiming at asceticism or proclaiming a new version of Rousseau's "return to nature." This would be not only naïve but also stupid. Such a consumer "paradise," with its immediate fulfillment of all wishes and whims, is likely to lead rather quickly to spiritual stagnation and the kind of "degeneracy" to which von Hoerner, in his statistics of extraterrestrial civilizations, ascribes the role of the "extinguisher" of psychozoics.[10] But if we are rejecting this false ideal, with what are we left? A civilization of creative labor? Yet we are doing everything we can to mechanize, automatize, and self-activate any kind of work. The marker of progress lies in the separation between man and technology, or the total alienation of the latter in a cybernetic sense, that is, also with regard to the mental sphere. Apparently it will only be possible to automatize noncreative mental labor. But where is the evidence? Let us be clear about it: there is none, and what is more, there can be none. Such a groundless statement of "impossibility" has no more value than the biblical claim that man shall always eat bread in the sweat of his face. It would indeed be a unique form of consolation if we were to claim that we will always have something to do—not because we consider work a value in itself but because the very nature of the world in which we live is coercing us, and will always be coercing us, to work.

Conversely, how can man be doing something that can be done the same way, or even better, by a machine? Today he is acting like this out of necessity because our world is constructed in an exceptionally imperfect way, and on many continents, human labor is cheaper and economically more viable than machine labor. But we are talking about future prospects, and very remote ones at that. Are people supposed to say at some point, "Enough! We will not be automatizing such and such areas of work anymore, even though we can; we shall stop Technology to save man's labor so that he does not feel redundant"? This would be a peculiar kind of freedom and a peculiar way of using it, after centuries of fighting for it.

Such questions, with all their apparent matter-of-factness, are actually very naïve because we shall never gain freedom in any absolute sense—either as an absolute freedom of choice with regard to our actions or as a freedom from all action (resulting from "total automatization"). We shall not gain the first type of freedom because what seemed like freedom yesterday is not the same anymore today. The state of having been liberated from the compulsion to undertake actions that have to satisfy elementary needs will allow us to select a further option, but this will not be a unique historical event. The situation of choice will be repeated at the higher levels we subsequently reach. Yet it will always be a choice from among a limited number of options, and thus the freedom acquired each time will be relative—because it seems impossible for all the constraints to depart man suddenly, leaving him just with the omnipotence and omniscience he has at last gained. The other, undesirable type of freedom is also a fiction. It is an alleged consequence of the full alienation of Technology—which, through its cybernetic power, is supposed to create a synthetic civilization that will remove mankind from all domains of activity.

The fear of unemployment as a consequence of automatization is justified, especially in highly developed capitalist countries. Yet we cannot consider as justified the fear of unemployment resulting from "excessive consumer prosperity." The vision of the cybernetic Land of Cockaigne[11] is false because it assumes the replacement of human labor by machine labor that forecloses all options for man, whereas in fact, the opposite is the case. Such a replacement will no doubt happen, but it will open up some new possibilities, about which we can only have a rather faint idea today. It will not happen in the narrow sense of laborers and technicians being replaced by programmers of digital machines because future generations and future types of these machines will not need programmers anymore. So this will not be a mere swap of one category of old jobs for another—that of new and different jobs, yet essentially similar to the old ones—but a fundamental transformation, perhaps equaling the transformation of anthropoids into humans. Man cannot enter into direct competition with Nature because Nature is too complex for him to measure himself against it on his own. Figuratively speaking, man has to build a whole sequence of intermediaries between himself and Nature, every one of which will be more powerful in its role as an amplifier of the Intellect than the previous one. This is a way of enhancing thinking rather than might, of enabling control over the material properties of the world in the future—properties that are inaccessible to the human mind. Surely these intermediaries will be in some sense "cleverer" than their human designer—but "cleverer" does not yet mean "rebellious." We shall also, by way of speculation, consider the domains in which man's enhanced activity of this kind will match Nature's work. Even then will man remain subject to limitations, the material aspect of which—conditioned as it will be by the technology of the future—we cannot predict but the psychological effects of which we can at least partially grasp because we are ourselves human. The thread of such understanding will only be broken when man, in a thousand or a million years' time, gives up his entire animal heritage, his imperfect and impermanent body, for the sake of a more perfect design, and when he turns into a being so much higher than us that it will become alien to us. Our preview of the future will thus have to stop at sketching out the beginnings of this autoevolution of the species.

3

CIVILIZATIONS IN THE UNIVERSE

THE FORMULATION OF THE PROBLEM

How exactly have we been searching for a direction in which our civilization is headed? By examining our civilization's past and present. Why have we been comparing technical evolution with biological evolution? Because the latter is the only process of improving the regulation and homeostasis of very complex systems that is available to us. This process remains free from human intervention—which could distort the results of our observation and the conclusions drawn. We have acted like someone who, in trying to find out about his future and the possibilities that await him, studies himself and his environment. And yet there is another way, at least in principle. A young man can decipher his fate from the fate of other people. Observing them, he will find out what paths lie ahead of him, what choices await him, and what limitations his choices are constrained by. Observing the mortality of Nature's creations—mussels, fish, plants—a young Robinson on a desert island would have perhaps guessed his own limitedness in time. But he would have learned more about the possibilities that await him from the lights and smoke of far-away ships or from the planes flying high above the island. He would have deduced from them the existence of a civilization that was created by beings similar to him.

Humanity is such a Robinson, marooned as it is on a lonely planet. Its inquisitiveness has certainly been put to a more stringent test by the living conditions, but is such a test not worth undertaking? Seeing some signs of cosmic activity coming from other civilizations would teach us something about our own fate. If we were to succeed at this, we would not have to rely just on guesswork, based on our limited terrestrial experience: facts drawn from across the Universe would provide us with a large plane of reference. Additionally, we would be able to map our place on the "curve of the factorization of civilization." We would thus be able to find out whether we are an average or a marginal phenomenon, whether we are something ordinary, a developmental norm within the Universe, or just an aberration.

We suspect that only a few years—a few dozen at most—separate us from obtaining information about biogenesis on the scale of the solar system. The probability that some other highly developed civilizations exist within it is almost zero. Attempts to let the inhabitants of Mars or Venus know about our existence, so common at the end of the nineteenth century, are not being undertaken at present—not because they would be impossible but rather because they would be futile. Either there are no forms of life on those planets, or, if they do exist, they have not created any technologies. Otherwise they would have by now discovered our existence—which can be observed on a planetary scale thanks to shortwave radiation. The presence of TV sets already makes the Earth's radio emission within the range of VHF waves (which can freely cross the atmosphere) match the total emission of the Sun within this range.

Any civilization within the solar system that would be able to compete with ours on the level of development would therefore notice our existence and would no doubt make contact with us—via light, radio, or matter. But there are no such civilizations within it. This problem, fascinating as it is, does not concern us at the moment because we are not inquiring about civilizations in general but only about civilizations that have exceeded the Earth's level of development. It is only from them, from their existence, that we could draw conclusions about our own future. The answer we would get, based on our observation of outer space, would make the majority of our current speculative analyses entirely futile. A Robinson who would be able to communicate with other intelligent beings, or at least observe their activity from afar, would not have to suffer the uncertainty of complicated guesswork anymore. Naturally, there is something dangerous about such a situation. Any too explicit and too definite answers would show us that we are slaves to developmental determinism rather than creatures exposed to ever greater freedom—which stands for an unlimited freedom of choice. The latter would be the more illusory the more convergence there would be between the paths of the different galaxies within a given group.

The idea of starting a new chapter in our investigations, one extended to the whole of the Universe, is thus appealing as well as dangerous. What distinguishes us from "lower beings," that is, animals, is not only our civilization but also an awareness of our own limitations—the greatest of which is mortality. We have no idea in what way beings higher than us exceed us. In any case, we must emphasize that what interests us are facts and their interpretation in accordance with the scientific principles, not fantasy work. This is why we shall not be taking into account any of those countless "futures" that have been prophesized for the Earth, and for some other celestial bodies, by writers who occupy themselves with this rather prolific genre called "science fiction." As we know, it is not literature's habit to use scientific methods or apply mathematical and methodological principles or probability theory—not even when it comes to science fiction literature. I am not saying this to accuse science fiction of betraying the scientific truth, only to emphasize how important it is for us to distance ourselves from any kind of arbitrariness. We shall be drawing on observational material from astrophysics and on the method followed by the scientist—which has very little in common with the artist's method. It is not because the latter would be more inclined to take risks than the former but only because the scientist's ideal—that is, the complete isolation of what he aims to represent from the world of his own experiences, the purification of the objective facts and conclusions from any subjective emotions—is foreign to the artist. In other words, the closer one gets to being a scientist, the more one silences one's humanity to allow Nature itself to speak, as it were. The more of an artist one becomes, in turn, the more this artist imposes himself on us, with all the greatness and frailty of his unique being. The fact that we do not encounter any pure positions of this kind testifies to the impossibility of their full enactment because every scientist has something of an artist about him, just as every artist is also a little bit of a scientist. But we are talking about the direction of development and not about some inaccessible final point.

THE FORMULATION OF THE METHOD

Scholarly works devoted to the area discussed previously have grown in number over recent years, but they are mainly available in specialist journals, which is why they are rather difficult to get hold of. This gap is filled by the book by the Russian astrophysicist I. Shklovsky titled

Universe, Life, Intelligence (1962). As far as I know, this is the first monograph that deals with the problem of extraterrestrial civilizations, whereby questions of their existence and development, of the possibility of communication between them, and of the frequency of their appearance in our Galaxy and in other stellar systems do not just appear on the margins of a cosmological treatise proper but are in fact the book's principal subject. Unlike the other experts, Professor Shklovsky deals with this problem at the greatest scale possible, only devoting one chapter of his work to biogenesis in the solar system. His book is especially significant since it presents the views and calculation results obtained from a large number of astronomers—radio astronomers, in particular. They have all used probabilistic methods to calculate the civilization "density" of the Universe and have attempted to match the results of their studies with current observations and theories.

Given our present concerns, we shall draw on the extensive material cited by Shklovsky only as far it relates to the problem of "cosmic technoevolution." We shall also discuss certain initial premises on which English, American, and German scholars have based their calculations. We are justified in doing so as those premises are to a large extent arbitrary and hypothetical.

Astronomy today is incapable of determining, either directly (e.g., in a visual way) or indirectly, the existence of planets around stars—unless they are the nearest stars, while the mass of those planets' bodies exceeds that of Jupiter. Only then can the existence of such bodies, situated dozens of light-years away, be deduced from the perturbations to stellar trajectories. It may sound surprising, to say the least, that under such circumstances, one can actually speak about the apparently scientific results of the search for "alien civilizations." Yet it is hard not to accept at least some basic elements of the reasoning that provides the foundation for scholarly work of this kind.

The existence of "aliens" in outer space can be confirmed in two ways: first, by receiving some signals emitted by those "aliens" (radio, light, or material signals, such as "alien" rocket probes), and second, by seeing some "miracles." Shklovsky uses this latter term to describe impossible phenomena, that is, phenomena that are inexplicable from an astronomical point of view, just as a highway crisscrossing the landscape of a planet is inexplicable from a geological point of view. And just as a geologist would deduce from its presence the existence of some intelligent beings that must have constructed it, an astronomer—on having discovered deviations from the expectations dictated to him by his knowledge, deviations that cannot be explained in any "natural" way—would have to conclude that there exist some signs of intentional activity in the field of vision of his apparatus.

"Miracles" would not thus be a form of purposeful signaling, intended to notify potential observers about the existence of life in outer space. They would just be a side product of the existence of a highly developed civilization, in the same way a glow lighting up the sky for miles accompanies the existence of a large metropolis. A simple calculation shows that such phenomena can be seen from dozens, if not hundreds, of light-years away, thanks to the expenditure of energy, which would equal stellar power. Simply put, only the signs of "astroengineering" are astronomically observable.

Almost all authors (Dyson, Sagan, von Hoerner, and Bracewell as well as Shklovsky himself) claim that the emergence of "astroengineering," in one form or another, at a particular stage of development is absolutely certain. If we accept that the energetics of the Earth is going to grow annually by 1/3 percent (a modest estimate, given its present growth), then in twenty-five hundred years, the global production of energy will exceed the current level ten billion times, reaching in the year 4500 energy equal to 1/10,000 of the entire solar power. Even if we were to convert hydrogen from the oceans into energy, it would only last for several thousand years. Astrophysicists can see a number of options here. Dyson postulates using up all the solar power through constructing a "Dyson sphere," that is, an empty sphere with very thin walls, whose radius would equal that of the Earth's orbit around the Sun. The building material would be provided by the large planets, principally Jupiter. The inner surface of this sphere, which would be turned toward the Sun, would receive all the solar emissions (4×10^{33} ergs per second). Shklovsky also considers the possibility of using solar energy in a different way or even of influencing the course of nuclear transformations inside the Sun according to the demands set by the engineers of the future. It goes without saying that we do not know whether energy use will continue to increase significantly in future millennia, but we do know today the potential recipients of such massive amounts of energy: the only conceivable vehicle for star and intergalaxy travel (the duration of which would equal man's lifetime), the photon rocket, would require this amount of power. This is, of course, just an example.

Since the Sun is quite an average star in many ways, also when it comes to its age, we can assume that the number of stars that are similar to it yet older, which belong to families of planets, is more or less the same as the number of stars that are younger than the Sun. This leads to the conclusion that the number of extraterrestrial civilizations that are more advanced in their development than us should be the same as the number of those that are lagging behind us.

The argument based on the conviction about our average standing has so far proved infallible: the very positioning of the Sun within the Milky Way is "average" (neither on its edge nor too close to its center), and the Milky Way itself, that is, our Galaxy, is a typical spiral galaxy, like billions of others that are represented in the massive catalog of nebulae. This is why we are justified to see Earth's civilization as typical, ordinary, and common.

Bracewell and von Hoerner have conducted independent statistical calculations of "civilization density" in the Universe, starting from the premise that only 1 out of 150 stars in our Galaxy has planets. As the Galaxy contains around 150 billion stars, there should be around a billion planetary systems orbiting within it. This is a rather modest estimate. If each of the billion planets were to experience an evolution of life at some point that would, after a certain time, reach a "psychozoic phase," the calculations show that, should the duration of this phase (i.e., of the technological era) only depend on the lifetime of home suns (i.e., if an average civilization could last for as long as it receives the energy it needs to maintain life from its star), then the average distance between two civilizations would be less than ten light-years.

This conclusion, mathematically flawless, does not find confirmation in the facts. With such civilization density, we should already be receiving signals from the star's vicinity—and not just via special instruments of the kind that have been used since 1960 by a group of astronomers headed by Drake in the Green Bank Observatory in the United States. Their instruments are capable of receiving signals of an efficiency as high as that which the transmitters from Earth would be capable of sending today, from a distance of ten light-years. Of course, the American radio telescope would have received signals even from a distance a hundred times greater, if only an adequately strong signal had been sent in the direction in which its twenty-seven-meter-long antenna was pointed. The silence of the instruments therefore immediately reveals the obvious fact of a "civilization vacuum" around the Epsilon Eridani and Tau Ceti stars but also the absence of any stronger signals traveling from within the Universe beyond these stars in our direction. A group of scientists led by Drake undertook the first attempt in the history of astronomy to conduct an interstellar "search for extraterrestrial civilizations," taking up an idea put forward by two other American astronomers, Cocconi and Morrison. The scientists used equipment that had been specially constructed for receiving "artificial" signals, thus enabling their differentiation from "galaxy noise"—as the whole of the Milky Way, that is, both its stars and interstellar matter, generates radio waves. This was a scientific experiment, which involved searching for some kind of regularity in the radio waves that reach us, a regularity that would mean that a packet of transmitted waves was being modulated, that is, that it was an information carrier sent by intelligent beings. This was the first but certainly not the last attempt, even though the astrophysicists' expectations had not been fulfilled. Day after day, week after week, their instruments registered nothing but the monotonous cosmic noise generated by dead matter.

THE STATISTICS OF CIVILIZATIONS IN THE UNIVERSE

As we have already said, if we assign to civilizations in the Universe a life span equal to that of their home stars—which practically means that once-formed civilizations will have to exist for billions of years—this will inevitably lead to the conclusion regarding the "civilization density" of the Universe, whereby any two inhabited worlds will have to be only several light-years away. This conclusion is contradicted by the entire body of observations, encapsulating the negative results of the radio search in the Universe, an absence of any other types of signals (e.g., "alien" rocket probes), and last but not least, a total absence of "miracles," that is, phenomena caused by astroengineering activity. This state of events has led Bracewell and von Hoerner, as well as Shklovsky, to accept the hypothesis about the short life span of civilizations when compared with the longevity of stars. If an average life span of a civilization is "only" one hundred million years, then the most probable distance between any two civilizations in statistical terms (as a result of the inevitable separation of their existence in time) is about fifty light-years. This also sounds extremely doubtful. Consequently, the preceding authors are more inclined to accept the hypothesis that estimates the average life span of a civilization at less than twenty thousand years. If this is indeed the case, then any two highly developed worlds are separated by a thousand light-years—which finally makes the failure of our attempts to search for and find them understandable.

And thus the greater the number of planets in the Galaxy that we perceive as being capable of biogenesis (which leads to the emergence of the "psychozoic"), the shorter the average life span of a particular civilization we must posit—if we are not to contradict the observations. It is currently accepted that out of 150 billion stars in the Galaxy, about a billion have planets capable of generating life. Yet even if we were to decrease this number tenfold, this would not change the results of probabilistic calculations in any significant way. This seems completely inexplicable: since the evolution of life in its precivilization form takes millions of years, it is difficult to understand why, after such a wonderful start, the "psychozoic" is then supposed to become extinct just after several dozen centuries. As soon as we realize that even a million years is only a tiny fraction of the time period during which an average civilization could be developing further (since its home star provides a constant supply of radiant power for many billions of years), we shall be able to accept the mysteriousness of this phenomenon, the explanation of which is not satisfying our curiosity at the moment.

In the light of our deliberations here, intelligent life seems to be a rare phenomenon in the Universe. Importantly, we are not talking about life in general, but about a life that is parallel to our own existence, since what concerns us is not the myriad civilizations that have appeared and disappeared during the whole life span of the Galaxy (the period of around fifteen billion years) but rather those civilizations that are cotemporal with us.

Taking the transience of "psychozoics" as a fact that needs to be explained, von Hoerner lists four possible reasons for it: (1) extinction of all life on the planet; (2) extinction of only highly organized beings; (3) mental or physical degeneracy; or (4) loss of scientific and technical interest.

Having assigned an arbitrarily chosen coefficient of probability to each of these causes, von Hoerner posits the figure of sixty-five hundred years as an average life span of a civilization and the figure of one thousand light-years as a distance between civilizations. His calculations also lead him to conclude that the most probable age of a civilization with which we shall establish our first contact will be twelve thousand years. The probability of the (first) contact with a civilization that matches the Earth's stage of development is rather slim: it is only 0.5 percent. Among other things, von Hoerner takes into account the possibility of the repeated emergence and disappearance of civilizations on the same planet.

In light of these results, it is easy to understand the lack of success of the American search for extraterrestrial civilizations. Information exchange does not look very likely either, should we actually manage to receive any signals, if, after posing our questions, we will have to wait two thousand years for an answer...

Von Hoerner considers the possibility of receiving some "positive feedback," if, given the statistical character of the distribution of life in the Galaxy, a local cluster of extraterrestrial civilizations was to emerge in it. When the period of waiting for a reply becomes (in such a local "crowd of psychozoics") a relatively small fraction of the whole of a given civilization's life span, a successful exchange of information between civilizations will be able to take place. This, in turn, will prolong their life span (thanks to the exchange of experiences).

Shklovsky compares this process to a rapid reproduction of organisms in a propitious environment. If it was to take place somewhere in the Galaxy, such a process, covering ever greater areas, could draw into its orbit an increasing number of galactic civilizations. This would lead to the formation of a "superorganism." What is actually most surprising, and, to be honest, completely inexplicable, is that this possibility has not been actualized yet. Let us just assume briefly that von Hoerner's catastrophic hypothesis applies widely across the Universe. The statistical character of this supposed rule makes the existence of a small handful of long-lived civilizations (rare as they may be) highly possible. To assume that no civilization can exist for up to a million years would amount to transforming a statistical rule into fatalistic determinism, a demonic inevitability of rapid extinction. And even if this was indeed the case, then at least several out of those long-lived, million-year-old civilizations should have conquered the star fields a long time ago—fields that are very far away from their home planets. In other words, a handful of those civilizations would have become a decisive factor in the development of a galaxy. The postulated "positive feedback" would then have become a reality. As a matter of fact, such feedback should have already been fully operational for centuries. Why, then, are there no signals from such civilizations? Why are we seeing no signs of their gigantic astroengineering activity? Why are there no countless information-gathering probes produced by them that would then populate the vacuum; no self-propagating machines penetrating the most remote corners of our stellar system?

Why, in other words, are we seeing no "miracles"?

A CATASTROPHIC THEORY OF THE UNIVERSE

The Milky Way is a typical spiral galaxy; the Sun a typical star; the Earth most probably a typical planet. But to what extent can we extrapolate civilization's phenomena that take place on Earth to the whole of the Universe? Are we really supposed to believe that, when looking at the sky above us, we are seeing an abyss filled with worlds that have already been turned to ashes by the power of their suicidal intelligence or that are headed directly toward such an end? This is what von Hoerner claims, assigning a 65 percent chance to the hypothesis about the "self-destruction of psychozoics." Once we realize that there are billions of galaxies similar to ours, and assume that, owing to the similarities in their building material and dynamic laws, similar planetary and psychozoic evolutions take place in all of them, we are faced with a trillion civilizations that develop only to become extinct after a short period of time—which, on an astronomical scale, amounts to a blink. I cannot accept such a statistical hell—not because it is too terrifying but because it is too naïve. We must not therefore accuse von Hoerner's hypothesis that the Universe is a machine generating swarms of atomic slaughter-houses of catastrophism, and it is not through moral repugnance that we should reject it because emotional reactions must not play any role in scientific analysis. Yet this hypothesis assumes quite an unlikely similarity between the trajectories of different planets. We do not believe that the Earth, with its bloody history of wars, and man, with all the immoral and dark aspects of his nature, constitute some kind of cosmic exception or that outer space has been populated by beings who are more perfect in their nature than us. However, an extrapolation of examined processes to unexamined ones—so useful in cosmology, astronomy, and physics—can easily turn an attempt to practice metagalactic sociology into its own reductio ad absurdum.

By way of example, let us point out that the fate of the world could have been entirely different if the genocidal politics of the Third Reich had excluded German Jews from its extermination plans, or at least if Hitler's dictatorship had recognized the value of certain physics experiments and the possibility of developing the "miraculous weapons"—so desired by the German rulers—from them. This could have happened as a result of some kind of "prophetic dream," of the kind Hitler used to have. At last, Einstein would not have had to be Jewish. In any case, one can easily imagine a situation in the 1940s in which the resources of the Nazi state would have been directed toward research into atomic energy. The German scientists would have surely hesitated before putting hydrogen bombs into the hands of the fascists, but we know from elsewhere that scruples of this kind can be overcome. (Given all the doubts regarding the accusations aimed at Heisenberg after the war, when looking into this matter more thoroughly, one cannot resist the feeling that he was actually trying to build the first microfusion cell and that this was not just connected with his academic ambitions.) The turn of the events was, as we know, quite different: the atomic bomb was first produced by the Americans—with the hands and minds of the emigrants from the Third Reich. If those individuals had remained in Germany, Hitler would have perhaps got hold of this terrible weapon about which he dreamed. We shall not be proceeding any further with such groundless speculation: the idea was to show how a given conjuncture of events led to Germany's rapid defeat and to the emergence, over its rubble, of the two remaining potential opponents: socialism and capitalism. Regardless of whether the Germans would have been able to gain world domination thanks to their nuclear lead, the nuclear factor as an important aspect of the technology of war would have changed the planet's equilibrium. Perhaps a whole number of wars would have taken place, which would have left humanity decimated but united. These suppositions, futile and irrelevant if seen as just a form of speculative armchair philosophy, gain significance when we extrapolate them to the Universe, since the emergence of one big hegemonic leader in the historical process of the unification of fragmented collectivities can occur as frequently as the coemergence of two equal antagonists can. The modeling of socioevolutionary processes in digital machines, which will be possible in the not-too-distant future, is likely to throw some light on this matter. I am referring here to the phenomenon of the planetary unification of societies, whose mutual antagonisms and isolationisms will be eliminated by the progression of technoevolution. Given that it is easier to tame Nature than to carry out acts of social regulation on a global scale, the overtaking of socioevolution by technoevolution may actually be a typical dynamic feature of such processes. It is difficult to accept that the possibility of delaying the regulation of social forces, as compared with the regulation of the forces of Nature, would always have to remain constant in the Universe and would therefore be a fixed quantity for all possible civilizations. Yet the scale of this delay—which, as an important parameter, is part of the Earth's social phenomena—has shaped the ongoing process of the planetary unification of humanity in a way that has led to the simultaneous emergence of two great antagonistic coalitions. Without even going so far as to say that this type of development does not necessarily have to end in total extinction, we are presumably allowed to believe that in a considerable number of "worlds" (remember that we are talking about models!), the distribution of forces will be so different from the one we have on Earth that the possibility of mutually destructive combat between opponents will even not arise. Such combat can also be less dramatic: after an initial decline caused by it, a unification of all the societies of the "planet" may take place.

What then? Then—a supporter of von Hoerner's hypothesis will say—some other factors will come into play that will shorten technological duration. For example, some "degenerative" tendencies will manifest themselves owing to the no doubt hedonistic and consumerist nature of the goals toward which a large part of the world is headed today. We shall speak about the possibility of the "hedonistic taming" of development later on, and also about the possible cyclical halting of "technological acceleration." But those other causes are only afforded a 35 percent chance by von Hoerner. We have in turn presented a definite possibility, one rooted in mathematics and modeling, of theoretically repudiating von Hoerner's hypothesis about self-destruction as an existential rule for the majority of civilizations in the Universe. Even if von Hoerner was closer to the truth than we think, the statistical aspect of his "laws" must allow—as we have already said—for some exceptions, precisely because of the probabilistic character of those laws. Imagine that 990 million planets out of their galactic billion are inherently characterized by the short duration of their technological eras. Imagine that out of the remaining ten million, only one hundred thousand, or even just one thousand, escape "the law of the impermanence of civilizations." Then that one thousand planets will develop their civilizations for hundreds of millions of years. In that case, we will be faced with a unique cosmic analogue of the Earth's bioevolution: as this is how its activity manifests itself. The number of animal species that became extinct during evolution is incomparably higher than the number of those that survived. Yet each species that survived gave rise to a great number of new ones. This is what allows us to postulate the existence of such "evolutionary radiation"— not biological but rather working at the scale of the Universe and civilization. Our hypothesis does not even need to include any "pastoral" elements. Sure, we can imagine those billion-year-old civilizations making contact with one another during their stellar expansion to fight, but in that case, their wars will look to us like disappearing acts of whole constellations, like enormous eruptions caused by beams of destructive radiation, or like "miracles" of some kind of astroengineering activity (either peaceful or destructive).

And thus we are returning once again to the question we posed earlier: why are we not seeing any "miracles"? Please note that in the preceding paragraph, we were prepared to accept an even more "catastrophic," as it were, idea of the development of civilizations than the one put forward by von Hoerner. He does not just claim that civilizations in the Universe destroy themselves but also that they do this in their developmental phase—a phase that is rather similar to the one reached by humanity (in that it is astronomically invisible). It seems to me that we are not applying probabilistic methods to sociogenetic phenomena anymore but simply dressing modern man's anxieties (a designation that, of course, also applies to the learned astrophysicist) in the masks of cosmic universality.

Astrophysics is unable to provide us with an answer to the question posed. Let us then search for this answer elsewhere.

A METATHEORY OF MIRACLES

What form could these "miracles," briefly described by us as manifestations of astroengineering activity, actually take? Among some "possible miracles" of this kind, Shklovsky lists artificially caused supernovae explosions or the presence of the spectral lines of the element technetium in the spectra of certain rare stars. Since technetium does not appear naturally (we create it artificially on Earth)—and in fact cannot appear naturally, as it disintegrates rapidly (within several thousand years)—we can conclude that its presence in the star's radiation can be triggered by nothing less than "sprinkling" it over the star's fire, a procedure that would, of course, have to be performed by astroengineers. Incidentally speaking, to visualize the spectral line of an element in stellar emission, an astronomically insignificant amount of this element is needed—some several million tons.

This hypothesis, together with the hypothesis about "the artificial explosion of supernovae," is presented by Shklovsky partly as a joke. Yet the reason for this is quite serious. One of the fundamental methodological principles in science is "Occam's razor," that is, a theory that postulates that *entia non sunt multiplicanda praeter necessitatem.* In constructing hypotheses, one must not multiply "entities" beyond what is necessary. "Entities" are understood here as basic concepts that have been introduced into the theory, concepts that are irreducible to any others. This principle is observed so widely that it is actually difficult to notice its presence in every scientific argument. A new concept can be introduced into the theoretical model of reality only in extraordinary circumstances: when the few theses that constitute the foundation of our knowledge are being jeopardized. When, in certain instances of atomic disintegration, the law of conservation of mass was put in jeopardy (it seemed that some of the mass was "disappearing" without trace), Pauli introduced the concept of the "neutrino"—initially a purely hypothetical particle whose existence was only later proved in an experiment—to save this law. Occam's razor, or the principle of the economy of thought, requires the scientist to attempt to explain every phenomenon in the simplest way possible, without introducing any "additional entities," that is, unnecessary hypotheses. The wide use of this principle results in a drive to unify all the sciences: this drive manifests itself in explaining diversity by constantly reducing it to elementary concepts such as those used by physics. Various sciences have at times opposed such reductionism: for example, biologists maintained for a long time that to explain the phenomena of life, one needed the concepts of "entelechy" or "vital force." A supernatural act of creation, aimed to liberate us from having to explain the origins of biogenesis or the emergence of consciousness, is also such an "additional hypothesis." After some time, such concepts turn out to be violations of Occam's principle and are therefore rejected as superfluous. The astronomer looking at the starry sky sees many phenomena that he is capable of explaining by referring to specific theoretical models (e.g., the model of the evolution of the stars, the model of their inner design) as well as many other, yet unexplained phenomena. The massive outflow of interstellar hydrogen from the nuclear region of the Galaxy, or the enormous radio emissions of some outer-galactic nebulae, have not yet found a theoretical explanation. Yet the scientist is reluctant to announce, "This is incomprehensible to us, which means it testifies to the activity of some intelligent beings." Such behavior would be too perilous because it would close the door to any attempts to explain such phenomena in a "natural" way. If, during a lone walk on a sea bank, we see symmetrically arranged clusters of rocks, and if we are struck by this symmetry, we are inclined to believe that it is a consequence of a phenomenon whose examination may turn out to be quite beneficial to science: perhaps we are faced here with some yet-unknown symptom of the working of hydrodynamic forces in a rising tide? But if we decide that another person must have walked along this path before us and arranged those stones in this way just because he felt like it, all our knowledge of physics or geology will have nothing to prove. This is why the scientist is inclined to consider even the most "nonnormative" galactic behavior of certain spiral nebulae to be a sign of Nature's work, not a consequence of the intervention of the Intellect.

We can multiply hypotheses about "miracles" at will. It has been said, for example, that cosmic radiation is a consequence of the ejection effect of massive "quantoplanes" distributed all over the Galaxy, the flight paths of which cross a vast amount of space in all directions. If we assume that over millions of years, photon rockets take off from various remote planets, we can consider the section of the radio emission that reaches us from the Galaxy to be a trace of their radiation, shifted as far as radio waves as a result of the Doppler effect (since the alleged sources of such waves, i.e., those rockets, are moving at near–light speed). The stars that suddenly "fly out," at the speed of hundreds of kilometers per second, from within certain clusters are capable of speeding like this as a result of a "catapult" effect caused by the natural explosion of their star companions—but those companions may have also been annihilated by some astroengineering activity. Finally, some of the supernovae explosions could themselves be of artificial origin. However, Occam's razor inexorably forbids us to accept such hypotheses. Incidentally speaking, one of the cardinal sins of science fiction consists in the multiplication of "additional entities," that is, hypotheses without which science can manage very well. A great number of sci-fi works accept as their foundational premise the belief that the development of life on Earth (or perhaps just the transformation of lower mammals into our human ancestors) took place as a result of some external influence: once upon a time, an "alien" rocket is believed to have landed on Earth. Having decided that the conditions for the "cultivation of life" were propitious under our Sun, the "aliens" planted the first seeds of life on our planet. Perhaps they thought that they were doing a good deed; perhaps it was an experiment; or perhaps it was just a "mistake" on the part of one of the visitors from the stars—who, on walking back to its rocket, dropped a test tube containing the germs of life . . . one can develop quite an infinite number of similar ideas. However, from Occam's point of view, they are banned as superfluous because biogenesis can also be explained without the "theory of extraterrestrial visitation." Yet that latter suggestion cannot be ruled out completely (Shklovsky mentions this in his book)—who knows if man himself will not be involved in the propagation of life on other planets one day? The American astronomer Sagan mentioned earlier postulates that we should make Venus open to colonization by disseminating some algae from the Earth onto it. The results of methodological analysis are therefore clear. The scientist, searching for signs of astroengineering activity in the Universe, may have actually been seeing it for a while, yet he is banned from classifying it as a separate phenomenon, that is, from isolating it from the natural world and ascribing its genesis to the Intellect, by the very science to which he is in service. Is there no way out of this dilemma? Can we not imagine some "clear-cut miracles" that cannot be explained in a nontechnological way?

No doubt we can. But what such miracles must have in common (apart from the use of large and thus astronomically observable powers, no doubt) is a course of action that resembles ours, at least broadly. What was it that drove our search for "miracles"? An attempt to find a multiplication of our current capabilities. In other words, we understood progress as moving along a horizontal line and the future as an era of Ever Greater and More Powerful Things. What would the cave man expect from the Earth's future or from the future of other planets? Large and well-made flints. And the ancient man, what would he hope to see on other planets? Galleys with extremely long oars, perhaps. Is this where the error of our thinking lies? Perhaps a highly developed civilization does not need the highest energy but rather the best regulation? Does the recently discovered similarity between microfusion cells and atomic bombs, on one hand, and stars, on the other, mean that we have now figured out the way? Is the highest civilization equivalent to the most densely populated one? Probably not. And if not, then its sociostasis does not need to amount to a growing appetite for energy. What did primitive man do around the fire that he had started with his own hands? He threw all that was flammable into it, dancing and shouting at the flames, astounded by such a manifestation of his own power. But are we not quite similar to him? Perhaps. Despite all our attempts to "explain things away," one should accept the existence of various developmental paths, including some expansive ones—which resemble our heroic idea of the eternal conquest of the ever wider stretches of matter and space. Let us thus be honest with ourselves: we are not looking for "any civilizations" but predominantly for anthropomorphic ones. We introduce the law and order of a scientific experiment into Nature, and then on the basis of such phenomena, we aim to see beings that are similar to us. Yet we are not seeing any such phenomena. Is it because they are nonexistent? There is something deeply saddening in the silence of the stars that awaits us in response to this question—a silence so absolute that it seems eternal.

MAN'S UNIQUENESS

The Russian scientist Baumsztejn adopts a contrary position to Shklovsky on the preceding matter. He claims that the life span of a once-formed civilization is almost infinite, that is, it expands to billions of years. At the same time, he considers biogenesis to be exceptionally rare. He justifies his argument as follows. The probability that a single cod egg will develop into a mature fish is extremely small. Yet because there is a great number of such eggs (about three million in one clutch), the probability that at least one or two of the eggs will develop into fish is close to 1. He compares this phenomenon, which is extremely rare in each singular instance but highly probable when we add up all such cases, with processes of bio- and anthropogenesis. Looking at calculation results—which we shall not be reporting here—he concludes that out of a billion planets in the Galaxy, only several of them, and maybe actually just one, the Earth, have produced the "psychozoic." Baumsztejn draws here on the theory of probability, which says that given such a low probability of the occurrence of a certain phenomenon, frequent repetition of the preliminary conditions is needed if this phenomenon is to take place at all. It is thus not very probable for one player throwing ten dice to roll ten sixes. Yet if a billion players throw dice simultaneously, the probability of at least one of them rolling ten sixes is much higher. The development of man was conditioned by a great many factors. For instance, a common ancestor of all vertebrates—fish—had to emerge; the hegemony of small-brain reptiles had to give way to the era of the mammals; after which primates had to develop from mammals. The development of man from primates must have been greatly influenced by the glacial period—which suddenly increased selection and presented the organisms with enormous expectations with regard to their regulatory capabilities. This led to the dynamic development of a "second-order homeostatic regulator": the brain.[1]

The preceding argument is correct but for one significant reservation. Its author did indeed prove that certain organisms could only have emerged on a planet that had a large and lonely moon (which leads to the emergence of ebbs and flows, which in turn creates propitious conditions for the emergence of vegetation in inshore areas), and that "cephalization," or the growth of primitive man's brain, must have been significantly accelerated by the influence of the glacial period—which disturbed and simultaneously enhanced selection pressure (the glacial periods having in turn been caused, it is believed, by a decrease in radiant solar energy, which occurs every several hundred billion years). Put briefly, the author did indeed prove the rare occurrence of anthropogenesis, but he proved it *literally*; that is, he demonstrated how improbable the hypothesis about the emergence of *anthropoid* organisms on planets of various suns would have been.

This argument does not in any way solve the problem of the frequency with which biogenesis and bioevolution occur in the Universe. The probabilistic model of the development of one cod from millions of scattered eggs does not apply here anymore. It is fine if one creature emerges from three million eggs; yet the nonemergence of a fish from an egg amounts to the annihilation of that egg. At the same time, the nonemergence of the *Homo sapiens* species from primates would not yet have annulled the possibility of the emergence of intelligent beings on Earth. Rodents, for example, could have given rise to them. The probabilistic model of the dice game does not apply to a self-organizing system such as evolution. This model only recognizes winning or losing, so it is a game based on the "all or nothing" principle. Evolution, in turn, is inclined to make all sorts of compromises: if it "loses" on land, it propagates its organisms in water or in the air; if a whole branch of animals dies out, thanks to evolutionary radiation, its place is soon taken by some other organisms. Evolution is not a kind of player that would recognize it has lost; it is not like an opponent who either overcomes the obstacle or dies, or like a hard missile that can only smash against the wall or go right through it. It is more like a river that avoids obstacles by changing its course. And just as on Earth there are no two rivers with analogous routes and shapes of their beds, it is highly likely that in the Universe there are no two identical evolutionary rivers (or trees). Thus Baumsztejn actually proves something different from what he had intended. He demonstrates the improbability of *repeating* the Earth's evolution within some other planetary systems, or at least repeating it faithfully, down to each detail that led to the emergence of man in his current form.

Yet we do not really know which formative influences in biology are accidental (e.g., the existence of the Earth's large Moon) and which are an inevitable consequence of the laws of the homeostatic system. What is most puzzling are those "repetitions," those instances of "unconscious self-plagiarism," that evolution committed in repeating, after millions of years, the process of the adaptation of organisms to the environment they had already left a long time ago. Whales resemble fish as a result of such secondary imitation, at least in their outer appearance; something similar has happened with certain turtles—which used to have a shell and then lost it completely, but then, after tens of thousands of generations, produced it again from scratch. The shells of "primary" and "secondary" turtles are quite similar, but the first type of shell developed from the inner skeleton, the second from callused skin tissue. This very fact indicates that the formative influence of the environment is a key factor in the emergence of forms that are similar in their design. The drivers of every evolution are thought to lie, first, in the transformation of the hereditary information passed on from one generation to another and, second, in environmental changes. Shklovsky emphasizes the influence of the cosmic factor on the transfer of hereditary information. He puts forward an extremely original hypothesis that states that the intensity of cosmic radiation (which regulates the number of mutations) is variable and depends on the life-generating planet getting closer to a supernova. The intensity of cosmic radiation can then exceed that of "normal" radiation, that is, its average intensity in the Galaxy, tens or even hundreds of times. The resistance that certain types of organisms show against such radiation, which destroys genetic information, is bewildering. For example, insects can cope with levels of radiation hundreds of times higher than what would be a lethal dose for mammals. Besides, in organisms that live longer, such radiation increases the frequency of mutations to a more significant degree that in those that only have a short life span (which could have influenced to some degree the "negative selection" of potential Methuselahs in the organic world). Shklovsky puts forward a hypothesis that the mass extinction of large reptiles in the Mesozoic era was caused by the Earth accidentally approaching a supernova that was just on the verge of erupting. And thus, as we can see, the environmental factor turns out to be more universal than we would have thought because it can determine not only selection pressure but also the frequency with which genetic traits mutate. Generally speaking, the speed of evolution is minimal, even approaching zero, when environmental conditions remain virtually identical for hundreds of millions of years. Such environments include, first of all, oceans—in which some animal forms (specifically, fish) have remained unchanged since the Cretaceous and Jura periods. Thus planets that have a more significant climactic and geological stabilization than the Earth—in other words, planets that we would be inclined to consider a "paradise" inasmuch as they "favor" life's phenomena—can actually turn out to be areas of homeostatic stasis. This is because life does not evolve as a result of an inherent tendency toward "progress" but only in the face of absolute danger. Conversely, too sudden fluctuations occurring, for example, around variable or binary stars, seems either to exclude the possibility of the emergence of life or pose the constant threat of its sudden extinction.

We can therefore expect evolutions to take place on numerous celestial bodies. A question then arises as to whether they always, or at least almost always, have to culminate in the emergence of intelligence, or whether intelligence is an accident that is to some degree external to the dynamic regularities of the process: more like an unplanned walk down a developmental path that has suddenly opened up. Unfortunately, the Universe cannot provide us with an answer to this question and probably will not be able to for quite a while. We thus find ourselves back on Earth with our problems, limited as we are to the knowledge we can draw just from the events that are taking place on it.

INTELLIGENCE: AN ACCIDENT OR A NECESSITY?

"Nonintelligent" animals and plants are capable of adapting to changes caused by environmental factors—for example, by seasons of the year. The evolutionary catalog of homeostatic solutions to this problem is enormous. Temporary loss of leaves, spore dispersal, hibernation, insect metamorphosis—these are just selected examples. However, the regulatory mechanisms, determined by genetic information, can only cope with the kinds of changes by which they themselves had been selected during thousands of previous generations. The precision of instinctive behavior becomes ineffective when the need to find new solutions arises, solutions that are not yet known to a given species and are thus not fixed genetically. A plant, a bacterium, or an insect, as "homeostats of the first kind," all have built-in ways of reacting to changes. Using the language of cybernetics, we can say that such systems (or beings) are "programmed in advance" when it comes to the range of the possible changes they should overcome through regulation if they are to continue their existence—as well as that of their species. Such changes are mostly of rhythmic nature (change from day to night, seasons of the year, high and low tides), or at least of temporary nature (being approached by a predator, which mobilizes the innate defense mechanisms: fleeing or freezing suddenly "as if one was dead," etc.). When it comes to changes that would knock an organism out of its environmental equilibrium by "programming" some unforeseeable instincts into it, the answer of the "first-order regulator" turns out to be unsatisfactory—which results in a crisis. On one hand, the mortality of nonadapted organisms suddenly increases, while at the same time, selection pressure privileges some new forms (mutants). This can eventually result in reactions that are necessary for survival being inscribed into "genetic programming." On the other hand, an exceptional opportunity arises for organisms endowed with the "second-order regulator," that is, the brain, which—depending on the situation—is capable of changing the "action plan" ("self-programming via learning"). There probably exists a particular type, speed, and sequence of changes (we could call this sequence "labyrinthine," after the mazes in which scientists study the intelligence of animals, such as rats) that cannot be matched by the evolutionary plasticity of genetically determined regulators or instincts. This privileges the processes of the expansion of the central nervous system as a "second-order" homeostatic device, that is, as a system whose task consists in *producing test models* of various situations. The organism then either adapts to the altered environment (the rat learns how to find the exit from the maze) or adapts the environment to itself (man builds civilization)—and it does this "by itself," without relying on any preprepared action plans. Naturally, there also exists a third possibility—that of losing, when, after having created an incorrect model of a situation, the organism does not achieve adaptation and becomes extinct.

Organisms of the first type "know everything in advance"; those of the second type still need to learn what to do. An organism pays for the comfort of the first solution with its narrowness, of the second one with risk. The "channel" through which hereditary information is transmitted has a limited capacity, as a result of which the number of preplanned activities cannot be too high: this is what we mean by regulatory "narrowness." One knowledgeably assumes the existence of a preliminary period, during which an organism is particularly prone to errors. The cost of such errors can be quite high and can even include the loss of life. This is probably why both of these types of regulators have survived in the animal world. There are environments in which typical behavior, learned "from the cradle," is a more economical option than having to cope with the difficulties and cost of learning from one's mistakes. This, incidentally speaking, is where the "wondrous perfection" of instincts comes from. All this sounds fine, but what does it mean for the general laws of encephalogenesis? Does evolution always eventually need to produce powerful "second-order regulators" such as large brains in primates? Or, if no "critical changes" take place on the planet, does this mean that no brains emerge on it—since they are not needed?

It is not easy to answer a question posed in this way. The cursory understanding of evolution usually results in a naïve idea of progress: mammals had "bigger brains" than reptiles, which means "greater intelligence," and this is why the former ultimately drove out the latter. Yet mammals coexisted with reptiles as marginal, minor forms for hundreds of millions of years, while reptiles reigned supreme. It has recently been confirmed once again what amazing intelligence dolphins have in comparison with other oceanic creatures. Despite this, they did not take control over the water kingdom. We are inclined to overestimate the role of intelligence as a "value in itself." Ashby comes up with a number of interesting examples here.[2] A "stupid" rat, which is unwilling to learn, carefully samples the food it encounters. A "clever" rat, having learned that food is to be found always in the same place and at the same time, seems to have a greater chance of survival. Yet if this food is poison, the "stupid," rat which "is incapable of learning anything," will beat the "clever" one in the survival stakes thanks to its instinctive lack of trust, while the "clever" one will eat its fill and then die. Not every environment thus privileges "intelligence." Generally speaking, the extrapolation of experience (its "transfer") is extremely useful in the terrestrial environment. There are, however, some other environments where this trait becomes a disadvantage. We know that an experienced strategist can beat a less experienced one, but he can also lose to a complete cowboy because the latter's actions are "unintelligent," that is, completely unpredictable. It makes one wonder how evolution, which is so "economical" in every area of information transfer, produced the human brain—a device with such a high degree of "excess." This brain—which, even today, in the twentieth century, copes very well with the problems of a large civilization— is anatomically and biologically identical with the brain of our primitive, "barbarian" ancestor from a hundred thousand years ago. In what way did this massive "potential of intelligence," this excessiveness which, from the early days, seemed geared to build a civilization, emerge in the course of the probabilistic evolutionary game between two vectors: mutation pressure and selection pressure?

Evolutionism lacks a firm answer to this question. Experience demonstrates that the brain of virtually every animal is characterized by significant "excess," which manifests itself in the animal's ability to solve the tasks it does not encounter in everyday life when it is presented with them by a scientist conducting an experiment. The universal growth of brain mass is another fact: modern amphibians, reptiles, fish, and, by and large, all representatives of the animal kingdom have bigger brains than their ancestors from the Paleozoic or Mesozoic eras. In this sense, all animals have "become cleverer" in the course of evolution. This universal tendency seems to prove that, provided the process of evolution takes a long enough time, the brain mass must eventually exceed a "critical quantity," which will initiate the rapid progress of sociogenesis.

We should nevertheless refrain from turning this "gravitation toward intelligence" into a structural tendency of evolutionary processes. Certain factors connected with the use of "materials," or with the initial stage of the "construction process," can limit evolution's future capabilities in its early days and determine its developmental threshold to such an extent that "second-order regulators" will not appear at all. Insects, which are one of the oldest, most vital, and most fertile animal strains, serve as a good example here. There are over seven hundred thousand species of insects on Earth today, compared with eight thousand species of all vertebrates. Insects take up over three-quarters of the animal kingdom as a whole—yet they did not produce intelligence. They have been in existence for approximately the same period of time as vertebrates, so—from a statistical point of view (if it was to be decisive)—owing to the tenfold size of their population, they should have ten times as much chance of producing "second-order regulators." The fact that this has not happened clearly demonstrates that probability calculus is not a determining criterion in psychogenesis. And thus the latter is possible yet not inevitable; it is one of the better solutions but not in all cases, and it is not the most optimal one for all worlds. To construct Intelligence, evolution must have at its disposal diverse factors, such as not too strong gravitation, the relatively constant strength of cosmic radiation (which should not be too powerful), environmental variability that is not just cyclic, and many other, probably still unknown ones. Their convergence on the surface of the planets is most likely not an exception. Despite everything, we can thus expect to find Intelligence in the Universe, though some of the forms in which it will manifest itself may defy all our contemporary ideas.

HYPOTHESES

The situation is thus paradoxical. In seeking support for our efforts to look into the future of Earth's civilization, we unexpectedly got some help from astrophysics, a discipline that, drawing on statistical analysis, examines the frequency with which intelligent life occurs in the Universe. Then we immediately questioned the results of such research. An astrophysicist could ask what entitled us to do this because his expertise in the key area of making distinctions between "natural" and "artificial" phenomena is incomparably higher than ours. This rather likely charge deserves a reply. To some extent, this reply was already dispersed through the previous sections of our argument. All we have to do now is systematize it.

We should point out that radio astronomy is still in its developmental stage. The attempts to carry out extraterrestrial search are being continued (in the USSR, among other places, they will be conducted by one of Professor Shklovsky's collaborators). If we do indeed detect some astroengineering or signaling activity in the next few years, this will certainly be of great significance. Yet a complete absence of any positive data will be of even greater significance—which will increase even further the longer such attempts take and the more sensitive the instruments used in them are. After a sufficiently long period of time, the absence of any such phenomena will prompt us to revise our theory of bio- and psychogenesis in the Universe. Today it is still too early for this. Yet, in making our hypotheses, we find ourselves restricted by our current state of knowledge. We accept the absence of "miracles" or outer space signals just as an astrophysicist does. We therefore do not question the material observed, only its interpretation. Each of the three hypotheses, which we shall list subsequently, provides an explanation for this "psychozoic vacuum."

I. Civilizations develop in the Universe infrequently, but they last a long time. Fewer than twenty civilizations emerge in each galaxy. There is thus only one planet containing a "psychozoic" per billions of stars. Together with astrophysicists, we reject this hypothesis because it contradicts the universally accepted theory of how planetary systems typically emerge and of how life develops in them. Yet we would like to point out that, however improbable, this hypothesis does not have to be wrong. Since the age of different galaxies, just like the age of different stars, varies, galaxies older than ours should have experienced some astroengineering activity, the manifestation of which would be observable once the equipment has been developed further. We are assuming here, just as astrophysicists do, that all or almost all civilizations, rare as they are, develop in a technical direction—a process that, after a long enough time, leads to the emergence of astroengineering.

II. Civilizations develop in the Universe frequently, but they last a short time. This is caused by (1) their "self-destructive" tendencies; (2) their "degenerative" tendencies; or (3) reasons totally incomprehensible to us, which become operational at a certain stage of a civilization's development. It is to these hypotheses that Shklovsky devotes most of his attention in his book. We would like to outline the founding premises of these hypotheses, which can be reduced to two principal ones: (1) the premise that the *direction* of development in the large majority of civilizations is the same as it is on Earth, that is, technical, and (2) the premise that the *speed* of development in different civilizations is similar, at least on an astronomical scale, whereby the deviation of around one million years is seen as insignificant. Thus the premise that almost all civilizations develop in an *orthoevolutionary* manner is foundational for almost all civilizations. It is silently assumed that the acceleration of the technological progress we have been observing on Earth for the last two hundred years or so is a dynamically constant process that can only by halted by some destructive influences ("degeneration," the "suicide" of a civilization, etc.). Exponential growth is a key dynamic characteristic of all civilizations. It leads, in a straight line, to the emergence of astroengineering activity. We can raise questions for both these premises. Yet we lack data that would allow us to determine whether a technical direction is in itself a manifestation of the developmental law of "psychozoics." It does not have to be. However, following Occam's razor, we do not want to introduce any "unnecessary entities," that is, hypotheses that would not be based on any facts. We therefore assume that this is a typical direction because we consider ourselves and our own history a standard, regular, and hence typical phenomenon in the Universe.

It is a different matter with the second premise. While history has so far shown a steady exponential growth of our civilization since the Industrial Revolution, some firm and significant facts indicate that this is likely to change. If we question the alleged constancy (on an astronomical scale) of the *speed* of technoevolution, we are presented with a possibility of a new solution. We can speak here about a third group of hypotheses, hypotheses that conform with the observed (or rather unobserved...) facts.

III. Civilizations develop in the Universe frequently, and they last a long time, but they do not develop in an orthoevolutionary manner. It is not their existence that is short-lived but only a certain phase within it, which is characterized by exponential growth. This phase of expansion lasts a short time in astronomical terms: less than twenty thousand years (or possibly even shorter, as it will turn out later). After this initial period, the dynamic characteristics of the developmental process change. This change, however, does not have anything to do either with "self-destruction" or with "degeneracy." From this moment on, the paths of various civilizations can significantly diverge. We shall discuss the influences that condition this diversification later on. Such a discussion will not be a sin against the prohibition on idle speculation because the factors that change the developmental dynamics can already be found, in an embryonic form, in the contemporary world. They are of an extrasocial and extrasystemic nature and are simply derived from the very structure of the world in which we live, that is, from the way it is. We shall discuss behavioral changes that a civilization may manifest once it has reached a certain stage of development. Given that, within certain limits, it has freedom of choice with regard to its future strategy, we shall obviously not be able to predict what will happen to that civilization. From many various options, we shall choose only those that comply with the facts, that is, those that reconcile the multiplicity of inhabited worlds that have existed for a long time with the fact that they cannot all be observed astronomically.

The theory we shall obtain in this way will, on one hand, satisfy the demands of an astrophysicist (i.e., it will accept the lack of "miracles" and outer space signals), and on the other hand, it will avoid the catastrophic formalism of von Hoerner's hypotheses. I believe we should look once again at what made us reject the "statistical inevitability of extinction" that can be derived from those hypotheses. If the directions and speeds of development of all the civilizations in the Galaxy are similar, and if an average life span of a civilization is several thousand years, this does not mean that there can be no civilization that would exist for millions of years—as an aberration from the norm. Von Hoerner's statistics resemble the statistics of a gas. The majority of particles in a gas at room temperature have speeds of around several hundred meters per second, but there are also some rare particles with much higher speeds. Yet the presence of a handful of fast particles does not in any way influence the behavior of the tepid gas. At the same time, the presence of just a few civilizations with an "abnormal" life span in a galactic set would influence the whole Galaxy because those civilizations would give rise to enormous expansive radiations into ever greater star fields. In that case, it would be possible to see some signs of astroengineering—which is not what happens. Von Hoerner thus silently assumes that his statistics contain phenomena that are as limited and short-lived as human lives. Even though statistical aberrations from the average life span of sixty years do exist, no one can live for two hundred or three hundred years. But the universal mortality of humans after they have been alive for several decades results from the characteristics of their organisms—something that cannot be said about social organisms. Every developing civilization can no doubt go through several "crises" (connected, e.g., with the discovery of atomic energy and also with some other transformations about which we do not know anything). Yet we ought to expect a reverse proportionality to the one we can observe in a biological population: in such a population, an individual who has a higher probability of dying is the one that has lived the *longest*. A long-lived civilization should in turn be "less mortal," less exposed to the risk of interference than a short-lived one, because it acquires ever greater knowledge—thanks to which it gains control over its own homeostasis. The universal extinction of civilizations is thus an additional premise, plucked out of thin air. Von Hoerner introduces it into his mathematical querns even before he gets busy with his calculations. We consider this premise to be rather groundless. It is therefore methodology rather than optimism (which can really be out of place in the Universe) that pushes us toward some other explanations of the "psychozoic vacuum" in outer space.[3]

VOTUM SEPARATUM

We were supposed to come back to Earth, but we are going to stay high in the sky for a little while yet, because I would like to present my own opinion about the preceding matter. This statement may evoke surprise because it may look like I have been outlining my own position all the time, while entering into a debate with various hypotheses. Yet I hasten to explain that I have been acting like a judge—a self-appointed one, true, but one that observes articles of the law. What I am referring to here is my adherence to the strict rules of scientific accuracy, a position that manifests itself in cutting off with Occam's razor all kinds of speculation. I think this was a sensible thing to do. But, despite the available evidence, one sometimes does not feel like being sensible. This is why I shall present here my own personal point of view, after which I promise to return to being a humble servant of methodology.

And thus back to extraterrestrial civilizations...As long as the questions asked of Nature by Science concerned phenomena on a scale parallel to ours (I am referring here to our ability to approximate the examined phenomena to what we can grasp directly through our senses, an ability we developed through everyday experience), Nature's answers sounded reasonable to us. Yet when the question, Is matter a wave or a particle? was posed during an experiment, thus setting up a clear alternative between the two, the answer turned out to be both incomprehensible and difficult to accept. Therefore, when, in response to the question, Are extraterrestrial civilizations frequent or rare...long-lived or transient? we receive incomprehensible answers that are full of apparent contradictions, these contradictions reveal not so much the status quo as our inability to ask Nature the *right* sort of questions. Man asks many questions of Nature that seem meaningless "from its point of view," hoping to get an explicit answer that would fit into his own precious frameworks. Put briefly, we do not really want to arrive at Order *as such* but rather at a particular kind of order: economic (Occam's razor!), clear-cut (so that it does not yield itself to various interpretations), universal (so that it applies to the whole of the Universe), autonomous (i.e., independent of whether anyone is observing it and who that might be), and immutable (one in which the laws of Nature do not themselves change in the course of time). Yet all of these are just scholarly postulates, not some revealed truths. The Universe was not created for us, nor were we created for it. We are a side product of stellar transformations and of the products the Universe has been creating in infinite amounts. We must certainly continue with the extraterrestrial search and observations in the hope of encountering an Intelligence that will be so similar to ours that we will recognize it through its signs. But this is just hope, nothing else, because the Intelligence we shall discover one day will possibly be so different from our ideas of it that we shall not even want to call it Intelligence.

Our dear reader may be running out of patience at this point. Perhaps, he will say, Nature is indeed providing us with unclear answers, but you, the writer, are not Nature! Instead of clearly outlining your position on extraterrestrial civilizations, you have just complicated the matter by speaking about the Laws of Nature, about Order, and so on, and by retreating to semantics in the end—as if the existence of those intelligent beings in outer space depended on what we understand by "intelligence"! This is pure subjectivism, or even worse! Would it not be more honest to admit that you simply do not know anything?

Naturally, I say, I do lack any reliable knowledge, because where would I have obtained it from? It is also possible that I am wrong and that the "sociocosmic" contact that will be made in the near future will expose me and my argument to ridicule. But please let me explain. I think the reason we shall not see the presence of Intelligence in outer space is not because it is not there but rather because its behavior defies our expectations. This behavior can in turn be explained in two ways. First we can assume that, instead of there being only one Intelligence, there exist various "forms of Intelligence." Alternatively, having assumed that there exists only one form of Intelligence, which is exactly like ours, we can investigate whether it does not perhaps change in the course of evolution to such a degree that, in its manifestations, it eventually stops resembling its original state.

The ensemble of humans who differ from one another in their temperaments, personalities, and so on, is an example of the situation of the first kind.

An example of the situation of the second kind can be found in a set of various subsequent states in which a human being finds himself: as a baby, a child, an adult, and eventually, an old man.

We shall consider the situation of the second kind separately because certain facts favor such an interpretation of the "cosmic status quo." As we shall be drawing on facts, we can expect to receive support from Methodology to conduct our investigations.

Unfortunately, the situation of the first kind is not rooted in any facts: it is pure speculation, a "what if" type of reasoning. Hence all the reservations I have raised when discussing it.

Let us thus go back to those various "forms of Intelligence." I would not even dare to suggest considering some different possible directions in a civilization's development, including nontechnical ones—because we could easily argue about the term "Technology," just as we could about "Intelligence." In any case, different forms of Intelligence do not mean "more stupid" or "cleverer" than human intelligence. By Intelligence we understand a second-level homeostatic regulator that is capable of coping with disturbances to its environment thanks to the activities in which it engages on the basis of the historically acquired knowledge. Human intelligence has led us to the Technological Era because the terrestrial environment has a number of unique characteristics. Would the Industrial Revolution have been possible if it had not been for the Carboniferous, a geological period during which the reserves of solar energy had been stored in the sunk forests that were undergoing carbonization? Would it have been possible if not for the oil reserves, which emerged in the course of some other transformations? "So what," I hear. "On the planets which did not go through their own Carboniferous period it is possible to use other kinds of energy, such as solar or atomic energy...In any case, we are getting away from the main issue. We were supposed to be talking about Intelligence."

But we *are* talking about Intelligence! Yet it would have been impossible to reach the Atomic Age without the Age of Coal and the Age of Electricity that preceded it. Or a different environment would have at least required a different sequence of discoveries, which would have involved more than just rearranging the calendars of Einsteins and Newtons from other planets. In an environment with a high degree of disturbance that exceeds the regulatory capacity of a society, Intelligence can manifest itself not in an expansive form, as a desire to take control over the environment, but rather as a desire to subjugate itself to that environment. I am referring here to the emergence of biological technology prior to physical technology: creatures inhabiting such a world transform themselves to function in a given environment, instead of transforming that environment so that it serves them—the way humans do. "But this is not intelligent behavior any more; this is not Intelligence!" we hear in response. "Every biological species behaves in this way in the course of evolution..."

A biological species does not know what it is doing, I reply. It does not rule itself but is rather ruled by Evolution, which throws its hecatombs onto the altar of Natural Selection. I was referring to conscious activity: a planned and directed autoevolution, a kind of "adaptation retreat." It does not look like intelligent activity to us because man favors a heroic attack on the surrounding matter. But this is just a sign of our anthropocentrism. The more different the conditions of the inhabited worlds are, the more diversity there must be between various types of Intelligence within these worlds. If someone thinks that coniferous trees are the only ones that exist, he will not find any "trees" even in the thickest of oak woods. Whatever positive things we can say about our civilization, we can be sure of one thing: its development has certainly not been harmonious. Capable of annihilating the biosphere of the planet within a few hours, this civilization is on the verge of collapse when facing a relatively severe winter! I am not saying this to "foul my own nest." Indeed, inequality of development is most certainly a norm in the Universe. If there is no "one single Intelligence" but rather its countless types, if a "cosmic intellectual constant" is nothing more than a fiction, then the absence of signals can be understood more easily, even if we take into account considerable civilization density. We thus have a multiplicity of Intelligence types, whereby they are all embroiled in their "own planetary matters" and moving along different trajectories, separated as they are by their different ways of thinking and acting and by different goals. As we know, man can be alone in a large crowd. Yet is this to say that this crowd does not actually exist? Is this loneliness merely the consequence of a "semantic misunderstanding"?[4]

FUTURE PROSPECTS

In 1966,[5] we still do not know anything specific about the existence of extraterrestrial civilizations. Yet this issue is increasingly becoming a subject of research and planning. Scholarly conferences focused on "aliens," and on making contact with them, have taken place in the United States and the USSR. Of course, the question of whether those "aliens" actually exist is still a fundamental one. It may seem that, given the lack of empirical data, the choice of an answer still depends on one's personal view, on a given scientist's "taste." Yet ever greater numbers of scientists are slowly realizing that a complete "psychozoic vacuum" in the Universe would contradict the totality of our natural knowledge. Though this knowledge does not *explicitly* postulate the existence of "aliens," it does so *implicitly,* because research conducted in the natural sciences leads us to see astrogenesis, planetogenesis, and biogenesis as normal, that is, average and "typical," phenomena in the Universe. Being able to prove empirically that there are no "aliens" in the Metagalaxy[6] that is perceivable by us (regardless of how such a demonstration would be accomplished and whether it would be possible to conduct it at all) would not just mean invalidating a certain isolated hypothesis (the one about the frequency with which life and intelligence emerge in the Universe). It would also pose a serious methodological threat to the very foundations of our natural knowledge. Acknowledging the existence of such a vacuum would amount to acknowledging that the continuous transition from selected material phenomena to others, a transition that constitutes an unshakeable foundation of all knowledge and is based on extrapolation widely accepted in science—that is, from the formation of stars through to the formation of planets, and then to the emergence of life and its evolution—is not allowed in the world. In other words, it would mean that there exists somewhere in the world a crack in the postulated general laws—a crack that lies beyond our grasp. Such an acknowledgment would call for a revision to a great many theories that are widely accepted as true today. To recall Shklovsky's statement from the 1964 conference in Byurakan, "For me, the biggest and most true 'miracle' would be the ability to prove that there are no 'cosmic miracles.' Only an expert astronomer can fully understand the significance of the [possible] fact that amongst the 10^{21} stars which form the observable part of the Universe (about 10^{10} galaxies, each containing between 10^{10} and 10^{11} stars), not a single one of them is surrounded by an adequately developed civilization, even though the percentage of stars which are surrounded by planetary systems is quite high."

At the previously mentioned conference, a young Russian astrophysicist called Kardashev divided the hypothetical civilizations into three types: (1) Earthlike civilizations, with annual energy use of 4×10^{19} ergs; (2) civilizations whose energy use approaches 4×10^{33} ergs; and (3) "supercivilizations," that is, civilizations that have dominated their galaxies with their energy use, which amounts to 4×10^{44} ergs. He estimated the time needed for the development of a civilization of the first type at several billion years (by drawing on the example of the Earth). The shift from the first to the second type would only take several thousand years (an estimate based on the pace of increase in energy production on Earth during the last few centuries); whereas the shift from the second to the third type would take between twenty and a hundred million years. The latter estimate met with criticism from the other experts because, with such a "pace of psychogenesis," practically all galaxies would already have had to have developed their own "supercivilizations." Consequently, the sky would be a space of some very intense "astroengineering" activity and would be heavily populated with "cosmic miracles"—something that most certainly is *not* taking place right now. We thus have to conclude that either the development of (any) civilization is a highly improbable and hence rare phenomenon, and that civilizations develop only in certain galaxies (which means we could be alone in ours), or that some phenomenon (a kind of barrier perhaps?), or even some multiple phenomena that remain a mystery to us, are hindering the growth of energy (and thus of technology).

Of course, there may exist a relatively trivial solution to this puzzle. As we have said, developmental paths are likely to be shared up to a certain stage (such as the one in which the Earth finds itself at the moment), after which they diverge through a radiation of developmental directions. Only a very small fraction of those who "started the race" are able to continue with the exponential pace of early development. This developmental barrier, which has a probabilistic character, is radically different from a mysterious "prohibition" marked by *fatalistic* determinism. A similar approach, rooted in statistics, sees the Universe once again as a zone of play and struggle for further growth—a difficult and dangerous struggle but one that is also worthwhile. A deterministic approach, in turn, would be represented by a mysterious suspended sentence that no cognitive or emotional effort on our part would be able to overcome.

The preceding solution to this issue, conducted in a probabilistic spirit (rather than just as "consolation"), seems to be the most apposite one today, also owing to its methodology.

We can posit the following rule as an almost irrefutable conclusion: starting from planetogenesis—which, as we know, *is* a rather typical phenomenon in the Universe—any further convergence of the processes of biogenesis (then of psychogenesis, and then of the development of a civilization and its direction) disappears at a certain point on this path. But we do not know whether we are dealing here with one clear "point" of the already ongoing divergence or whether there are a great number of stages in which subsequent deviations from the set directions of the Earth's "model" come together. Statistics seem to tell us that the number of planetary systems is much higher than that of life-generating ones. The latter are in turn more numerous than those that give rise to a civilization—and thus all the way until we reach a civilization whose technical achievements can be seen across the whole Universe.

Understandably, scientists do not spend much time on the preceding hypotheses, focusing instead on physical and technical aspects of communication between civilizations. With regard to this latter issue, it is worth briefly mentioning the following. First, predicting manned star flights, in, say, photonic ships, is not "fashionable," nor is it the subject of any theoretical studies at the moment, because balance analyses of energy (such as the one provided by von Hoerner) have demonstrated that even using annihilation as a source of propulsion does not solve the key problems of the energy that would have to be involved in such a journey. The amount of matter that would have to be annihilated to fly from one galaxy to another in a "reasonable" time span (within a human lifetime), and thus with near–light speed, almost equals the mass of our Moon. Such flights are thus considered impossible for the time being and are likely to remain impossible over the next several hundred years. It has been pointed out, however, that a "near-light" ship could cover at least part of its original mass deficit with cosmic matter—which, thinned down as it is, would become an attractive potential fuel for an equally fast vehicle. Perhaps some other energy sources will be discovered. In any case, the difficulties encountered by astronautics are different from those that, for example, make the construction of a perpetual motion machine impossible. It is because astronautics is not prohibited by the laws of nature, and even when the evidence shows that the original mass of a galaxy ship would have to be close to that of the Moon, attention is drawn to the extreme technical difficulties of achieving this but not to any fundamental impossibility as such. Yet the Moon does in fact exist, which is why if any of the Earth's future generations were stubborn enough, they would perhaps send our honorable satellite, which has so kindly been prepared for us by the planetogenesis of the solar system, on an appropriate journey.

Second, the issue that preoccupies the scientists the most, that is, that of radio (or possibly also laser) contact with "aliens," requires some serious material investment, as it turns out, if it is to be actualized. (It would require us to construct a great amount of equipment that would enable a "search for extraterrestrials"—and possibly also transmitting stations. If, for economic reasons, all the civilizations were only working on reception, then no one would ever hear anyone else, as many have already pointed out.) This kind of investment would have to be even more substantial than the current investment in research on nuclear energy.

No doubt scientists still need to "cultivate" a generation of rulers who will be inclined to dip deeply enough into the public coffers to finance issues that, worrying as it may seem, traditionally belong to science fiction territory. Apart from its material aspect, radio contact has some interesting information characteristics. The issue here is that the more precisely the transmitted message uses the capacity of the information channel (i.e., the higher the reduction of the excess of this transmission), the more it starts to resemble noise. In this situation, the receiver—who does not know the coding system—would actually have tremendous difficulties not just with deciphering the information he receives but also with recognizing it precisely *as information,* as opposed to cosmic background noise. We cannot thus exclude the possibility that already today, we are receiving, *as noise,* via our radio telescopes, fragments of "interstellar conversations" conducted by "supercivilizations." Such civilizations—if we were to discover them at all—would also have to be transmitting signals of an entirely different nature, signals that would not use the full capacity of the transmission channels and that would thus take a form of special "call signs" with a relatively simple, clearly ordered, and repeatable structure. Given that "call signs" of this kind can only be a fraction of the total information emission of those civilizations, constructing a sizeable amount of specialized receiving equipment once again proves to be an extremely important issue (and, as we said earlier, an extremely costly one).

And thus the only mystery we cannot still solve, even partially, is the absence of "cosmic miracles." This problem entails a certain paradox. What has so far been put forward as a "model" of such a "miracle," for example, a Dyson sphere, is most likely not being built anywhere, as we said earlier. Incidentally, we are aware that a great number of phenomena that take place in galaxies and stars are still awaiting an explanation; yet none of the experts are inclined to describe the unknown as a "cosmic miracle." It is one thing to invent phenomena (e.g., a Dyson sphere) that would create *for us* as observers propitious conditions for coming up with a dichotomous solution (the alternative between "natural" vs. "artificial"), but it is quite another to actually create phenomena that would be a side product to the already active energetics of stars, neutrinos, or even quarks.[7]

For this hypothetical civilization, its energetics is not a special device focused on signaling its existence to the Universe. This is why there can emerge, almost accidentally, a kind of "camouflage," which will lead us to interpret that which has been produced by "aliens" artificially as something created through the force of Nature, as long as its laws allow for such an interpretation. A nonexpert will find it hard to understand that some doubts might arise here. If he found a page from a letter written in an incomprehensible language and alphabet, he would then have no doubts as to whether this letter was produced by an intelligent being or whether it emerged thanks to some natural, "nonhuman" phenomenon. Yet it turns out that the very same sequence of cosmic "noise" can be considered either as a "sign" or as radiation of dead matter. This controversy arose with regard to the spectrum of some very remote objects—objects that Kardashev, contrary to the majority of astrophysicists, attempted to identify as transmitting civilizations. It was the other astrophysicists who were most probably right, not him.

And now for our final remark. For a great majority of people, including the scientists—and excluding a very small group of engaged experts—the whole issue of "aliens" smacks of science fiction, and more important, it is almost entirely devoid of any emotional aspect. Most people are used to the idea of the populated Earth and the deserted Moon (unless we are talking about fairy tales); they see it as obvious that this is the only option possible. This is why theories suggesting that we may be alone in the Universe do not present themselves as some kind of tremendous revelation to people, yet this is precisely the position conveyed by Shklovsky, as cited earlier. I am fully in support of it. Just to be loyal, let us add that our loneliness will seem rather monstrous, mysterious, and terrifying to a materialist and an empiricist but rather wonderful and perhaps even "calming" to a spiritualist. This even applies to scientists. On a daily basis, we are used to accepting the exclusive existence of humans in the class of "intelligent beings." The existence of aliens—which is not only in agreement with, but also strongly implied by, the natural sciences—seems to be an extremely abstract proposition to us. Such anthropocentrism cannot give way all too easily to some kind of galaxocentrism. This is quite understandable, as people still find it hard to coexist on Earth with one another. To postulate a universalism of a cosmic kind under those circumstances sounds like a dreamy, ironic, or even irresponsible fantasy, one that is being promoted among the ever-fighting Earthlings by a bunch of eccentrics.

I am perfectly aware of this, and so I am not making an appeal to correct school textbooks in the spirit of the argument presented previously. At the same time, I find it hard to accept that one can be a fully rounded human being in the second half of the twentieth century unless one thinks at least from time to time about this still unknown community of intelligent beings, of which we are allegedly part.

4

INTELECTRONICS

RETURN TO EARTH

In this chapter we aim to investigate whether intelligent activity that manifests itself in technoevolution is a dynamic and permanent process, one that does not alter its expansive nature during any period, or whether it must undergo a transformation until any similarity to its original state has disappeared.

Please note that this discussion will differ considerably from the cosmic debate that preceded it. Everything we said about extraterrestrial civilizations was not a product of vacuous speculation, yet the hypotheses we discussed had been based on further hypotheses, as a result of which the plausibility of our conclusions was at times rather low. In turn, the phenomena we shall be discussing here are predictions that are based on well-known and thoroughly researched facts. And thus the plausibility of the processes we shall outline in this chapter is many times higher than the plausibility that characterized our discussion of civilization density in the Universe.

We shall examine the future of civilization while taking into account the developmental potential of science. It is easy to say that science will "always" be developing and that the more we find out, the higher the number of new problems that we shall face. Will there be no limits to this process? It seems that the rapid pace of discovery also has its limit—which we are likely to reach soon.

The Industrial Revolution began in the seventeenth century. Its roots—or rather fuses, as it was more like an explosion than slow maturation—reach far back. Einstein answered the question about the "first cause" of science in a way that was both amusing and poignant:

"No one scratches unless they have an itch." Science as a driving force of technology was mobilized by social needs. It was mobilized, popularized, and accelerated—but not created—by them. The early origins of science go back to the Babylonian and Greek times. Science started with astronomy, with the exploration of the mechanics of the skies. The huge regularities of such mechanics brought to life first mathematical systems—systems that were considerably more complex than the first steps in arithmetic called forth by ancient technology (measurement of land, of buildings, etc.). The Greeks created formal axiomatic systems (Euclid's geometry), while the Babylonians came up with an arithmetic that was independent from geometry. A historian of science is very much aware of the first-born status of astronomy among the sciences. Experimental physics, the development of which was largely driven by questions posed by astronomy, came second. Physics, in turn, gave life to chemistry—while also belatedly freeing it from the mythological dream of the alchemists. Biology was the last one among the natural sciences to surface from the mist of unverifiable concepts at the turn of the twentieth century. I am highlighting here important but not sole reasons that lie behind the development of individual scientific disciplines, since the mutual crisscrossing of their findings accelerated their respective trajectories and the subsequent development of some new branches from them. All this clearly indicates that both the "mathematical spirit" of contemporary sciences and its material instrument—the experimental method—already existed, in embryonic form, before the Industrial Revolution. The Revolution gave impetus to the development of science, since it combined theoretical knowledge with manufacturing practice. As a result of this, over the last three hundred years, Technology has maintained positive feedback with Science. The Scientists pass on their discoveries to the Technologists, and, should the results prove successful, their research undergoes the process of "amplification." The feedback is positive, as negative attitude on the part of the Technologists toward a certain discovery by the Scientists still does not mean that theoretical research in that area will have to be terminated. I have simplified here the nature of the relationship between the two areas on purpose: this relationship is actually much more complex.

As science is a form of information gathering, the pace of its development can be estimated quite accurately on the basis of the number of specialist periodicals currently in print. This number has been growing exponentially since the seventeenth century. Every fifteen years, the number of scientific journals doubles. Exponential growth is usually a transitional phase that does not last very long, at least not in Nature. An embryo or a bacterial colony on the surface of a solid medium grows exponentially (i.e., its rate of growth must be expressed by an exponent) for a short time. It is possible to calculate how quickly a bacterial colony would turn the mass of the Earth into itself.

In reality, the environment soon limits such exponential growth, which then turns into linear growth or into stunted growth and stagnation. The development of science, determined as it is by an increase in the amount of scientific information, is the only phenomenon known to us whose astounding pace of development has not changed over the last three hundred years. The law of exponential growth says that the bigger a given set, the faster it grows. The manifestation of this law in science leads to each discovery resulting in a whole series of new discoveries. The number of such "births" is directly proportional to the size of the "population of discoveries" in a given period of time. We currently have around one hundred thousand scientific journals. If the pace of growth does not change, in the year 2000, we will have *one million* of them.

The number of scientists has also been growing exponentially. It has been calculated that if all the universities and colleges in the United States were to train solely physicists from now on, by the end of the next century, we would run out of humans (not just university candidates, but people in general, including the elderly, women, and children). And thus, given the current pace of scientific growth, in about fifty years, every inhabitant of the Earth would be a scientist. This is the "absolute threshold" that cannot be crossed, because if we did cross it, then one human being would have to stand for a few scientists at the same time.

The exponential growth of science will therefore be halted by the lack of human resources. The beginnings of this phenomenon can already be observed today. Several dozen years ago, Roentgen's discovery drew a considerable section of the world's physics to the exploration of X-rays. Today, discoveries of no lesser significance attract only a fraction of a percent of all physicists because, as a result of the immense widening of the scope of scientific research, the number of people focusing on any single section of it has decreased.

As theoretical knowledge is always ahead of the knowledge that has already been actualized in industry, then, even if the growth of theoretical knowledge were to stop, there would be enough of its already accumulated resources to help with further technological expansion for another hundred years or so. This "inertial" effect of technological progress (driven by the already accumulated but still unexploited results of scientific enquiry) would eventually come to a halt, after which we would have a developmental crisis. When we experience "scientific saturation" on the scale of the planet, the number of phenomena that need to be investigated but that are neglected owing to the lack of humans will increase. The growth of theoretical knowledge will not stop, but it will be hindered. Can we envisage the fate of a civilization whose science has exhausted all of its human resources, while still having a need for them?

Global technological improvements are estimated at 6 percent annually. At this rate, the needs of a large part of humanity are not being fulfilled. Given the current birthrate, the slowing down of technological growth through limiting the pace of scientific development would not just lead to stagnation—it would actually start a decline. The scientists on whose works I have drawn here[1] are anxious about the future. It is because they are predicting a situation when we will have to decide what kind of research must be continued and what kind of research has to be abandoned. Important as it is, the question of *who* will have to make such decisions—scientists themselves or politicians—pales into insignificance in the face of the issue that no matter *who* this is, such decisions may turn out to be wrong. The entire history of science demonstrates that technological progress is always a consequence of discoveries gained by means of "pure" research, which is not focused on any practical goals. The reverse process, in which new knowledge emerges from a technology that is already in use, has in turn been extremely rare and hence seems quite unusual. This unpredictability as to what kind of theoretical investigation will produce something technologically useful, tested over time since the days of the Industrial Revolution, is still with us. Let us imagine that a lottery issues one million tickets, a thousand of which are winning. If we sell all the tickets, the group that has bought them is bound to receive all the prizes. Yet if this group only buys half the tickets, it may turn out that no winning tickets will be included in this half. Science is a similar kind of "lottery." Humanity "bets" all of its "tickets" on scientists. The winning tickets stand for new discoveries that are valuable from the point of view of a technology and civilization. Should we be in a position of having to decide one day in an arbitrary manner which areas of research we must "bet on" and which ones to exclude, it may turn out that it is precisely those we are not "betting on" that would have turned out particularly successful in bringing about unforeseeable results. Actually, our world is going through the first phase of such a "gamble" right now. The concentration of experts in the areas of rocket ballistics, atomic physics, and so on, is so high that other fields of enquiry are suffering.

The preceding ideas are not to be seen as a prediction of the fall of civilization. This could only be accepted as true by someone who understands the Future as an exponentiated Present Time and who does not see any possibility of progress other than orthoevolution—someone who believes that any civilization must resemble ours, with its rapid growth over the last three hundred years, or it will not count as a civilization at all. The place in which a growth curve changes its steep ascent into the bend of "saturation" represents the change in the dynamic characteristic of a system, that is, science. Science is not going to disappear: the only thing that will disappear will be its current form, whose growth has no limits. The explosion phase is therefore only a stage in the history of civilization. Is it the only one? What will a "post-explosion" civilization look like? Will the omnidirectional nature of Intelligence—something we have considered its permanent trait—have to give way to a bunch of selective activities? We shall look for an answer to this question too, yet what we have discussed so far already throws some interesting light onto the problem of stellar psychozoics. Exponential growth can serve as a dynamic law of a civilization over thousands of years—but not over millions of years. On an astronomical scale, such growth takes a short while, during which the ongoing process of cognition leads to a cumulative chain reaction. A civilization that uses up its own human resources in a "scientific explosion" can be compared to a star that burns up its matter in a single flare—after which it reaches a state of alternative equilibrium—or to processes that have made many cosmic civilizations fall silent.

A MEGABYTE BOMB

We have compared an expansive civilization to a supernova. Just as a star burns its material resources in an explosion, a civilization uses up its human resources in the "chain reaction" of rapidly growing science. A skeptic may ask, Is such a comparison not a bit of an exaggeration? Have you not overstated the possible consequences of hindering scientific growth? When the state of "saturation" is reached, science will continue to grow at the limit of its human resources, yet it will no longer be growing exponentially but rather proportionally to the number of all living beings. When it comes to neglected phenomena that are being ignored in research, they have always existed in the history of science. In any case, thanks to intelligent planning, the main frontiers of science, the key directions of technological offensive, will still have armies of experts at their disposal. And thus the theory that a future civilization will be completely different from ours, because a highly developed Intelligence will not resemble its earlier forms, has not been proved yet. The "stellar" model of civilizations is particularly incorrect, as the depletion of material resources results in the star's death, while the "glow" of a civilization is not diminished by the depletion of the sources of energy used by it. It is because a civilization can switch to some other energy sources.

Incidentally speaking, the preceding approach lies at the foundation of a theory that predicts an astroengineering future for every civilization. It is true that the stellar model is a simplification: a star is just an energy machine, while a civilization is both an energy and an information machine. This is why a star is far more determined than a civilization when it comes to its development. This does not mean that the development of a civilization has no limits. Yet there are only differences in kind here: a civilization possesses energetic "freedom" until it hits an "information barrier." By and large, we have access to all the sources of energy that the Universe has at its disposal. But will we manage, or rather, *will we be quick enough,* to access them?

The shift from the type of energy sources that are being depleted to their new type—from water, wind, and muscle power to coal and oil, and then from them to atomic energy—requires obtaining the appropriate information *in advance.* It is only when the amount of such information has gone beyond a certain "critical point" that the new technology produced on the basis of such information will reveal some new reserves of energy and new domains of activity.

If coal and oil deposits had run out, say, at the end of the nineteenth century, we would have been highly unlikely to have produced atom technology in the middle of the twentieth century because enormous powers were needed to make it happen. Those powers were initially installed in laboratories, after which they became available on an industrial scale. And yet humanity is still not ready to switch to an exclusive use of "heavy" atomic energy (i.e., energy derived from the decomposition of heavy nuclei). Given the current increase in power use, this would result in "burning up" all the deposits of uranium and its cognate elements within a few centuries. The exploitation of energy from nuclear synthesis (of hydrogen into helium) has not been accomplished yet. The difficulties have turned out to be more substantial than expected. All this means that, first, a civilization should have considerable reserves of energy at its disposal to have enough *time* to obtain the information that will provide access to some new energy sources, and second, that a civilization must consider the primacy of this model of information acquisition over any other. Otherwise, its resources may run out before it has learned to exploit the new ones. Also, past experience indicates that during a switchover period from older to newer sources of energy, the energy costs involved in acquiring new information grow. It was much "cheaper" energy-wise to create coal and oil technology than to create nuclear energy.

Information is thus crucial to all sources of energy—and to all sources of knowledge. The dramatic increase in the number of scientists since the Industrial Revolution has been caused by a phenomenon that is well known to cyberneticists. The amount of information that can be transmitted via one particular channel is restricted. Science is such a channel, one that connects civilization with the external world (and also with its own, internal world, as it studies not only the material surroundings but also man and society). The number of scientists, which is growing exponentially, signifies a constant increase in the capacity of this channel. This increase is necessary because the amount of information that has to be transmitted is also growing exponentially. The increasing number of scientists has led to an increase in the amount of information being created, which has in turn necessitated the "widening" of the information channel through the "parallel connection" of some new channels, that is, the recruitment of some new scientists. This, in turn, has led to a further increase in the amount of information that has to be transmitted. This is a process with positive feedback.

Yet, eventually, a state must be reached when any further increase in the transmission capacity of science at a speed dictated by an increase in the amount of information will turn out to be impossible. There will be no more prospective scientists. This is a situation that can be described as a "megabyte bomb," aka, "information barrier." Science cannot traverse this barrier; it cannot absorb the avalanche of information that is moving in its direction.

Science uses a probabilistic strategy. We can hardly ever be sure what kind of research will pay off and what will not. Discoveries tend to be accidental, just like mutations in a genotype are. They can similarly lead to radical and dramatic changes. The example of penicillin, X-rays, and also "cold" nuclear reactions (i.e., reactions taking place in low temperatures—which, even though still impossible to carry out, may lead to a future breakthrough in energetics) confirm this random character of discoveries. And thus, because "nothing is known in advance," we have to "research whatever we can." This is the reason for this multidirectional expansion, which is so frequent in science. The probability of making discoveries is higher the higher the number of scientists engaged in a research project. But what are they researching exactly? Anything they are capable of researching. The situation when we are not studying x because we do not know whether x exists (the relationship between the amount of bacteria in a sick person's organism and the presence of penicillin in his blood can be one such x) is entirely different from the situation when we imagine that it might be possible to detect x only if we first examined a whole series of other phenomena: r, s, t, v, x, z—and yet this cannot be done because there is *no one* to do it. Having thus reached the limit of human resources, we will have to add all this neglected research that we will have to ignore consciously to the research that is not being undertaken because we are not even aware that it lies within our capabilities. With regard to the lack of scientists, the latter situation is represented by a line formation that, on entering an ever wider space, still maintains a fixed distance between any two members, as some new individuals are constantly joining in.

The second situation is represented by a line formation that becomes thinner the more it is stretched.

We should add that another adverse phenomenon can also be observed: the number of discoveries being made is not proportional to the number of scientists (where the doubling of the number of scientists would lead to twice as much research). The situation is rather as follows: the number of discoveries doubles every thirty years, whereas the number of scientists doubles every ten years. This may seem to contradict what we have said about the exponential growth of scientific information. Yet there is no contradiction here: the number of discoveries is also growing exponentially, but more slowly (its growth is expressed by a smaller exponent) than the number of scientists. All the discoveries taken together are just a fraction of all the information being acquired by science. It is enough to flick through the dusty piles of articles and dissertations produced with a view to obtaining an academic degree and now stored in university archives to see that not a single one of them has led to an at least partially useful result. Reaching the limits of the information capacity of science means significantly lowering the probability of making discoveries. What is more, as the curve of an actual increase in the number of scientists will be getting further away from the hypothetical curve of further exponential growth (which is not possible anymore) in its descent, the coefficient of such probability should be constantly decreasing from now on.

Scientific investigations resemble genetic mutations to some extent: valuable and groundbreaking ones are only a small part of the set of all mutations or of all investigations. And just as a population that lacks any significant reserves of "mutation pressure" is exposed to the risk of losing its homeostatic balance, a civilization whose "discovery pressure" gets weaker must mobilize all of its resources to try to reverse this gradient because it leads from a stable to an ever more unsteady balance.

And thus, remedies. But what remedies? Could cybernetics, this creator of "artificial researchers" or "Great Brains"—Generators and Transmitters of Information—be one of them? Or perhaps a development that transcends the "information barrier" could lead to the speciation of civilization? But what does it mean? Not much, because everything we shall be talking about is pure fiction. The only thing that is not fiction is the S-fold, that is, the descent of the curve of exponential growth that lies between thirty and seventy years away from us.

THE BIG GAME

What happens to a civilization that has reached an "information peak," that is, exhausted the transfer capacity of science as a "communication channel"? We shall present three possible ways out of such a situation. This will not cover all possibilities. We have chosen just three because they correspond to outcomes of a strategy game in which Civilization and Nature are set against each other as adversaries. We are familiar with the first level of the "game": civilization makes "moves" that lead to an expansive growth in science and technology. At level two, we have the information crisis. Civilization can either overcome it and thus win this level or it can lose. It can also score a "draw"—which will be a kind of compromise.

Without actualizing the possibilities offered to us by cybernetics, a win or a draw is impossible to achieve. A win means having to create channels of *any* capacity, no matter how big. Using cybernetics to create an "army of artificial scientists," promising as it sounds, is just a continuation of a strategy from the previous level. The very structure of science does not undergo any major transformation here; it is just that the scientific front is supported by some "intelectronic reinforcements." Contrary to what it may seem, it is a rather traditional solution. The number of "synthetic researchers" cannot be increased infinitely. By doing this, we can defer a crisis but not overcome it. An actual win demands a radical restructuring of science as a system that acquires and transmits information. We can imagine it in a form envisaged by many cyberneticists today—that of building ever greater "intelligence amplifiers" (which would not just become scientists' "allies" but which, thanks to their "intelectronic" supremacy over the human brain, would quickly leave scientists behind)—or in a form that would be radically different from any positions that are currently under discussion.

This would be a total rejection of a traditional approach to phenomena produced by science. The theory behind an "information revolution" of this kind can be summarized as follows: the idea is to "extract" information from Nature directly, without going through the brain, no matter if human or electronic, to create something like an "information farm" or "information evolution." This idea sounds entirely make-believe today, especially in such a unorthodox formulation when compared with the dominant position. We shall nevertheless look into it, but a little later and separately, as it requires some additional preliminary discussion. We shall do this not because of the confidence it can bestow on us (it is highly hypothetical) but rather because it is the only idea that could allow us to "break through the information barrier," that is, to achieve a full strategic victory in the game against Nature. For the time being, we shall only consider a natural process that contains a credible possibility of achieving such a solution. Genetics in its evolutionary dimension is looking into it. It is a way in which Nature *acquires* and *transforms* information, leading to its *growth* outside any brain—in the hereditary substance of living organisms. Yet, as we said earlier, we shall discuss such "molecular biochemistry of information" separately.

A draw is the second possible outcome of the game. Every civilization creates an artificial environment for itself while transforming the surface of its planet, its interior, and its cosmic neighborhood. Yet this process does not cut it off from Nature in any radical way; it only moves it further away from Nature. But the process can be continued so that an "encystment" of a civilization in relation to the whole Universe eventually takes place. Such "encystment," which could be enacted through a particular application of cybernetics, would facilitate the "tamponing" of excess information and the production of information of an entirely different kind. The fate of a civilization is first of all determined by its regulatory influences on its feedback with Nature. In placing various natural phenomena (oxygenation of coal, atomic disintegration) in the relation of feedback, we can get all the way to astroengineering. A civilization that is experiencing an information crisis and that already has access to feedback from Nature, and to sources of energy that will guarantee its existence for millions of years—while realizing that an "exhaustion of Nature's information potential" is not possible, whereas continuing with the current strategy may result in a defeat (because the constant march "inside Nature" will eventually lead to the dismantling of science as a result of its hyperspecialization and thus, possibly, to a loss of control over its own homeostasis)—will be able to construct an entirely new type of feedback, from within itself. Producing such "encystment" will involve having to construct "a world within a world," an autonomous reality that is not directly connected with the material reality of Nature. The emergent "cybernetic–sociotechnical" shell will enclose the civilization under discussion within itself. The latter will continue to exist and grow, but in a way that is not visible to an external observer anymore (especially one in outer space).

This may sound a little mysterious, but we are now capable of roughly sketching out the preceding idea in its different incarnations. We shall consider one or two of these incarnations in more detail later on; at the moment, we only want to highlight that such a compromise is not a fiction. This is because Nature does not stop us in any way from moving from our current state of knowledge to one required to score a "draw." Constructing a perpetual mobility machine or, say, introducing flights at faster-than-light speed are in turn fictions.

Let us now consider the possibility of a defeat. What will happen to a civilization that does not manage to overcome its crisis? It will become transformed from one that studies "everything" (as ours does at the moment) into one that only focuses on a few selected directions. With each one of those directions gradually beginning to experience the lack of human resources, their number will steadily decrease. Civilizations that are getting close to the exhaustion of their energy sources will no doubt focus their research activities on this area. Other more plentiful ones will be able to specialize in different areas. This is what I had in mind when I spoke about their "speciation," that is, the emergence of species, earlier on—although we are talking here about civilization species, not biological ones. In this view, the Universe is populated by numerous civilizations from which only a section focuses on astroengineering or, more generally, on cosmic phenomena (e.g., cosmonautics). It is possible that, for some, conducting research into astronomy is already a "luxury" they cannot afford—owing to the lack of researchers. Such a possibility does not seem very likely at first. As we know, the more extensive the development of science, the more connections there are between its various branches. One cannot constrain physics without adversely affecting chemistry or medicine. The opposite is also true: physics can draw new problems from outside its own field, for example, from biology. Put briefly, by limiting the speed of development of a research area that is considered less significant, we can adversely affect the areas on which we have chosen to focus. Besides, narrow specialization decreases homeostatic equilibrium. Civilizations that would be immune to perturbations of stellar nature but not to epidemics, or that would lack "memory" (having given up on studying their own history), would become crippled; they would become exposed to threats proportional to the size of those civilizations' one-dimensional character. This argument is correct. Yet "speciation" cannot be discounted as a possible solution. Does our civilization not show excessive hyperspecialization, even though it has not yet reached the "information barrier"? Does its military potential not resemble the enormous jaws and carapaces of Mesozoic reptiles, whose abilities in many other areas were so limited that they eventually determined their fate? Contemporary hyperspecialization has most certainly been caused by political rather than information or scientific factors. It could thus be reversed once the unification of humanity has taken place. Incidentally, this is where the difference between biological specialization and the specialization of a civilization would manifest itself: the former cannot be fully reversed, whereas the latter can.

A developing science is like a growing tree whose boughs divide into branches, which in turn divide into twigs. When the number of scientists stops growing exponentially, while the number of new "twigs," or new disciplines, continues to grow, a gap, or an inequality of information gain, will ensue. Research planning is only able to shift this process from one side to the other, back and forth. This is what we can describe as a shortfall. After thousands of years, the speciation of a civilization can develop in the three following directions: social, biological, and cosmic.

None of these appear in a pure form. The main direction of development is affected by a planet's conditions, the history of a given civilization, scholarly productivity or lack thereof in certain areas of knowledge, and so on. In any case, the reversibility of the changes that have already taken place and that are the consequence of decisions made earlier (with regard to abandoning, or continuing with, certain types of research) decreases with time, until we reach a point when those early decisions begin to have an enormous impact on life in its totality. Decreasing the number of degrees of freedom in a given civilization as a whole also decreases personal freedoms of its members. It may turn out to be necessary to introduce limitations to the birthrate or to professional choice. Speciation thus carries with itself unforeseeable dangers (because one necessarily has to make decisions, the consequences of which can reveal themselves after hundreds of years). This is why we consider it a defeat in the strategic game with Nature. Of course, the appearance of disturbances that are not subject to immediate regulation does not yet signify a downfall or a decline. The development of such a society could no doubt be pictured as a sequence of oscillations, rises, and falls, stretched over hundreds of years.

Yet, as we have already said, a defeat is a consequence of the non-application, or misapplication, of the possibilities opened up by the potential universality of cybernetics. Given that it is cybernetics that will ultimately decide about the results of the Big Game, we shall address it now with some new questions.[2]

SCIENTIFIC MYTHS

Cybernetics is twenty years old. It is therefore a young discipline, but it is developing at an astounding pace. It has its schools and directions, its enthusiasts and skeptics. The former believe in its universality, the latter are looking for limitations with regard to how it can be applied. Linguists and philosophers, physicists and physicians, communication engineers and sociologists have all taken it up. It is not uniform anymore because it has diversified into multiple branches. Its specialization is ongoing, just as is the case with the other sciences. And as every science creates its own mythology, cybernetics has one too. The mythology of science sounds like a contradiction in terms or like the irrationalism of empiricism. Yet every discipline, even the most scientific one, develops not only thanks to new theories and facts but also thanks to hypotheses and hopes of the scientists. Only some of those hypotheses pass the development test. The rest turn out to be illusions, and this is why they look like myths. The myth of classical mechanics was personalized in Laplace's demon—which, knowing the momentum and location of all the atoms in the Universe at a given point in time, was supposedly able to foresee its entire future. Of course, science rids itself of such erroneous convictions that accompany its development. But we only find out *post factum*, that is, from a historical perspective, which guesses were correct and which were badly framed presuppositions. The impossible becomes possible in the course of such transformations. Importantly, the goals themselves also change in the process. If you asked a nineteenth-century scientist whether the transmutation of mercury into gold, the alchemists' dream at the time, was possible, he would categorically deny it. A twentieth-century scientist knows that it is possible to transform atoms of mercury into atoms of gold. Does this mean that the alchemists were right, as opposed to the scientists? No, it does not, since that which was supposed to be a goal in itself, that is, gold burning in test tubes, has stopped being important for atomic physics. Nuclear energy is not only infinitely more precious than gold, it is also, more importantly, very new and very different from the alchemists' bravest dreams. It was the method used by the scientists that led to its discovery rather than the alchemical ways of their rivals.

Why am I speaking about this? Cybernetics today is haunted by the medieval myth of the homunculus, an artificially created intelligent being. The dispute about the possibility of creating an artificial brain that would show characteristics of a human mind has frequently engaged philosophers and cyberneticists. Yet it is a pointless dispute.

"Is it possible to transform mercury into gold?" we ask a nucleonic physicist. "Yes," he says, "but we are not really working on it. This kind of transmutation is not important to us and does not influence the direction of our work."

"Will it be possible to construct an electronic brain that will be an indistinguishable copy of a living brain one day?" "Most certainly it will, but no one is going to do it."

We thus have to differentiate between possibilities and realistic goals. In science, possibilities have always had their "negative prophets." The number of such prophets has at times surprised me, as has the passion with which they have been trying to prove the futility of constructing flying, atomic, or thinking machines. The most sensible thing we can do is refrain from arguing with those forecasters of the impossible— not because we have to believe in everything coming true one day but rather because, when drawn into heated debates, people can easily lose sight of what the real problems are. "Anti-homunculists" are convinced that in negating the possibility of a synthetic mind, they are defending the superiority of man over his creations—creations that, they believe, should never overtake the human genius. This kind of defense would only make sense if someone were really trying to replace man with a machine, not within a particular workplace but rather within civilization as a whole. But nobody intends to do this. The point is not to construct synthetic humanity but rather to open up a new chapter in the Book of Technology: one containing systems of any degree of complexity. As man himself, his body and brain, all belong to such a class of systems, such a new technology will mean a completely new type of control man will gain over himself, that is, over his organism. This will in turn enable the fulfillment of some age-long dreams, such as the desire for immortality, or even perhaps the reversal of processes that are considered irreversible today (biological processes in particular, especially aging). Yet those goals may turn out to be a fantasy, just as the alchemists' gold was. Even if man is indeed capable of anything, he surely cannot achieve it in just *any* way. He will eventually achieve every goal if he so desires, but he will understand before that that the price he would have to pay for achieving such a goal would reduce this goal to absurdity.

It is because even if we ourselves choose the end point, our way of getting there is chosen by Nature. We can fly, but not by flapping our arms. We can walk on water, but not in the way it is depicted in the Bible. Perhaps we will eventually gain a kind of longevity that will practically amount to immortality, but to do this, we will have to give up on the bodily form that nature gave us. Perhaps, thanks to hibernation, we will be able to travel freely across millions of years, but those who wake up after their glacial dream will find themselves in an unfamiliar world, since the world and the culture that have shaped them will have disappeared during their reversible death. Thus, when fulfilling our dreams, the material world requires us to undertake actions the realization of which can equally resemble a victory and a defeat.

Our domination over our environment is based on the feedback with natural processes, thanks to which coal emerges from mines, great weights can travel across enormous distances, while shiny cars leave the assembly line. All of this happens because Nature repeats itself in several simple laws, which are studied by physics, thermodynamics, and chemistry.

Complex systems such as the brain or society cannot be described in the language of such simple laws. Understood in this way, relativity theory and its mechanics are still simple, but the mechanics of thought processes is not simple any more. Cybernetics focuses its attention on those latter processes because it aims to comprehend and master complexity. Among the material systems known to us, the brain is the most complex one. Probably, or in fact almost certainly, it is possible to develop systems even more complex than that. We will get to know them once we have learned how to build them. Cybernetics is thus first of all a science of achieving goals that cannot be achieved directly.

"We've seen a model of a device consisting of eight trillion elements," we tell an engineer. "This device has its own energy center, locomotive systems, a hierarchy of regulators, and a timing belt that consists of fifteen billion parts. It can perform so many functions that we wouldn't even be able to list them all during our lifetime. Yet the formula that not only enabled the construction of this device but actually *constructed it* was fully contained within the cubic capacity of 8/1000 of a millimeter."

The engineer replies that this is impossible. He is wrong, because we were talking about the head of the human spermatozoon, which, as we know, contains all the information that is needed to produce a specimen of the *Homo sapiens*.

Cybernetics looks at such "formulas" not because of its "homunculist" ambitions but because it would like to be able to solve design tasks of a similar kind. It is still really far from being able to do this. However, it is only twenty years old. Evolution needed more than two billion years to come up with its own solutions. Let us say cybernetics will need another hundred or even thousand years to catch up with evolution: the difference in time scales is still to our advantage.

As far as "homunculists" and "anti-homunculists" are concerned, their arguments resemble the passionate debate between epigeneticists and preformists in biology. Such arguments are typical of the early stages of a new science. As this science develops further, there will be no traces of such arguments left. There will be no artificial people because it is unnecessary to have them. Nor will a "revolt" of thinking machines against man take place. At the core of this latter belief lies another old myth—a satanic one—yet no Intelligence Amplifier will turn into an Electronic Antichrist. All these myths have a common anthropomorphic denominator, to which machines' thinking activities are supposed to be reduced. This is a severe misunderstanding! True, we do not know whether, once they have crossed a certain "complication threshold," machines will not start manifesting signs of their own "personality." Should this happen, that personality will be as different from human personality as a human body is different from a microfusion cell. We can expect some surprises, problems, and dangers that we cannot even imagine today—but not a return of demons and monsters straight from the Middle Ages, dressed as technical larvae. We are certainly unable to imagine the majority of such future problems. Yet we shall still attempt to outline a few of them, in a series of thought experiments.

THE INTELLIGENCE AMPLIFIER

The general trend to introduce mathematical thinking into various sciences (including disciplines that did not previously use any math tools, such as biology, psychology, and medicine) is slowly extending to the humanities. For now, we have had some rare efforts in language studies (theoretical linguistics) and literary theory (the application of information theory to the study of literary texts, especially poetry). We are also beginning to see an unusual and unexpected phenomenon, that is, the inadequacy of (all kinds of) mathematics for actualizing some recent goals in the most advanced field, that is, tasks facing self-organizing homeostatic systems. By way of example, I want to list several crucial problems that have allowed experts to see this weakness of mathematics for the first time. They include attempts to construct an intelligence amplifier and a self-programming steering mechanism for industry. The biggest task was an attempt to construct a universal homeostat, whose complexity would be comparable to our own, human complexity.

The intelligence amplifier, proposed, I believe, as a realistic design project for the first time by Ashby,[3] is supposed to be equivalent as far as mind power is concerned to an amplifier of physical strength, where the latter term applies to every machine driven by a human. An automobile, a digger, a crane, a machine tool, and any other device in which a human being "is connected" to a control system as a source of its regulation rather than power is such a power amplifier. Contrary to what it may seem, deviations from the norm as far as the level of intelligence in an individual is concerned are not any more significant than similar deviations within the realm of physical fitness. An average intelligence quotient (measured with the most common psychological tests) is between 100 and 110; in extremely intelligent persons, it reaches 140–150, while its top limit, which is achieved very rarely, is around 180–190. An intelligence amplifier with a multiplier that equals the average of the worker's power multiplied by the machine he is operating in industry would give the intelligence quotient of 10,000. The possibility of constructing such an amplifier is not any less real than the possibility of constructing a machine that would be a hundred times more powerful than a human being. The chances of carrying out such a construction are in fact rather slim now, mainly because the current focus is on the construction of a different device: the control system for industry discussed earlier (a "homeostatic brain of an automatic factory"). However, I will pause by the amplifier example as it can demonstrate better the fundamental difficulty the constructor encounters here. Significantly, he has to construct a machine that is "cleverer than him." It is obvious that if he wanted to use traditional methods applied in cybernetics, that is, prepare an appropriate plan of action for the machine, he would not solve the set task, because this very program already defines the limits of the "intelligence" that the machine under construction can have. This problem may seem to be an irresolvable paradox—similar to the proposition to lift oneself off by one's hair (while also having a hundred-kilo weight attached to one's feet)—but this is only an illusion. Indeed, the problem seems irresolvable, at least according to today's criteria, if we are to postulate the need to put together a mathematics-based theory in advance of constructing the amplifier. Yet we can think of an entirely different approach to this task, which, for the time being, is only a hypothetical possibility. We lack any detailed knowledge about the inner design of the intelligence amplifier. We may not even need it. It may be enough to treat such an amplifier as a "black box," a device about whose inner workings and various states we do not have the faintest idea because we are only interested in the final outcome. Like every respectable cybernetic device, this amplifier is equipped with "input" and "output" points. Between them stretches the gray zone of our ignorance, but why should it matter if this machine was really to behave like an intellect with an intelligence quotient of 10,000?

As this method is new and has never been used previously, I admit it may sound like something from the theater of the absurd rather than like a technology-based manufacturing instruction. Yet here are some examples that, it is hoped, will make its application sound more probable. For instance, we can pour a little bit of powdered iron into a small aquarium containing a ciliate colony. (It has been done!) Together with their food, the ciliates are also consuming small amounts of the iron. If we now apply a magnetic field to the aquarium from the outside, it will influence the movements of the ciliates. Such changes to the field strength will actually be changes to the "input" signals of our "homeostat," while the "output" state will be determined by the ciliates' own behavior. It is not important that we do not know as yet how we could apply this "ciliate-magnetic" homeostat, or that, in this particular form, it is very far from our hypothetical intelligence amplifier in every way. The point is that, although we do not really know what the complexity of each individual ciliate is, and although we are unable to draft its construction diagram the way one drafts a machine diagram, we have nevertheless managed to construct a higher entity out of the elements that we know very little about, an entity that is subject to systemic laws and that has signal "inputs" and "outputs." Instead of ciliates, one can use, for example, certain types of colloids or run electric current through multiphase solutions. This can result in the precipitation of certain substances, which will change the conductivity of the solution as a whole. This, in turn, can lead to "positive feedback," that is, signal amplification. We should admit here that these trials have not yet resulted in any breakthrough and that many cyberneticists do not support such a radical departure from the traditional operations conducted on electronic parts, nor do they favor this search for some new substances or new construction materials that would partly resemble the building material of living organisms (a phenomenon that is by no means accidental!).[4]

Without deciding in advance on the result of such deliberations, we can now understand a little better how it is possible to construct systems that function in a way that suits us from "incomprehensible" elements. We are faced here, at an early design stage, with a significant methodological shift. Contemporary engineering behaves like someone who will not even try to jump over a ditch until he has determined on a theoretical level all the key parameters and the relations between them, that is, until he has measured local gravitation force and the strength of his own muscles and mastered the kinematics of his bodily movements and the characteristics of the control processes taking place in the cerebellum. An unorthodox technologist from the cybernetic school, in turn, is simply going to jump over this ditch, claiming, quite correctly, that if he manages to do it, then the problem itself will be solved.

He is drawing on the following fact here. Any kind of physical activity, such as the previously mentioned jump, requires some preparation and execution work on the part of the brain—which is nothing else than an extremely complex sequence of mathematical processes (as any activity in the brain's neural network can be reduced to such processes). Yet, at the same time, our jumper, who of course "carries in his head" all the mathematics connected with the jump, will not be able to write down its theoreticomathematical counterpart, that is, the appropriate scientific formulas and transformations. The reason for this is that, if it is to be conveyed in a way that is traditionally taught in schools and colleges, this kind of "biomathematics"—which is practiced by all living organisms, including ciliates—requires a repeated translation of the system-forming impulses from one language to another: from a word-free and "automatic" language of biochemical processes and of the transfer of neuron stimulations to a symbolic language. The formalization and constitution of the latter is performed by completely different sections of the brain than those that directly supervise and actualize the "innate mathematics" we have discussed. And thus the key issue is to make sure that the intelligence amplifier does not have to formalize, constitute, or verbalize anything but rather that it acts automatically and "naïvely," yet at the same time as efficiently and infallibly as the neural processes of our jumper, that is, that it does not do anything else apart from transforming "input" impulses in a way that provides a ready-made solution at the "output." Neither he, nor the amplifier, nor its constructor, nor anyone else will know how it is doing this, but we will obtain the only thing that is of importance to us: results.

THE BLACK BOX

In the old days, people understood both the function and the structure of their tools: a hammer, an arrow, a bow. The increasing division of labor has gradually narrowed down such individual knowledge, as a result of which in modern industrial society we have a clear distinction between those who operate devices (technicians, manual workers), those who use them (a person in an elevator, in front of a TV, or driving a car), and those who understand their design principles. No living person today understands the design principles of all the devices at the disposal of our civilization. Yet there is someone who has such understanding: society. Partial knowledge possessed by individuals becomes complete when we take into account all the members of a given social group.

The process of alienation, the separation of the knowledge about various devices from social consciousness, carries on. Cybernetics furthers this process, moving it to a higher level—since it is theoretically capable of producing things the structure of which will not be understood by anyone. A cybernetic device thus becomes a "black box" (a term frequently used by experts). A "black box" can be a regulator involved in a particular process (one that involves the production of goods, their economic circulation, the coordination of transportation, curing an illness, etc.). The important thing is for given "inputs" to correspond to given "outputs"—that is all. The "black boxes" constructed at the present moment are still quite simple, which is why an engineer–cyberneticist is able to understand the relationship between such pairs—which is represented by a mathematical function. Yet a situation may arise when even he will not know a mathematical representation of this function. The designer's task will be to build a "black box" which performs the necessary regulation. But neither the designer nor anyone else will know how this "black box" is performing it. He will not know the mathematical function representing the correlation between "inputs" and "outputs." The reason he will not know it is not because it will be impossible to find out but, first of all, because it will not be necessary.

By way of a rather useful introduction to the problem of the "black box," we can recall the story of a centipede who was asked how it was able to remember which leg to raise after it had lifted the eighty-ninth one.[5] The centipede, as we know, pondered the question, and not being able to find an answer, died of hunger since it could not move anymore. The centipede is in itself such a "black box," which fulfils designated functions even though it "has no idea" how it does this. The operating principle of a "black box" is very general and usually quite simple, such as "centipedes walk" or "cats catch mice." A "black box" contains an appropriate "inside program," which its various actions must follow.

A technologist today will begin building something by preparing the necessary designs and calculations. We can say he creates a bridge, a locomotive, a house, a jet, or a rocket twice: at first theoretically, on paper, and then in real life, a stage during which the symbolic language of his equations and designs, that is, his algorithm for action, "is translated" into a series of material activities.

A "black box" cannot be programmed with an algorithm. An algorithm is a ready-made program that predicts everything in advance. It is commonly said that an algorithm is a scientific, repeatable, and reproducible prescription that shows how to solve a particular task step by step. Any formalized proof of a mathematical thesis and a computer program that undertakes translation from one language into another are all algorithms. The idea of an algorithm comes from mathematics: this is why my use of this term with regard to engineering is a little unconventional. A math theorist's algorithm never lets him down: if someone has worked out an algorithm for a mathematical proof, he can be sure that this proof will never "collapse." But an applied algorithm used by an engineer can fail because it only seems to "foresee everything in advance." The strength of bridges is calculated with reference to particular algorithms, but this does not guarantee its robustness. A bridge may collapse if it is subjected to forces that are greater than those considered by its designer. In any case, if we have an algorithm for any process, we can learn, within certain limits, about all the subsequent phases, that is, all the stages, of this process.

Yet when it comes to very complex systems such as society, the brain, or the yet nonexistent "very large black boxes," it is not possible to gain such knowledge, as systems of this kind do not have algorithms. How are we supposed to understand this? We will no doubt agree that every system, including the brain and society, acts in a certain determined way. Its actions could therefore be represented symbolically. Yet this would not mean a lot because an algorithm must be repeatable; it must allow us to predict the system's future states, while a society that finds itself in the same situation on two different occasions does not have to behave in the same way. This is precisely the case with all highly complex systems.

How can one construct a "black box"? We know that it is possible to do it and that a system of any complexity can be constructed without any prior designs, calculations, or algorithms—because we ourselves are such "black boxes." Our bodies are controlled by us; we can give them specific orders, even though we do not know (i.e., we do not have to know) their inner mechanics. We are faced here once again with the problem of the jumper, who is capable of jumping but does not himself know how he does it, that is, he has no knowledge about the dynamics of nerve and muscle circuits that result in a jump. Every human being is thus an excellent example of a device that can be used without knowing its algorithm. Our own brain is one of the "devices" that is "closest to us" in the whole Universe: we have it in our heads. Yet even today, we still do not know how the brain works exactly. As demonstrated by the history of psychology, the examination of its mechanics via introspection is highly fallible and leads one astray, to some most fallacious hypotheses. The brain is constructed in such a way that while it enables our actions, it remains "hidden." Of course, this is not a consequence of any maliciousness on the part of our Designer, Nature—only a consequence of natural selection. The latter equipped us with the capacity for thinking when this capacity was useful for evolutionary purposes, which is why we are thinking beings—even though we do not know how we do this, since providing us with an understanding of this process was not in evolution's "interest." Evolution did not hide anything from us; it only eliminated from its activity any knowledge that was considered unnecessary "from its point of view." If this knowledge is *not* unnecessary—from our point of view—then we must gain it ourselves.

The uniqueness of the cybernetic solution, whereby a machine is completely alienated from the domain of human knowledge, has actually already been used by Nature for a long time now.

This may be true, someone will say, but man has been given his "black box," his body and brain, focused on finding optimal solutions to life problems, by Nature—which constructed them through a trial-and-error period lasting billions of years. Are we supposed to try to copy its results? And, if this is indeed the case, then how are we supposed to do it? We cannot, of course, seriously consider another evolution (one that would take place on a technical level this time). Such a "cybernetic evolution" would probably take billions, millions, or at least hundreds of thousands of years. How are we supposed even to begin this process? Should we approach this problem from a biological or an abiological perspective?

We do not know an answer to this. We will no doubt have to try all possible avenues, especially those that, for one reason or another, have remained closed to Evolution. Yet it is not our ambition to fantasize about some possible, that is, thinkable, "black boxes" that would become creators of Technology. We just want to pose the problem as such. We know that only a very complex regulator will cope with a very complex system. We therefore have to look for such regulators anywhere we can: in biochemistry, in living cells, in the molecular engineering of the solid body. Indeed, we know what we want and what we are looking for; we also know—thanks to the private tuition we got from Nature—that this task *can* be solved. We already know so much that we are halfway there!

THE MORALITY OF HOMEOSTATS

It is time to introduce moral issues into our cybernetic deliberations. But it is in fact the other way around: it is not we who are introducing questions of ethics into cybernetics; it is cybernetics that, as it expands, envelops with its consequences all that which we understand as morality, that is, a system of criteria that evaluate behavior in a way that, from a purely objective perspective, looks arbitrary. Morality is arbitrary just as mathematics is, because both are deduced from accepted axioms by means of logical reasoning. One of the axioms of geometry that we may want to agree on is that through a point outside a straight line, we can only draw one other line that is parallel to the first one. We can also reject this postulate: then we will have non-Euclidean geometry. The important thing is that we are aware that we act in a way that has been agreed in advance, as is the case with the approved geometric axioms, because this agreement and this decision are up to us. We can agree on a moral axiom saying that children born with a disability must be killed. We will then have a so-called "Tarpeian" morality known to us from history, the discussion and ultimate rejection of which was evoked by the recent infamous thalidomide scandal. It is often said that there exist transhistorical moral pointers. From this point of view, "Tarpeian morality,"[6] even in its most benign form (manifested, e.g., in postulating euthanasia for people experiencing agony as a result of incurable diseases), is immoral, criminal, and evil. What happens here is that we evaluate one moral system from the standpoint of another system. It goes without saying that we opt for this other, "non-Tarpeian" system, but if we agree that it emerged over the course of man's social evolution rather than via an act of revelation, we must take into account the fact that different moral systems are at work in different historical periods. The issue of the divergence between the morality we preach and the morality we practice complicates the matter, but such complications are of no interest to us because we shall only limit ourselves to presenting actual behavior and not its dissimulation—which is no doubt possible but which simply amounts to misinformation. The person who misinforms others practices a different kind of morality in his speech than in his actions. The very need for misinformation shows that the moral axioms he preaches are widely accepted in society—otherwise, he would not have to distort the facts. Yet these very same facts can have a diametrically opposed evaluation in different civilizations. Let us compare the moral aspects of contemporary and Babylonian prostitution. Babylonian "sacred prostitutes" gave themselves to men not for individual profit but rather because of some "higher ground": such behavior was accepted by their religion. They functioned in full compliance with the morality deduced from the religion of the time. Therefore, in their times and in their society, they did not deserve condemnation—unlike contemporary courtesans, since, according to today's criteria, prostitution is morally evil. The very same behavior thus finds two diametrically different evaluations in two different civilizations.

The introduction of cybernetic automation leads to some rather unexpected moral dilemmas. Stafford Beer, one of the American pioneers[7] of introducing cybernetics into large capitalist production plants, postulates the construction of a "homeostatic corporation." To illustrate how it might work, he offers a detailed discussion of the theory of regulation as applied to the working of a large steel mill. Its "brain" is supposed to optimize all the processes involved in the production of steel in a way that will make the production process as efficient and reliable as possible, free from disturbances to supply (of workforce, ore, coal, etc.) and to market demand and from inner systemic malfunctions (irregular production, undesirable increase in one's own costs, maximum efficiency per worker). According to Beer, such a production unit is supposed to be an ultrastable homeostat, which responds to every deviation from the state of equilibrium with inner reorganization, thus restoring this state. When presented with this theoretical model, some experts pointed out that it lacked "religion." Beer had consciously modeled his homeostatic steel mill on the working principles of a living organism. And the only criterion of "value" for a natural organism is its capacity for survival—at any cost. This means possibly also at a cost of annihilating other organisms. A biologist who understands that Nature lacks any "systems of moral judgment" does not consider the actions of hungry predators "morally evil." This then leads to the following question: can a "steel mill-organism" "devour" its competitors if it has to; that is, does it "have the right" to do it? There are more questions of this kind, perhaps some of them a little less drastic. Is such a homeostatic unit supposed to maximize production or profit? And what if, after a certain period of time, the production of steel will turn out to be superfluous as a result of ongoing technological changes? Should the "survival tendency," built into the "brain" of this production system, result in its complete redesign so that it reorganizes itself into, say, a producer of plastic? But why plastic, of all things? What is it supposed to be driven by in this process of its total reorganization: a chance to maximize its social usefulness? Or, again, profit?

Beer avoids answering these questions when he declares that above the "brain" of the steel mill, there is also a board of private owners. This board makes decisions of a general kind—which are decisions of the highest level. The brain only actualizes those decisions in an optimal way.

Beer thus reneges on the "autonomous–organic" principle of his own theory and shifts all the "moral" issues outside the system of the "black box"—onto the board of directors. But this avoidance is only apparent. The "black box," even with those limitations, will be making moral decisions, by, for example, laying off workers or reducing pay when the principle of the optimal organization of the steel mill as a whole requires it. It is easy to imagine a "struggle for life" erupting between Beer's "homeostatic steel mill" and some other homeostats, designed by different cyberneticists and employed on behalf of some other corporations. Either the competences of those latter homeostats will be so limited that they will always have to turn to their human manager to make a decision (as to whether a competitor should be bankrupted because an opportunity has presented itself, etc.) or their activities, loaded will moral consequences, will become more extensive in their scope. In the first case, the self-regulation principle of the homeostatic producer will be violated. In the second, homeostats will start to influence human lives in a way that has not been fully envisaged by their creators—which can then lead to an economic collapse in the country as a whole just because one of the homeostats will be performing its set task too well, by bankrupting all its competitors...

Why is the operational principle of the "black box" violated in the first case? Because such a "box" as a regulator is not like a human being, in the sense that one cannot ask questions (about the social consequences of its actions) of it at every stage of its decision-making process and expect it to be able to answer them. Incidentally speaking, even a human manager does not often know the far-reaching consequences of his decisions. The "black box" that is to "maintain the life" of the steel mill by responding to any fluctuations of "input" (price of coal, ore, and machines; salary costs) and "output" (market price of steel; demand for its different varieties) and the "black box" that also takes into account workers', and maybe even competitors', interests are two completely different devices. The first one will be a more efficient producer than the second. If we introduce some work legislation that applies to all the producers present on the market into the initial program, that is, the "axiomatic kernel" of the box's behavior, it will constrain the harmful behavior of the homeostat toward the workforce, but it may also increase such behavior toward, say, some competitor companies or toward steel producers in some other capitalist countries. The most significant issue is that a "black box" does not actually "know" when exactly it is acting harmfully and toward whom. We cannot expect it to inform us about such consequences of its decisions because, by definition, no one knows about its inner states, not even its designer–constructor. It is precisely such consequences of introducing homeostatic regulators that Norbert Wiener had in mind when he devoted a separate chapter of the new edition of his magisterial work, *Cybernetics*,[8] to the unpredictable results of their actions. It might seem that dangers of this kind will be nipped in the bud by constructing a "black box" of a higher type, as a kind of "ruling machine"—one that would not rule over people but rather over the "black boxes" of individual producers that would be subordinate to it. The discussion of the consequences of this step will prove extremely interesting.

THE DANGERS OF ELECTROCRACY

And thus, to avoid the socially harmful consequences of actions performed by "black boxes" that function as regulators of individual production units, we have placed The Black Box: The Regulator of the Highest Kind on the throne of economic power. Let us imagine its aim is to limit the freedom of the regulators of production by imposing on them, through programming (which amounts to legislation), the observance of the articles of labor law, the principle of loyalty toward one's competitors, the desire to liquidate the reserve workforce (i.e., to get rid of unemployment), and so on. Is something like this possible? In theory, yes. In practice, however, such conduct will be burdened with a great number of what we could euphemistically call "inconveniences."

The Black Box as a very complex system is indescribable. No one knows its algorithm—and no one *can* know it. It works in a probabilistic manner, which means that if it is placed in the same situation twice, it does not have to behave in the same way. Most important, the Black Box is also a type of device that, in the course of the actions it performs, learns from its own mistakes. The fundamentals of cybernetics tell us that it is impossible to construct one kind of Black Box: The Economic Ruler that would immediately become omniscient and that would be able to predict all the possible consequences of the decisions it makes. Yet, over time, the regulator will be getting close to such an ideal. We cannot determine how quickly this will happen. It will perhaps first lead the country to a series of terrible crises and then gradually pull it out from them. Perhaps it will announce that there is a contradiction between the axioms that are being introduced by means of the Operating Program (e.g., one cannot conduct an economically viable program of the automation of production and simultaneously aim to reduce unemployment unless one does many other things at the same time, e.g., introduce a state- or capital-supported reskilling program for those who are being made redundant as a result of the automation, etc.). What then? It is difficult to offer an analysis of such a complex issue. We can only say that The Black Box, either as a regulator of production in one of its lower sections or as a universal regulator on the scale of the state, always acts from the position of partial knowledge. There is no other way. Even if, after much trial and error—which would make millions of people's lives miserable—the Black Box: The Economic Ruler was to gain an enormous amount of knowledge, incomparably larger than the sum total of the knowledge possessed by all the economists of capitalism put together, there is still no guarantee that our Black Box will not try to prevent future fluctuation caused by some new factors in a way that will shock everyone, including its own designers. We have to consider such a possibility by looking at a specific example.

Imagine that the prognostic part (the "subsystem") of the Black Box: The Economic Regulator notices that the state of homeostatic equilibrium—which, after many oscillations, has at last been achieved—is facing some kind of danger. This danger results from the fact that population growth is higher than the ability to satisfy human needs that the current state of our civilization presents us with—in the sense that, given the current birthrate, the standard of living will begin to decrease from next year on, or thirty years from now on. Also, it suddenly turns out that some information about a chemical agent that has been detected has entered the "black box" through one of its "inputs." This chemical agent is entirely harmless as far as our health is concerned. Yet, when taken regularly, it causes a decrease in the speed of ovulation in women so that they are capable of conceiving on just several days a year rather than on more than a hundred days. The "black box" then makes a decision to introduce this chemical, in required microscopic doses, into the drinking water in all of the water supply systems across the country. Of course, to be successful, this operation needs to remain a secret, otherwise the birthrate parameter will once again start to show a growth tendency because many people will no doubt try to drink the water without this chemical, for example, from rivers or wells. The "black box" will thus be posed with the following alternative: either to inform society about the situation and meet with opposition on its part or not to inform it and thus save the existing state of equilibrium, for the common good. Let us assume that to protect society again the move of the "black box" toward such "cryptocracy," its program includes the publication of all intended changes. The "black box" also has an inbuilt "safety brake," which becomes operational every time the situation like the preceding one arises. The "advisory body" of the regulator, consisting of humans, will therefore thwart the plan to introduce this fertility-reducing substance into the water. Yet there are not going to be many situations as straightforward as the one described here, which means that in the great majority of cases, the "advisory body" will not know whether the "safety break" needs to be activated at a given moment. In any case, its too frequent activation can make all the regulatory activity of the "box" rather illusory and hence throw society into total chaos—not to mention the fact that it is highly unclear whose interests this "advisory body" will actually be representing. In the United States at the present moment, it would, for example, prevent the introduction of free health care and of a pension system (something that had actually been undertaken there by Congress, with the role of the "box" that had proposed such a change played by President Kennedy—who was then halted when the "safety break" was activated). Indeed, we must not underestimate the "black box." If its activities are "halted" once, twice, or three times, it will most likely develop a new strategy. It will, for example, aim to ensure that marriages take place relatively late in people's lives or that having a small number of children brings particular economic benefits, and if all of this still does not bring the required results, it will try to reduce population growth in some other, even more roundabout way. Let us imagine that there exists a medication that prevents tooth decay and that, at a certain point during the course of treatment, causes gene mutation. The new ("mutated") gene does not decrease fertility on its own but only after encountering another mutated gene—one that developed as a result of using a different medication that had been in circulation for quite a while. Imagine that the latter medication has liberated the male half of the population from the worry of premature hair loss. The "black box" will thus use all possible ways to spread the use of the medication that prevents tooth decay and, consequently, get the result it wants: after a certain period of time, the two (recessive) genes, which have undergone mutations within the population, will increase in number to such an extent that they will be frequently encountering each other. This will lead to a decrease in population growth. One might ask why the "black box" will not have informed anyone in due time about its proposed move—did we say not that, according to a principle that had been built into it, it *must* provide information about any changes it intends to undertake?

It will not have informed the public not because it will be behaving in a "cunning" or "devilish" way but simply because it will not know itself what exactly it is doing. It is no "electronic Satan"; it is not an omniscient being capable of reasoning the way a human, or a superhuman, is, but merely a device that is constantly looking for connections, for statistical correlations between particular social phenomena—of which there are millions and thousands of millions. Given that, as a regulator, it should optimize economic relations, the high standard of living of the general population is a state of its own equilibrium. Population growth is a threat to such equilibrium. One day the "box" will notice a positive correlation between a *decrease* in population growth and the use of the medication that prevents tooth decay. It will notify the "advisory board" about this, which will then undertake some research and conclude that this medication does *not* lead to a decrease in fertility (the scientists who are members of this "board" will be conducting experiments on animals, and animals, of course, do not use any medication to prevent hair loss). The "black box" will not be able to hide anything from the public because it itself does not know anything about genes, mutations, or the causative relation between the introduction of these two types of medication and the decrease in fertility. The "black box" will only detect the desired correlation and then try to use it. This example also smacks of primitivism, but it is not improbable (as the thalidomide affair testifies). In reality, the "black box" will be acting even more indirectly and gradually, "not knowing what it is doing," because it is aiming at a state of ultrastable equilibrium, while the correlations between phenomena it detects and then uses to maintain this state are a consequence of the processes it does not investigate—processes whose causes it does not know (or does not have to know). It may eventually turn out, in a hundred years' time, that the price that had to be paid for increasing the standard of living and decreasing unemployment is a short tail grown by every sixth child or the general lowering of intelligence in society (since more intelligent people cause more problems for the regulatory activity of the machine, which is why it will aim to reduce their number). I hope it is now clear that the "axiomatics" of the machine will not be able to consider all the options in advance, from the "short tail" to the general increase in idiocy.

We have thus conducted a reductio ad absurdum of the theory of the Black Box: The Regulator of the Highest Kind in the human population.

CYBERNETICS AND SOCIOLOGY

The failure of the "black box" as a regulator of social processes is caused by several things.

First, there is a difference between trying to regulate a system that has been given *in advance,* that is, to demand a regulator that will maintain the homeostasis of a capitalist society, and trying to regulate a system that has been *designed* by using suitable sociological research.

In principle, any complex system can be regulated. Yet neither the methods used nor their consequences will necessarily find acceptance among the regulated if this regulated entity is society. If a system such as the capitalist one has a tendency to fall into self-induced oscillations, aka economic upturns and downturns, the regulator may decide to perform maneuvers that will meet with some strong objections if it wants to eliminate these oscillations. We can easily imagine the reaction of the owners of Stafford Beer's "homeostatic steel mill" if its "brain" were to decide that, to maintain homeostasis, it is necessary to nationalize the means of production or at least to reduce profits by half. If a particular system is given in advance, its rules of behavior within a particular range of parameters are also given. No regulator can suspend the working of these laws because this would amount to performing miracles. A regulator can only choose between various states of the system that can be actualized. Biological regulation—evolution—can increase either the size of organisms or their mobility. It is impossible to have a whale as agile as a flea. A regulator thus has to look for compromise solutions. If certain parameters are "untouchable"—such as, for example, private ownership—then the choice of possible maneuvers is reduced, which can lead to a situation in which the only way to maintain "stability" within the system is by using force. We have placed stability in quotation marks here because we are talking about the stability of a collapsing building, which is being held together with iron clamps. In trying to contain self-induced oscillations of a system by force, one is abandoning the principle of homeostasis, since the system's self-organization is replaced with violence. This is how historical forms of power, such as tyranny, absolutism, fascism, and so on, came about.

Second, from the regulator's point of view, the elements of the system only need the knowledge that is absolutely essential for it to operate. This principle—which does not meet with any opposition from a machine or a living organism—goes against human desires. As elements of the social system, we wish to have information that does not just concern what we ourselves are supposed to do but that also applies to this system as a whole.

Given that the "nonhuman" regulator (the "black box") that is connected to society aims at various forms of cryptocracy, any form of social homeostasis that uses a "ruling machine" is undesirable. The reason for this is that, if the second of the examples mentioned earlier was to take place, that is, the regulation of the system designed on the basis of some sociological research, there will still be no guarantees that the state of equilibrium that has been achieved will not find itself under threat in the future. The goals that society sets itself are not the same in any given period. Homeostasis is not an "existence for the sake of existence" but a teleological phenomenon. Initially, in the design phase, then, the goals of the regulator and of a society that remains under its control will overlap, but then antagonisms may appear. Society cannot give up the burden of having to decide about its own fate by sacrificing this freedom for the sake of the cybernetic regulator.

Third, the number of the degrees of freedom that a society under development shows is higher than the number of such degrees of freedom within the realm of biological evolution. A society can undertake a sudden systemic change; it can gradually enhance various domains of its activity by introducing "cybernetic rulers"—who will be granted limited yet extensive power. All such revolutionary changes are impossible within bioevolution. The inner freedom of action in a society is not just greater than the freedom of a living organism seen in isolation (to which a society has sometimes been compared) but also than the freedom of all the organisms within evolution taken together.

History has taught us about different systems of power. When it comes to their classification, they function as "types" or superior units. The dynamics of systemic feedback is informed—but not determined–by the economics at work. And thus the same system can realize different economic "models," within a certain range of parameters. Besides, a given type of system is not determined by individual values of those parameters. In the capitalist system, cooperativeness can blossom, yet this does not stop it from being capitalist. Only the simultaneous change to a series of significant parameters can change not just the economic model but also the systemic model that is superior to it because this is when the totality of social relations will undergo a transformation. However, a distinction needs to be made once again between a regulator of a *given* system and a regulator that can transform a given system into a *different* one (should it decide such a change is required).

Given that people wish to make decisions with regard to what kind of system they will live in, what kind of economic system they will implement, and last but not least, what kinds of goals their society is supposed to aim for—as this very same society can either *rather* develop space exploration or *rather* focus on biological autoevolution—it is possible yet undesirable to resort to the machinic regulation of social systems.

It is a different matter when we consider applying such regulation to solving individual tasks (economic, administrative, etc.) and to modeling social processes in digital machines or other complex systems to learn as much as possible about their dynamic laws. There is a difference between using cybernetic methods to study social phenomena to improve the latter and awarding ruling power to cybernetic progeny. What we need is cybernetic sociology, not a theory of how to design and engineer ruling machines.

How should we approach the subject of cybernetic sociology? This is too broad a topic for us even to begin to sketch it out here. Yet, so that it does not remain just an empty concept, let me present a few introductory remarks.

The homeostasis of a civilization is a product of man's social evolution. Since times immemorial, all societies through history have practiced regulatory activity aimed at maintaining systemic equilibrium. Of course, people were not aware of the nature of their collective actions, just as they did not realize that their economic and manufacturing activities gave shape and form to political systems. Societies that had the same level of material development and analogous types of economics saw different structures emerge within this domain of postproductive life that we call superstructure. We can say that, just as a given degree of group cooperation on a primitive level leads to the development of speech—as an articulated and diversified system of communication—but does not determine what kind of speech it is going to be (a language from the Finno-Ugric group or a different one), a given degree of development with regard to the means of production leads to the development of social classes but does not determine what kinds of human ties will be established within them.

A given type of language, just like a given type of social ties, develops randomly (in a probabilistic way). What to an observer from a different culture may look like a most irrational type of social ties, obligations, imperatives, and prohibitions has practically always aimed at the same goal: reducing the individual spontaneity of action and its diversity—which is a potential source of disturbance to the state of equilibrium. An anthropologist is primarily interested in the content of beliefs and in social and religious pragmatics—initiation processes, the nature of family relations in a given society, the relations between the sexes, intergenerational relations, and so on. A sociologist–cyberneticist needs to disregard to a large extent the content of such rituals, regulations, and rules of behavior and search instead for the key aspects of their structure, as it is the latter that organizes the mechanism of feedback, that is, the regulatory system whose nature determines the degree of freedom for an individual, together with the degree of the stability of the system understood as a dynamic whole.

We can move from this analysis to an evaluation because, thanks to the plasticity of his nature, man can get used to functioning within diverse cultural models. Yet we tend to reject the majority of such models because their regulatory structures meet with opposition on our part. Such opposition is, of course, entirely rational, manifesting as it does some objective evaluative criteria, and is not merely based on what appeals to us as members of a particular culture. Sociostasis does not actually require the reduction of the diversity of action and thought, and thus of personal freedom, that has been practiced for a long time. We can say that the majority of regulatory systems, especially in primitive societies, were characterized by a significant restriction overload. An excess of such restrictions in family, social, personal, or erotic life is as undesirable as their deficit. There must exist an optimal regulatory stage of imperatives and prohibitions for a given society.

This is a very brief outline of one of the many topics with which a sociologist–cyberneticist occupies himself. His field is concerned with examining historical systems and offers a theory of how to construct optimal models of sociostasis—"optimal" with regard to freely chosen parameters. As the number of key factors is enormous, it is impossible to create a mathematically driven "ultimate formula for a society." We can only approach this problem by getting closer and closer to it, via studying ever more complex models. And thus we return in our conclusion to "black boxes"—not as future "electronic omni-rulers," or even superhuman sages, passing verdicts about the fate of humanity, but just as an experimental training ground for the scientists, a tool for finding answers to questions that are so complex that man will not resolve them without such help. Yet the decision itself, as well as the plan of action, should always remain in human hands.

BELIEF AND INFORMATION

For hundreds of years, philosophers have been trying to prove logically the validity of induction: a form of reasoning that anticipates the future on the basis of past experience. None of them have succeeded in doing so. They could not have succeeded because induction—whose beginnings lie in the conditioned response of an amoeba—is an attempt to transform incomplete into complete information. It thus constitutes a transgression against the law of information theory, which says that in an isolated system, information can decrease or maintain its current value, but it cannot increase. Yet induction—be it in the form of a conditioned response of a dog (a dog "believes" that it will be fed after the bell has rung because it has always been like this up until now and conveys this "faith" by salivating) or in the form of a scientific hypothesis—is being practiced by all living beings, including man. Acting on the basis of incomplete information, which is then completed through "guesswork" or "speculation," is a biological necessity.

Thus it is not as a result of some kind of anomaly that homeostatic systems manifest "belief." The reverse is true: every homeostat, that is, regulator, which aims to maintain its significant variables within the parameters whose transgression threatens its existence, must show "belief" or act on the basis of incomplete and uncertain information as if this information was both complete and certain.

Every action starts from a position of knowledge that contains gaps. In the light of such uncertainty, one can either refrain from action or undertake actions that involve risk. Refraining from action would mean stopping life processes. "Belief" stands for expecting that what we hope will happen is going to happen; that the world is the way we think it is; that a mental image is equivalent to an external situation. "Belief" can only be manifested by complex homeostats because they are systems that actively react to a change in their environment—unlike nonliving objects. Such objects do not "expect" or anticipate anything; in Nature's homeostatic systems, such anticipation precedes thought by far. Biological evolution would not have been possible if it had not been for that pinch of "belief" in the effectiveness of reactions, aimed at *future* states, that has been built into every particle of the living substance. We could represent a continuous spectrum of "beliefs" manifested by homeostats—from protozoa all the way to the human, with his scientific theories and metaphysical systems. A belief that has been confirmed many times by an experiment becomes more and more credible and is thus transformed into knowledge. The induction method by no means offers certainty, yet it is justified, since in a great number of cases, it is crowned with success. This is due to the nature of the world, to the fact that it is subject to various laws that can be detected through induction, yet sometimes the conclusions drawn via inductive reasoning turn out to be wrong. In that case, the model produced by a homeostat is at odds with reality, and the information obtained is false, which means that the belief based on it (that the world is like this) is also false.

Belief is thus a transient state—until it gets verified empirically. If it becomes too independent from such verification, it is transformed into a metaphysical construct. The peculiarity of such belief stems from the fact that realistic actions are being undertaken to achieve unrealistic goals, that is, goals that either cannot be realized at all or that can be realized but not via the action that is being performed. An attainment of a realistic goal can be verified empirically, an attainment of an unrealistic goal only via the kind of reasoning that matches external or internal states with dogmas. And thus we can check by means of an experiment whether the machine we have constructed is working but not whether man is going to be saved. Actions aimed at achieving salvation are realistic (particular behavior, fasting, performing good deeds, etc.), while the goal itself is unrealistic (because in this case, it is situated "in the next world"). Sometimes this goal is situated "here," for example, when one is praying for a natural disaster to be averted. An earthquake can stop; thus the goal has seemingly been achieved, yet the link between the prayers and the avoidance of a disaster does not emerge from the relations within Nature we discovered empirically but is rather a consequence of a reasoning that matches the state of prayer with the state in which the Earth finds itself at a given moment. Belief thus leads to an excessive use of the inductive method because the conclusions drawn by means of induction are either directed toward "the next world" (i.e., an empirical "nowhere") or supposed to establish connections within Nature that do not exist in it (every evening when I start to make scrambled eggs, stars appear in the sky; yet the conclusion that there exists a connection between my preparation of an evening meal and the appearance of stars is a case of false induction, which can become the subject of belief).

Like every science, cybernetics is unable to determine the existence of transcendent relations. Yet belief in such entities and relations is an entirely genuine phenomenon in the world. It is because belief is a form of information—which is sometimes true (I believe that the center of the Sun exists, even though I am never going to see it) and sometimes false. With this, we have been aiming to show that false information, as a directive for action undertaken within a certain environment, usually leads to failure. Yet the very same false information can fulfill other important functions within a homeostat. Belief is useful both on a psychological level, as a means of achieving spiritual equilibrium (as it manifests the usefulness of all kinds of metaphysical positions), and on the level of corporeal phenomena. Interventions that change the material state of the brain (the introduction of certain substances into it via blood) or its functional state (prayers, meditation practices) help to invoke subjective states that have been known across all times and religions. There is freedom of interpretation with regard to such states of consciousness, but those interpretations then become solidified as dogmas within given metaphysical systems. We hear references, for example, to "metaconsciousness" or "cosmic consciousness," to the merging of the individual "I" with the universe, to the annihilation of this "I," or to states of grace. Such states are genuine phenomena in an empirical sense because they are repeatable and can be evoked again through the performance of certain practices. Psychiatric terminology deprives those states of their mystical character, which of course does not change the fact that their emotional content can be more valuable than that of any other experiences for someone who is experiencing them. Science does not question either their existence or their significance for someone who is experiencing them; it only claims, contrary to metaphysical theories, that such states cannot be considered acts of cognition because cognition means increasing the amount of information about the world, and no such increase takes place during those states.

Significantly, the brain as a very complex system can adopt states that are more or less probable. The states with a very low degree of probability are those in which, during the brain's combinatorial activity, statements of the kind "energy equals mass times the velocity of light squared" are formulated within it on the basis of the information already possessed by it. This theorem can then be verified, and numerous consequences can potentially be drawn from it—which will eventually lead to astronautics, to the construction of devices that generate artificial gravitational fields, and so on and so forth.

States of "metaconsciousness" also result from the combinatorial activity of the brain. Even though the experience of such states can evoke the most sublime spiritual feeling, their information value amounts to zero. The process of cognition involves increasing the amount of information in one's possession. The information value of mystical states amounts to zero, which is evident in the fact that the content of such states cannot be transmitted and that it cannot increase our knowledge about the world in any way (so that it can then be applied, as in the preceding example).

The opposition discussed previously is not aimed to support triumphant atheism; this is not what we are concerned with. The only thing that matters to us is that the previously mentioned states are accompanied by the sensation of experiencing some ultimate truth, which is so intense and overwhelming that, having experienced it, one then looks with contempt or pity at the empiricists, miserably bustling around some trivial worldly matters. Two things thus need to be said. First, the divergence between "experiential truth" and "scientific truth" would perhaps be irrelevant if the first one did not claim some kind of superiority for itself. And if this is the case, then we should point out that the person undergoing the experience would not exist at all if it had not been for such lowly empiricism, which was initiated a long time ago by the Australopithecus and cavemen. Indeed, it was empiricism rather than any states of "higher cognition" that enabled the construction of civilization over a period of several hundred thousand years—which in turn turned man into the dominant species on Earth. If the opposite had been the case, primitive man—having experienced some of those higher states—would have been driven out by other animal species in the process of biological competition.

Second, the previously mentioned states can be evoked by taking various chemical compounds, that is, psilocybin (which is an extract from certain types of mushrooms). The person under observation, while realizing that his state has been caused in a nonmystical way, is nevertheless experiencing an exceptional intensity of emotions and sensations, so that the most common external impulses are being received as earth-shattering revelations. One can experience such a state even without psilocybin, for example, in a dream, when one wakes up deeply convinced that he had existential truth revealed to him in his dream. Having come around, he realizes that this was a sentence of the kind "bippos are pallowing in terpentine."

And thus a physiologically normal brain can achieve sublime sensations that are called mystical only after it has dutifully followed the course of treatments prescribed by a given ritual, or—in exceptional situations and infrequently—in a dream. Without having to believe in their extrasensory character, the same states can be "facilitated" through psilocybin, peyote, or mescaline. Such facilitations are so far only provided by pharmacopoeia, but, as we shall discuss later on, we can expect neurocybernetics to open up some new possibilities in this area. Let me make it clear that we are not discussing here whether states of this kind *should* be evoked; we are only saying that, even when there is no "mystical emergency service," it is entirely possible to evoke them.

Bodily consequences of belief are not any less far-reaching than psychological ones. So called miraculous healings, salutary effects of healers' therapies, and the curative power of suggestion where there is no evidence of mystification are all consequences of the working of a particular belief. It often happens that to achieve a particular effect, no introductory practices are necessary. For example, we know of some cases in which, having painted his patient's warts with some neutral dye, the doctor assures the patient authoritatively that the warts are going to disappear in an instant—which is indeed what often happens. What is key here is that it would be futile for the doctor to apply this procedure to himself or to one of his colleagues because the knowledge about the superficial character of this procedure, *the lack of belief* in its curative properties, would prevent the nerve mechanisms that lead to the shrinking of the vessels nourishing the wart in the "believer," and to its eventual disappearance, from being activated. In certain situations, then, *false* information may prove to be more effective than true information—with one significant reservation. The working of such information stops at the limits of an organism and, obviously, does not take place outside it. Faith can heal the believer, but it cannot move mountains—contrary to what has been said about it. On mountaintops in Ladakh, specially chosen lamas are attempting to bring rain on their country—which has been suffering from permanent droughts—through prayers. The prayers are not really working, but the believers remain convinced that some spiritual forces are stopping the lamas from carrying out their task. It is a beautiful model of metaphysical reasoning. I, too, can assure you that, thanks to a certain demon, I am capable of moving mountains, and it is only the influence of another demon, or perhaps an antidemon, that is preventing me from shifting this particular mountain.

Sometimes an act of faith itself is enough to enact the desired changes within a system (i.e., to cure warts). Sometimes, as with mystical states, some prior training is needed to achieve such results. Indian yoga is one of the most thoroughly codified and extensive kinds of training systems. Apart from bodily exercises it also contains spiritual ones.

As it turns out, man can gain mastery over his organism to a much higher degree than happens normally. He can regulate the amount of blood supply that reaches different body parts (which is actually what lies at the heart of the "disappearance" of warts) and also the working of the organs innervated by an autonomous system (heart, bowels, urinary and reproductive system) by slowing down, accelerating, or even changing the direction of the physiological action in the entrails area (reversing peristaltic bowel movement, etc.). Yet even such penetration of the will into the autonomous behavior of an organism, surprising as it is, has its limitations. As a superior regulator, the brain only has partial control over the body that is subordinated to it. It cannot, for example, slow down the aging process or an organic disease (atherosclerosis, cancers) or influence the processes that take place in the genotype (e.g., mutations). It can decrease cellular transformation, but only within a relatively small range of parameters. Thus stories such as those about the yogi who were able to survive buried in the ground for a long time turn out to be exaggerated or false on their verification—the suspension of life functions similar to that experienced by animals in hibernation (bats, bears) does not occur in the yogi.

Biotechnology allows for a significant widening of the scope of regulations available to a human organism: hypothermic states or even states that are close to clinical death have already been achieved via pharmacological and other interventions (e.g., by cooling down the body). Thanks to an "enhanced" biotechnological method, it will no doubt be possible to obtain results that used to be achieved through extreme perseverance and with much effort and sacrifice. It will thus be possible to achieve states (such as reversible death) that lie outside the reach of yoga or some other nonscientific method.

Put briefly, in the two areas discussed earlier, technology is clearly capable of successfully competing against belief as a source of spiritual equilibrium or intervention into the otherwise inaccessible realm of the organism's life functions, or even as a facilitator of "states of metaconsciousness" or "cosmic wonder."

To return to the problem of belief and information, we can now summarize our conclusions. The influence of the information entered into the homeostat depends not so much on whether this information is *objectively* false or true but rather, on one hand, on how predisposed the homeostat is to consider it true and, on the other, on whether the regulatory characteristics of the homeostat allow it to react in response to the information entered. To make it work, both postulates need to be fulfilled. Belief can heal me, yet it will not make me fly. It is because the first activity lies within the regulatory realm of my organism (although not always within the realm of my conscious will), the second one outside it.

The relative autonomy of the subsystems that make up the organism can lead a cancer sufferer who believes in the success of the therapy he is undergoing to feel better on a subjective level, despite the material failure of his treatment. Yet it is a subjective conviction, which results from an anticritical and selective working of belief (the sick person will not notice certain deterioration symptoms, e.g., the growth of a palpable tumor, or will somehow "explain them away," etc.). It cannot last a long time, and it will lead to a sudden physical collapse when the disparity between the real and the imaginary states of the organism becomes too significant.

It is interesting why true information can sometimes turn out to be less effective than false. Why cannot the medical knowledge of the physician—who is familiar with the belief-induced mechanism (the shrinking of vessels, which leads to the disappearance of the wart)—compete with the entirely false assessment of the situation by the patient, an assessment that does actually lead to him being cured? We can only suspect why this is so. There is a difference between knowing something and experiencing something. We can have knowledge about what love is, but it does not mean that we will be able to experience it on the basis of this knowledge. The neural mechanisms of cognitive acts are different from the mechanisms of "emotional involvement." The former only function as a transmission station for belief—which, through the simultaneous activation of the latter mechanisms, opens an information channel that subsequently leads to the unconscious shrinking of the vessels in the skin. We do not know the exact mechanism of this phenomenon. It is because we do not know enough about how the brain works. It is not only a "gnostic machine" but also a "believing machine"—something that neither psychologists and physicians nor neurocyberneticists should forget about.

EXPERIMENTAL METAPHYSICS

By metaphysical information, we understand a kind of information that is not subject to empirical verification—either because such verification is impossible (we cannot verify empirically the existence of purgatory or nirvana) or because such information by its very nature defies the criteria of experimental verification (i.e., in everyday understanding, religious truths cannot or should not be verified empirically).

If this is the case, then the phrase "experimental metaphysics" seems to be a contradiction in terms, because how can we analyze something by means of an experiment if it, by definition, does not yield itself to experimentation and is impossible to be judged on its basis?

It is only an apparent contradiction, as the task we have placed before ourselves is relatively modest. Science cannot say anything about the existence or nonexistence of transcendent phenomena. It can only examine or create conditions in which *belief* in such phenomena manifests itself—and this is what we shall be discussing.

The emergence of metaphysical belief within a homeostat is a state in which any further changes to its input, no matter how contrary to the produced model of the existential situation, are unable to disturb it any more. Prayers may remain unheard, reincarnation may be abolished on the basis of an internal logical contradiction, religious writings may contain some obvious untruths (in an empirical sense), yet such facts do not really shake belief as such. Of course, a theologian will say that if someone has lost his faith on encountering those facts, he only had "weak" or "small" faith in the first place, because true faith means that nothing—that is, no future "inputs"—will be able to abolish it. In practice, a form of selection often takes place. A metaphysical system is never truly consistent. Desperately trying to confirm it with empirical facts leads to a situation in which the kinds of inputs that seem to confirm the validity of the belief are accepted as its additional verification (someone is praying for someone else to be cured, and then this other person gets better; during the drought, a sacrifice is being made, and then it starts to rain). At the same time, the inputs that remain in contradiction with the belief are ignored or "explained away" with the help of an extensive arsenal of arguments that the metaphysical system has created in the course of its historical development.

Significantly, empirical verifiability is the only necessary and required characteristic of scientific propositions that differentiates them from metaphysical ones, yet the nature of a proposition is not determined by the presence of unverified information within it. Thus, for example, unified field theory, which Einstein developed toward the end of his life, does not bear any consequences that could be tested by means of an experiment. The information contained within field theory has not therefore been verified, yet it is not metaphysical, since its (so far unknown) consequences will be subject to an experiment, as long as we manage to draw such consequences from the formula. The information contained in Einstein's formula is therefore to some extent "suspended," "latent," or awaiting an opportunity to be verified. The formula itself needs to be seen as an *attempt* to present a general law about material phenomena, an attempt whose truth or falsity we have not yet managed to determine. Of course, there is a difference between expressing a conviction when making such an attempt about the possible behavior of matter and *believing* that it certainly behaves in this way. A scientist's proposition may be the product of a flash of intuition, and its validation with facts can be rather weak at the moment of its formulation. What is decisive, however, is the scientist's *willingness* to subject this proposition to empirical verification. And thus it is not the amount of information in his possession but rather the attitude or stance toward it that differentiates the scientist from the metaphysician.

The division of labor in a given civilization is accompanied by a phenomenon we may call "the division of information." It is not only that we do not do everything ourselves; it is also that we do not learn about every single thing *directly*. We learn at school that there is this planet called Saturn, and we believe it, even if we are never to see it ourselves. Statements of this kind can actually be verified by means of an experiment, although not always directly. We can see Saturn, but we cannot currently experience the existence of Napoleon or of biological evolution. Yet directly unverifiable scientific propositions have logical consequences that can be verified empirically (the historical consequences of Napoleon's existence; facts that testify to the evolution of life). A scientist should adopt an empirical stance. Every change to inputs (some new facts) that contradicts the model (the theory) should influence this model (putting into doubt its adequacy with regard to the situation modeled). Such a stance represents an ideal rather than a reality. Many theories that are rather widely accepted as scientific today have a purely metaphysical character, for example, the majority of psychoanalytic theories.

We cannot enter into a proper debate about psychoanalysis here as this would divert us from our main path, but we should still say a few words. The unconscious is not a metaphysical term for a variety of reasons; it is rather a nominal concept—such as, for example, a nuclear shaft. Such a shaft can be neither seen nor measured directly: we can say that the assumption with regard to its existence allows us to match theory with empirical facts. Equally numerous premises support the existence of the unconscious. There are some significant differences between the two concepts, which we are nevertheless unable to analyze here. We shall just say that the existence of the unconscious can be discovered by means of some empirical methods; yet it is impossible to determine whether a child is really scared during birth and whether its crying signifies the anxiety resulting from the suffering on its journey through the birth canal or rather the wonder at seeing the light of this world. The interpretation of dream symbols is equally arbitrary. According to the Freudian pansexual school, they exclusively signify various kinds of copulation and the organs that are necessary for it; Jung's students have a different "lexicon of dream symbols." It is truly inspirational to learn that patients of Freudian psychoanalysts dream according to the Freudian theoretical principles, whereas patients of Jungian psychoanalysts have dreams that are "aligned" with Jung's explications of them. This explicatory monomania, which "the interpretation of dreams" is, turns the otherwise valuable aspects of psychoanalysis into oases of sobriety in the desert of most arbitrary fabrications.

It should not surprise us that the majority of people tend to "shift" from an empirical to a metaphysical standpoint, when even scientists, who in a way are professionally obligated to remain loyal to empirical principles, at times sin against the scientific standpoint. According to our definition, superstitions, old wives' tales, fixed yet groundless convictions all belong to metaphysics; yet this kind of metaphysics is typical of small groups or just individuals. Metaphysical systems distributed across society in the form of religions are particularly significant. It is because every religion—no matter whether such tendencies played a role in its emergence—is a social regulator of interpersonal relations. Naturally it is not an exclusive regulator, as other regulators (those of economic and systemic nature) are more dominant, but every religion aims at performing a similar function. The question of the pragmatic value of religion for individuals, its ability to create spiritual equilibrium as a tool for coming to terms with one's existence, recedes into the background when compared with the (sometimes unintended) consequences of its social influence.

Such domination of religion within the spiritual culture of a given society used to be particularly strong in the past. This is why we can sometimes identify certain cultures with particular religions. The magic of ancient mysteries, the charm of metaphysical systems that inspired people to construct the most amazing temples and to create the most timeless works of art, the beautiful myths and legends derived from those systems and for them, tend to affect even otherwise rational scholars. And thus, for instance, Lévi-Strauss, who remains quite close to Marxism, considers practically all civilizations to be more or less equal (or incomparable, which amounts to the same thing) in his works. He believes that the values of ancient Asian civilizations—civilizations that had remained almost totally stagnant (in an economic and business sense) until their continent was invaded by imperialist capitalism—are equal to the values of our civilization, with its technological acceleration.

Similar opinions—such as those which, for example, consider Buddhism laudable in its contempt toward purely material values and in its neglect of empiricism—can sometimes be found among other Western scholars. Lévi-Strauss explicitly states that any judgment in this area must be relative because the person who is passing such a judgment is doing so in accordance with his own cultural tradition and thus is inclined to view as "better" or "worse" what to a large extent resembles the aspects of the civilization that has produced him.

We are talking about this issue because it is in Asia, in India in particular, that religion replaced any ideas of scientific or technological progress for the longest period of time. Through implementing such ways of thinking in subsequent generations, it arguably prevented any possibility of an autonomous revolution of action and thought arising in this country.

If it had not been for the Greek and Babylonian discovery of the deductive method and the turn toward empiricism, especially during the European Renaissance, science in its current form would no doubt have not been able to develop. At the same time, logical thinking (the law of excluded middle, of the equivalence of propositions, of the substitution of equivalents, etc.) as well as technical empiricism are held in utter contempt by the mystical religious doctrines of the East. The point here is not to enter into a dispute with this standpoint or to provide an apology for science; it is only to show the undeniable consequences of such a state of events. Whatever evils science has caused, it is only thanks to science that a significant part of humanity has been liberated from its hunger-stricken existence. It is only contemporary industrial technology and biotechnology that are able to cope with the problems of mass civilizations, while the proud catastrophic indifference toward mass problems, that is, problems facing the constantly growing human groupings, constitutes the foundation of the Asian religious doctrines. It is enough to look quickly into what thinkers of such religions have to say today to see the shocking inappropriateness and the terrifying anachronism of their teachings and instructions. The conviction that it is enough for individuals to practice the most beautiful kind of ethics in their lives, an ethics drawn from the most harmonious religion, for the whole of society or perhaps even humanity to automatically arrive at perfect equilibrium, is as tempting as it is fallacious. A society has to be understood as both a *human* collectivity and a *physical* system. To treat it as only a collection of individuals is erroneous, as is to see it as just a collection of molecules. Different things are beneficial for the human, and different ones for society as a whole, which is why, in this case, we need a compromise solution based on thorough and complex knowledge. Otherwise, if everyone does whatever he feels like, the final result can easily turn out to be rather terrifying. The (by now obvious) failure of Vinoba's religious and philanthropic campaign in India,[9] through which he attempted to raise fifty million acres of donations for the homeless and hungry of his land by setting off on a pilgrimage and knocking on people's hearts, is for some overshadowed by the astonishing courage and spiritual beauty of this man, who tried to solve the radically burning social problems in such a way. The point is not that he did not manage to get the millions his calculations told him he needed; it is just that, even if he had managed to achieve his goal, it would have only brought about short-term relief, because population growth would have quickly annihilated any temporary improvement in living conditions.

The conviction that Western civilization, with its standards set by mass culture and the ongoing mechanical facilitation of life, annihilates these potential spiritual riches—the cultivation of which should be at the core of our existence—regularly leads various people, among them even Western scientists at times, to turn to ancient Asia, especially India, in the hope that Buddhism will offer a panacea for the spiritual dry rot of technocracy. This is an extremely mistaken view. Individuals may become "saved" in this way, while those who are looking for consolation can apparently find it in Buddhist monasteries, yet this is pure escapism, an act of running away or even of intellectual desertion. No religion can do anything for humanity because it is not an empirical knowledge. It does reduce the "existential pain" of individuals, but at the same time, it increases the sum total of the calamities affecting whole populations precisely owing to its helplessness and idleness in the face of social problems. It cannot thus be defended as a useful tool even from a pragmatic point of view because it is a wrong kind of tool, one that remains helpless in the face of the fundamental problems of the world.

Religions in the West are increasingly shifting from the social sphere to the realm of individuals' private lives. However, the metaphysical hunger remains enormous because its emergence was not just a consequence of social phenomena. Metaphysical systems, be they the vague, aphoristic, and ambiguous ones from the East or those that use European logic, such as scholasticism, are always simple, at least when compared with the actual complexity of the world. It is precisely thanks to this simplicity, and also thanks to the definitive finality of their explanations (and avoidances), that they attract people. Each one of those systems can immediately tell us (although each one differently) that this is how the world came into being, this is who created it, and this is what man's destiny is.

The logical construction of the Judeo-Christian system is implied by its "mechanical determinism." It proclaims that *every* soul is immortal, *every* sin will be punished, and so on. Theology is not inclined to innovate on a methodological plane through introducing indeterministic types of relations between the "two worlds." The fact that prayers are not being answered would not mean anything within such "probabilistic" metaphysics because it would be governed only by probabilities: souls would be immortal, but not all of them, sins would be punished, but not always. Yet religion tends to establish a bookkeeping types of relations between here and now and the afterlife, rather than imitating the relations existing in Nature.

It is only fair to acknowledge that—when compared with Buddhism, with its various branches—the European religions, all branches of Christianity, are models of rationally constructed and logically coherent systems. Ever since Europe came across the term *nirvana,* religious studies experts have been fighting over its correct translation. It is not "nothingness," we hear, but neither is it "being." We are being referred to various parables, aphorisms, Buddha's remarks, and profound statements contained in holy books. Death is the end of existence, yet it is not the end, and so on. Even the mind of a theologian who is well versed in medieval scholasticism would feel tortured in the face of such pronouncements. The mystical content is to be contained precisely in a paradox, in a logical contradiction. Such associations are also to be found in Christianity, but their role is different.

I have just realized with horror how far we have departed from our main topic. We were supposed to be talking about experimental metaphysics, but we are almost practicing religious studies. I will only add, just to ease my conscience, that it has not been my intention to cast aspersions on Buddhism—one of the most beautiful religions I know. My critique derives simply from the fact that I am looking for something that does not really exist within it, that is, an answer to the kinds of questions that have never been posed within this system. We have to be clear about what our goals are. If the future of humanity does not concern us at all; if it is not the world but rather ourselves that we want to change, and only in a way that will allow us to adjust ourselves to the present one in the best way possible, for the rather short period of existence that awaits us, Buddhism will probably not be such a bad choice. But if we place Bentham's thesis about "the greatest good for the greatest number of people" above everything else, then neither ethical nor aesthetic aspects of any religion will be able to hide the fact that it is as outdated a tool for perfecting the world, for straightening its paths for the worthless, as a call for a "return to Nature" used to be.

We should really explain what Bentham's "good" actually stands for, yet we shall avoid this by declaring that the most important thing is to allow every person to live and to ensure that satisfying our needs is not a problem that has to be solved by heads of empires and scientists. This absence of hunger, poverty, diseases, anxiety, and uncertainty represents a very modest idea of good, yet there is still not enough of this good in our imperfectly designed world.

On to experimental metaphysics then...We shall not busy ourselves with translating the language of metaphysical models into their cybernetic equivalents because this would not take us very far, even though it is achievable. To a believer, an attempt to translate his own faith into the language of information theory would seem to be folly at best, blasphemy at worst. We could, of course, demonstrate how, in heading toward equilibrium, practically every homeostat experiences a "short circuit," as a result of which the system actually gains *eternal* equilibrium—though achieved thanks to unverifiable or false information. From this perspective, belief would stand for the compensation for all the gnostic and existentialist weaknesses manifested by homeostats that would triumphantly grant them permission to exist. There is injustice here, you say? "There" everything will be compensated for. There are a lot of things we cannot accept here? "There" we shall grasp everything, and then we shall accept it, and so on. Yet this whole exegesis does not lead anywhere because revealing the compensatory genesis of belief does not abolish its claims. Even if we were to demonstrate, by means of the mathematical apparatus of information theory, how the homeostat creates superficial metaphysical models of existence and how theogony develops within it, this argument would not solve the question of the actual existence of the designates of those terms (i.e., God, eternal life, providence). Since we were able to discover America while looking for India, and discover China while yearning for alchemists' gold, why would we not be able to discover God while looking not for an explanation— this is what science provides—but rather for a justification of our own existence? What is a cyberneticist to do then? There is only one thing left: to construct the kinds of homeostats that, while not being human, will be capable of spontaneously "creating" metaphysics. We are referring here to experimental metaphysics, that is, the modeling of the dynamic process through which belief emerges in self-organizing systems. (What we have in mind is its spontaneous, and not preprogrammed, emergence, based on the potential those homeostats have at their disposal and aimed at an optimal adaptation to the living conditions on Earth.)

Given that it is impossible to prove empirically the existence of the designates of belief, its adaptive value as a source of *universal* information is beyond all doubt. As we have already found out, the adaptive value of information does not always depend on whether such information is true or false. We can expect different homeostats to produce different "types of belief." It is only such comparative cybernetic metaphysics that will concern us here.

THE BELIEFS OF ELECTRIC BRAINS

Our idea for future research involving the construction of homeostats that will be capable of creating metaphysical systems, that is, "believing machines," is not just a bit of fun. It is not that we want to mockingly re-create within a machine the genesis of transcendental concepts. The aim of this task is to help us discover the general principles that govern the way in which metaphysical models of the world come into being. We can imagine (and, for now, *just* imagine) a group of colloid, electrochemical, or some other homeostats that are driven to develop certain beliefs in the course of their evolution. Those beliefs will emerge not because the homeostats have purposefully been programmed in this way. An experiment of this kind would be pointless. The homeostats will be capable of self-programming; that is, they will have the variability of goals—which is a cybernetic equivalent of "free will." Just as man consists of a series of subsystems that are hierarchically "connected" to the brain, each one of those homeostats will have various reception subsystems (inputs, "senses") and execution subsystems (inputs, effectors, e.g., a hidden locomotive system) as well as the "brain" proper—which we are not going to predetermine or limit in any way. We shall not introduce into it any instructions for action (apart from the crucial tendency to adapt to the environment, which emerges spontaneously in a homeostatic system). During the early stages of its activity, the homeostat will be empty; it will be like a blank slate. Via its "senses," it will then start observing its environment, while also being able to influence it thanks to its effectors. We shall introduce some limitations just to its effectors (executive subsystems, i.e., its "body" or "soma") to find out to what extent aspects of such "corporeality" affect the metaphysics generated by the "brain." Such metaphysics will no doubt have compensatory character with regard to those limitations. How are we supposed to understand this? Having learned about its limitations, that is, its "imperfect mortal life," the homeostat will no doubt create "eternal perfection," in the form of supplements and extensions to the former, as a result of which it will be able to achieve optimal internal equilibrium or, colloquially speaking, accept the current state of events. Yet there are also other "generators" of metaphysics, alongside compensatory reasons. Apart from equalizing, that is, "egoistical," factors, "gnostic" and genetic factors will also come into play. A homeostat will realize that its knowledge can only be approximate and incomplete. A natural drive to gain full and comprehensive knowledge will lead it to a "metaphysical model" that will allow it to believe that it "already knows everything." Yet, because such empirical knowledge is impossible to gain, the homeostat will shift the possibility of achieving such knowledge beyond the limits of its own material existence. In other words, it will become convinced that it possesses a "soul," most probably an immortal one.

"Genetic" factors, in turn, manifest themselves in looking for the "cause" of oneself and the surrounding world. Such a task becomes particularly interesting at this point since cybernetic modeling allows us to take into account not just the creation of homeostats but also the "creation of the world" for them. A digital machine (but one that is much more complex than those we have currently available) in which two cross-dependent processes are taking place is the simplest example. We could call them "process" and "antiprocess." "Process" stands for the system's self-regulation, whereby, after a period of time, the system becomes an equivalent of a living organism. "Antiprocess" is its "environment," that is, its "world." Of course, under those circumstances, both "intelligent beings" and their "world" are not material equivalents of our everyday living conditions but only an enormous set of (electric, atomic) processes that take place in the machine. How can we envisage such a state of events? It can be compared to a "transfer" of reality into the brain of a sleeping person. All the spaces, gardens, and palaces this person visits exist in his head, as do all the people he meets in his dream. His brain is thus an equivalent and an approximation of a "machine-world," because thanks to certain (biochemical, electronic) processes, phenomena from both places are divided into the "environment" and the "organisms" that live in it. The only difference is that a dream is a private matter for an individual, whereas what takes place in a machine can be controlled and examined by any expert.

We thus have a process and an antiprocess. Our task is to ensure that "organisms" become adapted to the "environment." We can now freely change not only the design principles of an "organism" but also those of its "world." For example, this can be a world of strict determinism or quite a static world. Finally, it can be an in-between world, which emerged as a result of an overlap between phenomena of both kinds. Consequently, a "machine" world would be very similar to ours. We can have a world in which "miracles," that is, phenomena that contradict observable laws, take place. We can have a world without "miracles." We can have a "reducible" world, a fully "mathematical" one, and a "finitely uncognizable" one. Also, such a world can manifest different forms of order. We shall be particularly interested in this issue because drawing conclusions about the Designer on the basis of the existence of order in the physical world is one of the key characteristics of metaphysics as professed by scientists (this type of argument about the existence of the Creator was typical, e.g., of Jeans and Eddington).[10]

The homeostats that would exist in those worlds would most probably also produce empirical knowledge. Some of them would no doubt turn into "materialists," "agnostics," or "atheists." The "spiritual" ones would then experience various schisms. A schism stands for a transformation of the axiomatic core of the postulated transcendence. In any case, what is most significant is that, by introducing particular modifications into homeostats' subsystems, that is, by limiting their *material* capabilities (but never spiritual ones, or their freedom to perform various mental operations), it is possible to bring about various kinds of metaphysical systems. By changing the characteristics of a "world," in turn, and comparing the results obtained, it will be possible to discover whether particular "worlds" prefer the emergence of a given structure of metaphysical beliefs and also to determine what kinds of worlds manifest such a preference. It is, I think, rather obvious that a thinking homeostat (yet still a "regular" one, and thus a form of "robot") that had not been brought up among other homeostats but rather among people—who were in turn religious—will adopt their "metaphysical model." This will lead to some rather unusual conflicts because the homeostat will demand equal rights with the practitioners of the religion with which it will identify from then on. The "transfer of a metaphysical model" to an individual from the social group into which he was born and in which he now lives is so common that the extrapolation presented earlier is very likely to occur. Yet such a demand for the same "metaphysical rights" that the religious practitioners have will be of more interest to theologians (who need to take a position in relation to this demand) than to empirical researchers.

Discussions such as the one presented earlier can be expanded on in various ways. And thus, for example, in a social group which consists of "higher" homeostats—that is, those that are more developed psychologically—as well as "lower" ones, a situation may emerge in which the "metaphysical solidarity" of the leading group will not extend to the "lower" homeostats. The attitude of the thinking machines toward their less complex companions will thus correspond exactly to the relationship between man and the rest of the living world. We often hear an argument in favor of metaphysics that reduces its role to giving meaning to our countless frailties, calamities, and sufferings—all of which lack earthly payback. Such solidarity does not extend to any other beings apart from man (in Christianity and similar religions). To a biologist, who knows the bottomlessness of this pit of misery that the history of life on Earth is, such a position is both amusing and terrifying. In this way, the billion-year-long history of all species is shifted beyond the limits of our myth-forming loyalty. This loyalty is now to apply only to a microscopic part of all the species, to a several-thousand-old primate branch—just because we represent this particular branch.

Another interesting possibility involves depriving homeostats of the knowledge about the finitude of their existence. This will perhaps reduce the probability of the development of metaphysics, but it will not eliminate it completely. The theory of homeostats lists their two types: a finite one (the only one that can be actualized, be it by man or Nature) and an infinite one (a so-called universal Turing machine). Of course, an infinite machine, that is, one whose transition from state to state has no limit, is only an abstraction (it would require both eternity and an infinite amount of material). Yet homeostats from our world can already live long enough, which is why the possibility of their own eternal existence will seem plausible to them. Each homeostat of this kind is free from "the gnostic conditioning of metaphysics by its own finitude" because it can express hope about being able to learn "everything" in the course of its eternal existence. However, given that this only eliminates cognitive reasons for metaphysics but not compensatory ones, such a homeostat can conclude that its limitless existence is an impediment, preventing it from entering a "better world"—access to which would only be offered by suicide.

THE GHOST IN THE MACHINE

"The ghost in the machine" is what some philosophers (such as Ryle)[11] call a belief that man is a "dual" being, consisting of "matter" and "soul."

Consciousness is not a technological problem because an engineer is not interested whether a machine has feelings, only whether it works. "The technology of consciousness" can thus only emerge accidentally, so to speak, when it turns out that a certain class of cybernetic machines is equipped with a subjective world of psychological experiences.

But how can we find out about the existence of consciousness in machines? The significance of this problem is not only abstractly philosophical, as the belief that a machine that is supposed to be sent to a scrap yard—because renovating it would cost too much—has consciousness transforms our decision from an act of destruction of a material object, such as a gramophone, to an act of an annihilation of personhood, and a conscious one at that. Someone could equip a gramophone with a switch and a record so that, after we have started it, we would hear it cry, "Oh, please, save my life!" How can we distinguish such a no doubt soulless apparatus from an intelligent machine? Only by entering into a conversation with it. In his paper "Computing Machinery and Intelligence" (1950), English mathematician Alan Turing offers "The Imitation Game" as a determining criterion, in which Someone is being asked random questions. On the basis of answers received, we are supposed to decide whether this Someone is a human being or a machine. If we cannot tell a machine from a human, then we should conclude that this machine is behaving like a human, that is, that it has consciousness.

We should add that this game can be complicated further; that is, we can think of two types of such machines. The first one is a "regular" digital machine, which is as complex as the human brain. We can play chess with it, talk with it about books, about the world, about all sorts of general things. If we were to open it, we would see a great number of feedback loops just like the neural feedback loops in the brain; we would also see its memory blocks, and so on and so forth.

The other machine is completely different. It is a Gramophone, enlarged to the size of the planet (or the Universe). It contains very many, say, one hundred trillion, recorded answers to all possible questions. And thus when we ask a question, the machine does not "understand" anything at all; it is only the form of the question, that is, the order of vibrations in our voice, that starts the transmitter which plays a record or tape containing a prerecorded answer. Let us not worry about technical details for now. It is obvious that such a machine will be inefficient, that no one is going to build it, because, first of all, this is not actually possible, but mainly because there is no need for it. Yet we are interested in the theoretical aspects of the problem. Given that it is behavior rather than internal design that determines whether a machine has consciousness, will this not make us jump to a conclusion that "the space gramophone" does have consciousness—and thus talk nonsense (or rather pronounce untruth)?

Yet can we actually program all possible questions in advance? No doubt an average person does not even answer one trillion questions during his lifetime. But we would record many times more of them, just in case. What is to be done? We have to play our game using the right strategy. We ask the machine (i.e., that Someone, as we do not know with whom we are dealing because the conversation is taking place on the phone) whether it likes jokes. The machine answers that yes, it does like good jokes. We thus tell it a joke. The machine is laughing (i.e., the voice at the other end of the receiver is laughing). Either it had this joke prerecorded, which allowed for the appearance of the correct reaction, that is, laughter, or it really is an intelligent machine (or a human being, because we do not know this either). After we have been talking to the machine for some time, we suddenly ask whether it remembers the joke we have told it. It should remember the joke if it is indeed intelligent. It will then say that it does remember the joke. We shall ask it to repeat it in its own words. Now, this is very difficult to program in advance, because it would mean the Designer of the "Cosmic Gramophone" would have to record not only particular answers to possible questions but also whole sequences of conversations that can potentially take place. This, of course, requires memory, that is, records or tapes, which probably cannot be contained even within the solar system. Let us say the machine is unable to repeat our joke. We thus expose it as a gramophone. The Designer, whose pride has been hurt, starts working on perfecting the machine by adding the kind of memory to it that will allow it to repeat what has been said. He has thus taken the first step on the way from a gramophone machine to an intelligent machine. As a soulless machine is unable to consider as identical questions that have analogous content but that have been formulated with even just slight formal variations, questions such as "Was it nice outside yesterday?" "Did you have beautiful weather yesterday?" "I wonder, was it nice on the day before today?" and so on, look different to a soulless machine yet identical to an intelligent one. The designer of the machine that is being constantly exposed for what it is has to keep working on it all the time. Eventually, after a long series of redesigns, he will introduce the skills of deduction and induction, the ability to link facts and to grasp the matching "form" of differently formulated yet identical content, into the machine, until he arrives at what is simply a "regular" intelligent machine.

We are faced here with an interesting problem: when exactly did consciousness arise in the machine? Let us say the designer was not adjusting those machines but rather took each one to a museum and then built a new model from scratch. There are ten thousand machines in the museum as this is how many subsequent models had been made. It is in fact a *fluid transition* from a "soulless machine" such as a jukebox to a "thinking machine." Should we consider machine no. 7852 as conscious, or only machine no. 9973? The difference between them is that the first one was unable to explain why it was laughing at the joke it had just heard but could only say that it was funny, whereas the second one was able to provide such an explanation. Yet some people laugh at jokes even though they are unable to explain what is funny about them, because, as we know, a theory of humor is a hard nut to crack. Does this mean that those people also lack consciousness? Not at all; they are probably not very bright, not that intelligent, their brain is not trained in thinking analytically about problems, yet we are not asking whether a machine is bright or somewhat stupid, only whether it has consciousness.

It seems we have to assume that model no. 1 has no consciousness, model no. 10,000 has full consciousness, and all the in-between models have "ever more" consciousness. This statement demonstrates how pointless it is to try to locate consciousness with exactitude. If we separate the individual elements ("neurons"), this will only result in insignificant quantitative changes to (or the "weakening" of) consciousness, which will be occurring in a similar way to a disease or a surgeon's knife progressing in a living organism. This issue has nothing to do either with the construction material used or with the size of such an "intelligent" device. An electric thinking machine can be built from individual blocks, which correspond, for example, to the brain's folds. Let us separate these blocks and place them all over the world so that one of them is in Moscow, another one in Paris, yet another in Melbourne, in Yokohama, and so on. When separated from one another, these blocks are "psychologically dead," yet when connected (e.g., by telephone wires), they become one integral "personality," one "intelligent homeostat." The consciousness of such a machine is, of course, not situated either in Moscow, or in Paris, or in Yokohama, but, to a certain degree, in all these cities, yet, at the same time, in none of them. We cannot say that, like the river Vistula, it stretches from the Tatra Mountains to the Baltic Sea.[12] In fact, the human brain represents a similar problem, although not as explicitly, because blood vessels, protein molecules, and connective tissue are contained within the brain yet not within consciousness, but at the same time, we cannot say that consciousness is situated right under the dome of the skull, or that it is situated a little lower, above the ears, on both sides of the head. It is rather "disseminated" across the whole of the homeostat, across its activity network. We cannot say anything else on this matter if we want to remain both sensible and cautious.

THE TROUBLE WITH INFORMATION

We are approaching the end of a section in our discussion devoted to the various topics that occupy cybernetics, topics that are rather remote from its core activities. In one of its most revolutionary phases, cybernetics formulated laws that regulate the transformation of information and, in this way, built a bridge, for the first time in science, between traditional humanities disciplines (such as logic) and thermodynamics (which is a branch of physics). We have already discussed various applications of information theory, but understandably, we have done it in a very general way and somewhat vaguely, because this book regrettably does not include any mathematical equations and formulas. Let us now consider what information actually is and what role it occupies in the world.

Information is currently being championed in disciplines that are very remote from physics (whose progeny it is) such as poetry or painting. We should make it clear right from the start that this kind of championing is somewhat in excess of the status quo of those disciplines, although not necessarily in excess of their future possibilities. We often hear references to the amount of information, yet before we start measuring it, it is certainly worth examining a more basic problem: the uniqueness of information—which, being a material phenomenon, is neither matter nor energy.

If there was not a single living being in the whole Universe, stars and stones would still exist. Would information exist then? Would *Hamlet* exist? In a way it would, as a series of objects covered with little spots of print, known as books. Does this mean that there are as many *Hamlets* as there are copies of those books? This is not the case. A large number of stars is still a large amount, no matter whether someone is experiencing their presence. With reference to such a number of stars, we cannot say that it is one and the same star repeated many times, even if those stars are very similar to one another. A million books titled *Hamlet* amount to a million *physical* objects, which all constitute only *one Hamlet*, repeated a million times. This is where the difference between a symbol, that is, a bit of information, and its material carrier lies. The existence of *Hamlet* as a series of physical objects that are information carriers does not depend on the existence of any intelligent beings. Yet for *Hamlet* to exist as information, there must exist someone who is capable of reading and understanding it. This leads to a rather shocking conclusion that *Hamlet* is not part of the material world, or at least not *as information*.

We could say that information exists also when there are no intelligent beings. Does an impregnated egg of a lizard not contain information? There is even more information in it than in *Hamlet*; the difference is that a book called *Hamlet* is a static structure that becomes dynamic only in the act of reading, that is, thanks to the processes that take place in the human brain, while an egg is a dynamic structure because it "reads itself," that is, it initiates the appropriate mechanisms of development, which lead to the emergence of a mature organism. As a book, *Hamlet* is a static structure. Yet it can "become dynamic." Let us imagine that an astroengineer connects *Hamlet*'s text to a large star via a coding device, after which this engineer dies, and so do all the intelligent beings across the whole Universe. The device is "reading" *Hamlet,* that is, transforming its text, letter by letter, into impulses, which cause some strictly determined transformations of that star. Bursting into protuberances, shrinking and expanding, the star is now "transmitting" *Hamlet* through fiery pulsation. In a way, *Hamlet* has now become its "chromosome apparatus" because it controls its transformations in the way the chromosomes of an egg control the development of a fetus.

Will we still say that *Hamlet* is not part of the material world? Yes, we will. We have created a large *transmitter* of information, the star, as well as its *transmission channel*—which is the whole Universe. Yet we still have no addressee and no receiver for such information. Imagine that the beams of radiation emitted by the star in the process of "transmitting" the scene of Polonius's murder prompt eruptions in the neighboring stars. Imagine that a number of planets are formed around the stars as a result of those eruptions. And when Hamlet dies, imagine that there already exist first germs of life on those planets. Sent as "text transmitted by a star," the last scenes of the play, in the form of very hard radiation, will increase the frequency of mutations in the protoplasm of living beings, which, in time, will give rise to proto-apes. It is no doubt a very interesting sequence of events, but what does it have to do with the content of *Hamlet*? Nothing. Perhaps this only concerns semantic information? Such information is of no interest to information theory. That's fine. There is much information in *Hamlet*, right? Its amount is proportional to the degree of probability of arriving at the other end of the communication channel, where the addressee is waiting. But who is this addressee? And where does the transmission channel end? In the Andromeda Nebula? Or perhaps in the Messier Catalog? Imagine we choose as an "addressee" a star that is close to the one that is "undertaking transmission." How are we now supposed to calculate probability? As the reverse of entropy? Absolutely not; entropy only functions as a measure of information when the system in which we measure it finds itself in a state of thermodynamic equilibrium. But what if this is not the case? Then everything depends on the frame of reference. But where is this frame?

It existed in Shakespeare's head, conditioned as it was by the structure of his brain, and in the whole civilization that had raised and formed Shakespeare. Yet this civilization does not exist anymore, nor does any other; there only exists a pulsating star, "connected" via a "translation" device to a book called *Hamlet*. This star is actually just an amplifier; the information is contained in the book. What does all this mean then?

Language is a system of symbols that refer to extralinguistic situations. This is why we can say that there is the Polish language but also the language of heredity ("the language of chromosomes"). Human language is an artificially created information code, constructed by biological evolution. Both have their addressees and their significance. A particular gene in the lizard's egg stands for a particular trait of the organism (it is currently a symbol of this trait but also its potential constructor, in the process of embryogenesis). If an egg "signifies" (i.e., contains a design recipe for) a lizard in the same way printed paper "signifies" *Hamlet* (i.e., contains a design recipe for the staging of the play), then we can perhaps go so far as to say that a nebula "signifies" a star that will develop from it in the future (i.e., contains a description listing a set of conditions that are necessary for its construction to take place).

But if that is the case, then a falling bomb will be a symbol of an explosion, lightning a symbol of a thunderclap, and a stomach ache a symbol of diarrhea. This approach is unacceptable. A symbol can be a thing, yet it does not refer to this thing but to something else. When carriers are taking ivory out of a warehouse, a Negro is putting aside pebbles. These pebbles are things, yet they refer to something else: in this case, they are numerical symbols referring to the elephant's tusks. A symbol is not just an earlier stage in the development of a phenomenon, at least not in the domain of human information technologies. The assignment of a symbol to what it signifies is arbitrary (this does not mean it is purely random, only that it does not involve creating a causative link between a symbol and its designate). Genes are actually not symbols because they represent a unique case in which an information carrier is simultaneously an earlier stage in its future "signification." Of course, we could decide that they *are* symbols: it is a question of definition, not empiricism, because no empirical research is going to demonstrate whether a gene is a "symbol" of blue eyes or whether it is only "a carrier of information about them." This would not, however, be convenient, because the word *gene* would then be a symbol of a symbol; besides, in our understanding, symbols are not capable of spontaneous transformation (letters in a chemical equation do not react with one another). This is why we should rather describe the gene as an information-carrying sign (one that is capable of autonomous transformation). A sign is therefore a more general term.

A sign assumes the existence of information (it is part of its code), while information exists only when it has an addressee. We know who the addressee of *Hamlet* is, and also that a nebula has no addressee—but who is the addressee of the chromosomal information contained in the lizard's egg? Not a mature organism for sure, as it is only the "next step" in the transmission of information. That organism, in turn, also has an addressee—but where? Lizards cannot live on the Moon or in the Sahara, only in rivers with muddy banks, whose waters provide them with nutrition, and where they can find partners and procreate. Thus an addressee of the lizard's genetic information is the environment itself, together with the whole population of its species and the other organisms—which it will either devour or itself be devoured by. In other words, an individual's biogeocenotic environment is the receiver of genetic information. It will beget other lizards in it, and in this way the circulation of genetic information as part of the evolution process will be maintained. Similarly, the human brain is an "environment" that enables the existence of *Hamlet*.

But if this is indeed the case, then why cannot we say that the Galaxy is the addressee of the information contained in a nebula? Even if not the Galaxy, then perhaps the planets which the star that emerged from the nebula will one day beget? Life will develop on these planets; it will reach its last stage of development there in the form of reason. Perhaps this "reason" is the addressee of the information contained in the nebula?

As we know from thermodynamics, the amount of information (i.e., of entropy) in a closed system cannot increase. The fact that we ourselves emerged from star dust and that the Universe is a closed system (because there is nothing "outside" it) clearly means that both *Hamlet* and everything else man has created, thought up, or lied about already existed as information in that primeval nebula from which galaxies, stellar systems, planets, we ourselves, and this book has developed. Thus we have happily reduced the whole thing to absurdity.

"Information in general" does not exist. There is no addressee for it. Information only exists in the context of a given set within which one is making a selection. This selection can produce a genus of lizards (natural selection) or a genre of plays (a selection which occurs in Shakespeare's brain).

When the police are trying to arrest a criminal about whom they only know that his name is Smith and that he lives in a certain town, the amount of information obtained as a result of knowing his name will depend on how many Smiths reside in this town. If there is only one Smith, then there is no selection, and the information equals one. If all the town's inhabitants are called Smith, the clue that the criminal's name is Smith contains zero information *for this particular set*. Incidentally, some claim that there is such a thing as negative information: in our case, negative information would consist of a denunciation made to the police informing them that the criminal's name is Brown.[13]

Measurements of information are thus relative and depend on the original assumption about the set of all possible eventualities (states). A given phenomenon can be a symbol, that is, an information carrier, because of the preselected set of the potential states of this phenomenon, and it can also not be a symbol if we change this set, that is, this frame of reference. In fact, it is very rare for Nature to preselect once and for all the set of all possible states. With or without his full awareness, it is man who chooses the frame of reference according to the *goal* he has set himself. This is why the information obtained is not a reflection of the actual state of things (the world) but rather a function of this state—a function whose values depend both on Nature (i.e., its examined part) and on the frame of reference provided by man.[14]

DOUBTS AND ANTINOMIES

The brave "total program" outlined by the creators of cybernetics has been subject to a critique, at times really severe, over the recent years, through which it has been considered a utopia or even a myth—as evidenced, for example, by the subtitle of Mortimer Taube's book *The Myth of Thinking Machines*. Taube writes,

> It can be remarked parenthetically that all the great mechanical brains, translating machines, learning-machines, chess-playing machines, perceiving machines, etc., accounts of which fill our press, owe their "reality" to a failure to use the subjunctive mood. The game is played as follows: First, it is asserted that except for trivial engineering details, a program for a machine is equivalent to a machine. The flow chart for a program is equivalent to a program. And finally, the statement that a flow chart could be written for

a nonexistent program for a nonexistent machine establishes the existence of the machine. In just this way Uttley's "Conditioned Reflex Machine," Rosenblatt's "Perceptron," Simon, Shaw, and Newell's "General Problem Solver," and many other nonexistent devices have been named in the literature and are referred to as though they existed.[15]

And then, a little further, he says that the relationship between "man" and "machine" has found itself "mired in the classical vicious circle":

(1) A machine is proposed or constructed to simulate the human brain which is not described.
(2) The characteristics of the machine which are carefully described are then stated to be analogous to the characteristics of the brain.
(3) It is then "discovered" that the machine behaves like a brain. The circularity consists in "discovering" what has already been posited.[16]

Given that progress in the area of design has undermined some of Taube's propositions, the debate about his book, published in 1961, is now superfluous. It is not just that perceptron exists; there also exists effective software for playing chess—admittedly, only at the level of an average player, but why should we not be able to acknowledge that machines that can play chess exist until the last invincible world champion has been checkmated by a digital machine, if the great majority of people cannot even play at an average level (including, unfortunately, the author of these words—even though this in itself does not constitute a proof)?

In his polemical and at times even nihilistic book, Taube raises some significant reservations in a way that is typical of scholars belonging to a particular school of thought. He brings up once again the classical dilemma of "whether a machine can think" by looking at it from two different angles: that of semantic actions and that of intuitive actions. It seems that there are indeed limitations to proceeding in a formal manner, which result from Gödel's proof about the incompleteness of deductive systems, and that it is impossible to translate successfully from one natural language into another by means of purely algorithmic methods, because there is no relation of mutually unequivocal correspondence between the two. We shall look at this issue in more detail later on. Before we get to consider the rather nebulous concept of intuition, let us add that Taube is also right when he points out how often the consequences of man's action and those of machine's action can be identical in their results, yet different in the processes that have led to those results. This brings me to the conclusion, which is also a warning, that one should not frivolously extrapolate the large amount of observation made while researching devices that had been programmed to solve particular tasks into human psychology. There is much more to such a comparative approach because very different brain activities are likely to lead to identical results in different persons. Last but not least, the very same person, when posed at different times with tasks that belong to the same class on the level of algorithm (i.e., a class of tasks for which a solution algorithm is known), sometimes approaches them via different methods. This nonuniformity of human behavior makes life difficult, no doubt, for all those involved in modeling brain processes.

When it comes to intuition, the possibility of automatizing it, that is, of imitation taking place outside the brain, does not seem as hopeless as Taube thinks. Some interesting research has been conducted into comparing man's heuristic with the heuristic of a machine, using chess as an example. It is because chess is not "semantically loaded," and thus problem solving during the game takes place independently, to a large extent, from the question of "meaning"—which tends to occlude, and introduce confusion into, the domain of psychological operations. But we should establish first of all what heuristic actually means. The Russian scholar Tichomirov, who conducted the preceding experiment,[17] understands by it some general laws a person uses in trying to solve the set task, when it is impossible to examine in a systematic manner all the potential alternatives (as is the case with the game of chess, when the number of possible moves amounts to 10^{99}). There were attempts to analyze the player's heuristic in the past by insisting he should think out loud during the whole game. It turned out, however, that the majority of "tracking" operations (i.e., those focused on finding an optimal solution) took place at a sublinguistic level—which is something the player did not in fact realize. Tichomirov therefore registered the movements of the chess player's eyes. It then turned out that the tracking heuristic of the player, at least partially reflected by those movements, had a rather complex structure. The width of the orientation sphere—that is, the section of the chessboard with the figures placed on it, which the player perceives most actively by signaling through the movement of his eyeballs that he is undertaking some kinds of "trial" moves, which are be-ing constantly extended (and which constitute the "interiorized" elements of the game; interior models of the subsequently considered sequential operations)—is changing dynamically. When his opponent's moves cor-respond to the player's inner expectations, that is, to his predictions, this sphere narrows down to a minimum, while every unexpected move that poses a surprise leads in turn to a considerable widening of the sphere of orientational tracking and to the far more extensive examination of the alternatives to the current situation. What is particularly interesting is that certain forms of "inspiration," a "sudden occurrence" of tactical ideas, an analogy to the "creative inspiration" that the classical anecdote sums up with the "Eureka!" cry, are preceded by a series of very fast eye movements at the time when the chess player remains completely unaware that an idea is about to "pop into his head." This leads to a conclusion that the alleged suddenness and the "arrival from nowhere" of entirely new ideas, which are greeted with a subjective sense of "rev-elation," or "enlightenment," is only an illusion or a delusion, resulting from limited introspective self-knowledge. In reality, every such idea is preceded by collecting information at maximum speed (in this case, from the chessboard), while the "suddenness" with which an idea appears is a consequence of information (now subliminally organized and at least schematically processed) reaching our consciousness and shifting from the lower levels of integration to the highest one, where the plan for the most effective action is ultimately formulated.

Of course, we still do not know what is going on at those lowest levels of brain dynamics. However, those experiments confirm the hypothesis about the multilevel structure of the processing of transmission signals that the brain is receiving. Assuming it is correct to talk about algorithms with regard to its working, many such algorithms are simultaneously involved in solving tasks—sometimes mutually connected to a certain degree, sometimes independently. The brain constitutes a system of sub-units, which are working relatively independently. With this, what we call "consciousness" can, figuratively speaking, be "pulling" in one direction, while at the same time, a person has a vague sense that "something" is pulling him away from the path he has already chosen consciously—even though no specific alternative for action presents itself as yet in his consciousness. Metaphorically speaking, we could say that, being still unable to deliver the final results of information processing to conscious-ness, nonconscious spheres notify it "somehow"—via the "channels" of emotional tension perhaps?—about the fact that a "surprise" is forthcoming. But we should abandon this manner of speaking as soon as possible, as it can only drive a designer who is intent on modeling intuitive heuristics insane. Indeed, he will not manage to accomplish anything in his studio by using even the most elaborate language of introspection—which is limited to reporting things and emoting about them.

A machine for playing chess (i.e., one programmed to do this) practices the kind of heuristics that its software (which, incidentally, is capable of learning) predisposes it to. We can state without any exaggeration that much depends on how gifted its programmer is (since programming is no doubt a gift). A machine can process an infinite number of times more operations in a single time unit than man can (it works more or less a million times faster than he does), yet man still beats it since he is capable of a unique kind of dynamic integration. If he is good at chess, he perceives each distribution of the pieces as a coherent system, a unity that is characterized by some clear developmental tendencies, which then diverge and "branch out." A machine uses tactics, which is why it can resort to certain moves to prepare subsequent ones; it can use gambits, and so on, but every time, it must also "quantize" the situation of the chessboard. It does not, of course, do this by making predictions many sequences ahead because this is physically impossible even for a machine. Man's chess heuristics allows him to make shortcuts that a machine is incapable of making. Acquiring some emotional and formal characteristics, the chessboard is perceived by him as an individualized whole. It is only the level of integration that turns chessboards with very similar distributions of pieces on them into completely different ones that allow a champion to play a few dozen games at the same time.

For the time being, our understanding has to stop at just recognizing such a phenomenal skill—phenomenal at least from the "machinic" point of view. In any case, human heuristic is a derivative of the "heuristic" of all living beings, since from their early days they have always had to act on the basis of incomplete and imprecise information, often settling on approximate constants and vague findings. A device that works on the basis of the purely logical premises in its possession, by deciding unambiguously what is true and what is false, would not be a modeling ideal, at least in the early stages; a device that works "more or less" well, "so so" and "approximately," would be a much better model. Given that evolution, at the level of all organisms, produced such latter "devices" *first*, it must have been easier for it to produce them than to construct systems that would explicitly use logic. Every human being, even a small child, uses logic (contained in the unconscious rules of language) "by accident," while the formal study of logic requires quite a lot of intellectual effort. The fact that a single neuron can be analyzed as a miniature logical element does not change the situation. We should add that even though the number of such elements is more or less the same in all the brains, there are some considerable differences between them which turn one human being into a great arithmetician but a poor mathematician, another one into a perfect mathematician who nevertheless has problems with arithmetic calculations, yet another one into a composer who is only capable of grasping the basics of mathematics, and yet another one into someone who lacks both creative and representational skills. While we know very little about what all these differently functioning brains have in common, we know nothing at all about the material reasons for such a high degree of diversification. This, in turn, complicates our issue even further. In any case, a cyberneticist is happy to welcome the arrival of devices that are less competent, at least on an elementary level, at certain kinds of differentiating operations, that is, devices that "more or less" work, even though there exists no general formal theory of such differentiation. We are thinking of perceptrons here.

Perceptrons are systems equipped with a "sight receptor" (which constitutes a rough analogy of the eye's retina) and with pseudo-neuron elements that are connected in an accidental, random manner. They are capable of recognizing images (simple planimetric configurations, e.g., digits or letters) thanks to the learning process controlled by a relatively simple algorithm. Perceptrons that are currently being constructed are still primitive and cannot recognize, say, human faces; neither can they, of course, "read texts." But they are an important step on the way to constructing machines that will be able to read such texts. This will greatly simplify all the initial procedures that are needed if we are to introduce information about the task to be solved into a digital machine. Today any such task has to be "translated" into a machinic language: this activity, which has not been automatized, takes a great amount of the machine operator's time. The possibility of constructing ever more complex and ever more efficient perceptrons thus sounds very promising. It does not mean that they offer more "accurate" models of the brain than digital machines do (especially that the working of a perceptron can also be modeled on a digital machine); it is also difficult to claim that a perceptron is "more similar" to the brain than such a machine is. Every one of those devices models, in a very fragmentary way, some elementary aspects of the brain's functioning—that is all. Future perceptrons will perhaps take us closer to understanding "intuition." We should add that there is terminological confusion, or conceptual lack of clarity, in the literature on the subject because some refer to a "heuristic behavior" as "nonalgorithmic," yet such a conclusion depends on whether we consider an algorithm to be an ultimately determined instruction that does not change in the course of its realization or an instruction that, thanks to the feedback that is restructuring it, in the course of its action "itself" becomes transformed into a form different from its original one. We are faced here in some instances with "self-programming," and the confusion that prevails results from the fact that this "self-programming" can also entail various states of things. In classical digital machines, programming is clearly separated from the operating systems that are subordinate to it, while in the brain, such a clear-cut distinction does not always apply. As soon as the functioning of a complex system becomes "plastic," that is, it is only conditionally and probabilistically determined, rather than being a one-way realization of some fixed "rules," the concept of an algorithm is not applicable anymore in a form that is taken directly from the deductive disciplines. It is possible to imagine a behavior—which will also be determined, but only to a certain extent—with which, after a certain number of steps, the system is "notified" about the fact that it now has to start "searching freely" for its next moves within the whole set of alternatives. The system thus starts working "by trial and error" until it hits an "optimal" value—for example, a minimum or a maximum of a certain function—which is when the "rigid" instruction switches on again for some time. Yet it is also possible for the whole algorithm to be "uniformly" probabilistic, which means that none of its future steps are assigned "apodictically": only certain brackets, or permitted ranges, are chosen within which can operate either algorithms of a different kind ("locally determined" ones) or "comparative" operations aimed at "finding similarities" (such as "recognizing images" or "shapes" or just looking for representational similarity). Thus control operations that have been "decided a priori," "search" operations, "comparative" operations, and last but not least, operations of induction can all be intertwined. The decision as to whether we are still dealing with an "algorithm" or already with a "heuristic" based on "intuition" will be made in an arbitrary manner to some extent—just as the decisions that see a virus in its crystallized form as "dead," but one that has been inserted into a bacterial cell as "alive," tend to be arbitrary.

How can we thus answer the question of whether products of "machinic thinking" can go beyond the limits of man's intellectual capacity?

We should probably list different versions of a possible answer, yet we do not know whether this will cover all possibilities or which one of those answers will be true.

1. Machinic thinking cannot go beyond the "limits of man's intellectual capacity" owing to some fundamental reasons: for example, no system can be "more intelligent" than a human being. We have arrived at these limits ourselves, yet we do not know about it. Also, there is only one way that leads to thinking systems of a "human" type, the way of natural evolution, and it can only be "repeated" if we take the whole planet as a ground for experimentation. Another reason is that nonprotein systems are always intellectually "weaker" (as information transformers) than protein ones, and so on.

All of the preceding sounds very improbable, even though we cannot discount it for the time being. In saying this, I am drawing on pointers from heuristic that suggest that man is in fact a rather ordinary intelligent being because he was formed through an elimination process on the basis of a relatively small number of parameters about a million years ago, that some more intelligent beings than him can actually "exist," that natural processes can be imitated, and last but not least, that different routes can be taken to reach states that Nature had arrived at via a sequence of some other states. Future developments in ergodic theory should be able to explain many phenomena in this field to us.

2. Machinic thinking can go beyond the "limits of man's intellectual capacity" just as a math teacher can be more "intelligent" than his pupils. Yet because man is capable of understanding what he is unable to arrive at himself (children understand Euclidean geometry, even if they do not come up with it themselves), he is not threatened with losing control over machines' "cognitive strategy" because he will always be able to understand what they are doing and why. This position, in turn, seems unacceptable to me.

What does it actually mean to say that "machinic thinking can go beyond the limits of man's intellectual capacity"? If it means the same as in the relationship between the teacher and his pupils, then this is a bad example, because the teacher did not invent geometry himself. We are referring here to the relationship between scientists and other people as it is analogous to the relationship between "man" and "machine." A machine is thus capable of creating theories, that is, detecting class constants, to a wider extent than man is. An intelligence amplifier as originally proposed by Ashby would not replace the scientist because it is a selector of information, while the scientist's work cannot be reduced to just making selections. Ashby's machine would actually be able to turn a much larger number of alternative options into elements of a situation in which selection has to be made than a human being would. Such a system would be possible and useful, but only in situations when we are facing a dilemma and have to choose a further path, but not when we first have to guess that such a path exists (e.g., the path of "process quantification"). Such an amplifier cannot thus even be the first approximation of a machine that would automatize the scientist's creative work. We are still unable to sketch it out even briefly, but at least we more or less know what a gnostic machine is supposed to do. If we are to create a theory of complex systems, this theory must take into account a great number of parameters—which the algorithms of contemporary science are unable to cope with. One can isolate various levels of phenomena in physics (atomic physics, nuclear physics, physics of solid bodies, mechanics). This is impossible in sociology because various levels (singular–individual, plural–mass) can turn out to be dominant, that is, alternately decisive, in choosing the system's dynamic route. The difficulty lies in the number of variables that have to be taken into account. If a "gnostic machine" was capable of creating a "theory of social systems," this theory would have to consider a great number of variables—which is what would differentiate it from the formalisms in physics that are known to us. And thus, at the "gnostic creator's" output, we get a theory that has been coded, say, as a system of equations. Will humans be able to do anything at all with those equations?

We shall perhaps understand the situation better if we use an example drawn from biology. If the information capacity of an egg cell equals the amount of information contained in an encyclopedia, then it will only be possible to read such an encyclopedia into which a genotype will one day no doubt be "translated" owing to the fact that the reader will be familiar with physics, chemistry, biology, the theory of embryogenesis, the theory of self-organizing systems, and so on. In other words, the reader will know the language and the rules of its application. In case of a theory "generated" by a machine, the reader will not know in advance either the language or its rules: he will still have to learn both. The ultimate question is thus as follows: will he be able to learn them?

The time factor enters into our deliberations at this point because it is presumably clear that more time is needed to read all the information contained in a bacterial cell and coded in the language of amino acids or nucleotides than for a cell to divide. During a single reading of the text of a "formalized and transcoded bacteria," a reading we undertake "with our eyes and brain," the bacteria will meanwhile divide hundreds of times because it "reads itself," in its subsequent divisions, incomparably faster. In the case of a "theory of society," or, generally, of any extremely complex system, reading time can turn out to be such that the only reason the reader will not understand what he is reading will simply be because he is unable to work with elements of equations on a mental level: they are too large; they escape his attention and go beyond the limits of his memory. It is indeed a Sisyphean task. The problem that then emerges is as follows: can a theory, in a form in which it has been issued by a machine, be reduced to a simple enough formula so that man can grasp it? I am afraid this will not be possible. This is to say that, yes, reduction as such is possible; it is only that every subsequent stage of a theory that emerges from such a reduction will still turn out to be too complicated for man, even though such a theory is already impoverished in comparison with the original one, as it has already lost some of its elements.

In undertaking a reduction, a machine will thus be doing what a physicist is doing when explaining gravitational wave theory to a wide audience by taking recourse to a modest arsenal of school-level math. Or it will be doing what a sage is doing in the fairy tale, when he is fetching a king—who is hungry for knowledge—with a library carried on the humps of a herd of camels; then, with a hundred volumes in the saddlebags of a mule; and eventually, with some large books carried by a slave—since all these "reductions" have still proved "too complicated" for the king.

We can thus see that we need not look into the following (third) possibility: (3) that a machine can go beyond man's intellectual capacity both with regard to what man is still capable of grasping and with regard to what he cannot grasp anymore. It is because this possibility presents itself as a conclusion after the second one has been invalidated.

It is probable that in situations when man will be able to figure something out himself, he will not need a machine in any other capacity than that of a slave—one that would undertake laborious subsidiary operations in his place (such as counting, delivering the required information, and thus serving as "prosthetic memory" and providing "assistance in step-by-step operations"). When he is unable to figure something out himself, he will be provided with ready-made models of phenomena, that is, ready-made theories, by a machine. We are thus posed with an antinomic question: "how can one control what cannot be controlled?" Perhaps one should create "antagonistic" machines that would control one another (i.e., control the outcomes of their actions)? But what should we do if they present contradictory results at output? Because it is up to us, after all, what we are going to do with theories generated by machines: in a situation of conflict, they will turn out to be useless. It is a different matter with ruling machines, that is, those that are the most probable incarnation yet of Ashby's amplifier. Robots endowed with quasi-human personalities are unlikely to be built—unless it is for the kinds of purposes imagined by Fritz Leiber in his novel *The Silver Egghead* (1961). There are even some amazing brothels with electronic ladies there, who hum Bach while doing "this" or who have tails like Chimeras. But machinic control centers that will manage the production and exchange of goods, their distribution as well as research (involving the coordination of scientists' efforts, "symbiotically" supported in the early phase by subsidiary machines), will emerge and grow. Such local coordinators will require some superior ones—at the scale of the country, at least, or even the continent. Is it possible that some conflicting situations will arise between them? Yes, absolutely. Conflicts will take place on the level of decisions concerning investment, research, and energy because it will be necessary to determine the order of various actions and steps owing to the plethora of mutually intertwined factors. Such conflicts will have to be solved. Of course, we quickly respond, this will be performed by humans. Very well then. Yet decisions will have to be made with regard to problems of extreme complexity. To find their way in the mathematical abyss opening up before them, people who are the Coordinator's controllers will have to turn for help to some other machines—machines that will optimize their decisions. There is a global economic aspect to all this: economy will also have to be coordinated. The Planetary Coordinator is also a machine, together with a "personal advisory board" consisting of people who check the local decisions of the "controller machine" systems on different continents. How do they do this? They have their own machines for optimizing decisions. And thus, is it possible for their machines—which, for the purposes of control, duplicate the labor of continental machines—to produce different results? This is entirely possible because, while performing a given sequence of steps involved in solving a task (through the method of, say, subsequent approximations, as the number of available variables is enormous), every machine becomes somewhat partial—or what the English philosophical discourse terms "biased."[18] We know that a human being cannot be entirely unbiased, but why should a machine be biased? Bias does not have to result from emotional predilections; it results from assigning different weightings to the conflicting elements of an alternative. Are diverse "evaluations" of these elements by several independently yet simultaneously working machines possible? Yes, they are, because these machines—which are probabilistic systems—do not work in an identical way. Algorithmically speaking, management process is a tree or a system of "decision trees": one has to reconcile contradictory needs, different kinds of supply, demand and interest. It is also impossible to set a "price list" for all the possible conflicting situations in advance so that only on the basis of using its items and the valuations assigned to them will one get exactly the same results every time one tries to solve this management problem—despite using probabilistic methods. The extent to which the results have been diversified is, of course, a function of the complexity of the problems solved. The situation will perhaps become clearer if we realize that it can also be expressed in the language of game theory. A machine is like a player who is playing against a "coalition" that consists of a great number of various production and market associations, transport and service associations, and so on. Figuratively speaking, its task is to ensure that an optimal equilibrium within the coalition is maintained and that none of its "members" are disadvantaged in relation to others or rewarded at their expense. From this perspective, the coalition simply refers to the whole planetary economy, which should be developing homeostatically but at the same time "fairly and evenly." "The machine's game against the coalition" stands for the systematic maintenance within the dynamically developing economy of a kind of equilibrium that either brings benefits to all or at least causes the least amount of damage, should such damage be impossible to avoid. And now, if a similar "game" is to be played "against" our "coalition" by a number of various machinic partners (which means that, at the outset, every one of them will face exactly the same situation within the coalition), it is highly unlikely for all those various games to be identical, both when it comes to their moves and to their final result. Suggesting such a thing would be like expecting all the people who are taking turns to play chess against the same player to play in exactly the same way, only because they all have the same opponent. What are we thus to do with the contradictory "valuations" of the machines that were supposed to support the person whose task it was to solve the argument between the local Coordinators? Infinite regress is not possible; something has to be done. But what? The situation is as follows: either electronic coordinators are incapable of considering a higher number of variables than man is, which means that there is no point in building them, or they *are* capable of doing it, which means that man himself cannot "find his way" among all the results; that is, he cannot make a decision independently from a machine by only relying on "his own opinion about the situation." The coordinator is managing the task well, but a human "controller" is not really controlling anything; he only thinks he is. Is this not obvious? The machine on whose help the human controller relies is, in a way, the Coordinator's double. At this point, man becomes a go-between, transporting the information tape from place to place. And if two machines give nonidentical results, the only thing left for man to do is to toss a coin to make a choice: he turns from the "supreme coordinator" to a random mechanism of selection! And thus again, even in the case of simple management machines, we are faced with a situation when they become "cleverer" than man. It may seem that they could be stopped from becoming like this—for example, on the basis of the following decree: "it is forbidden to construct and utilize coordinating machines whose potential for transforming information disables a human controller from assessing factually the results of their activity." This is pure fiction, though, because when the objective economic dynamics of regulatory processes demands a further increase in the number of Coordinators, the limits of man's capacity will have to be crossed—and here we are faced with an antinomy yet again.

One could ask whether I have not mystified the problem here. We do, of course, manage today without any machines whatsoever! Yes, but we are living in a society that is relatively simple in comparison with the society of the future. The difference between a civilization like ours, which is relatively primitive, and a highly complex future one is like the difference between a classical machine and the machine of a living organism. Classical machines and "simple" civilizations demonstrate various kinds of self-induced oscillation, uncontrolled vacillation of parameters, which lead to an economic crisis in one place, to hunger in another, and to poisoning by thalidomide in yet another. To understand how a complex machine functions, we should consider that the reason we move, walk, talk, and are generally alive is because, during every millisecond, masses of blood cells are running in file, in billions of our body parts at the same time, carrying tiny amounts of oxygen that control the continuous Brownian motion of particles heading toward an anarchic thermostatic chaos. The number of such processes that have to be constantly kept within a very narrow range of parameters is enormous—if that were not the case, then the system's dynamics would start to fall apart immediately. The more complex a system is, the more overall regulation is needed, and the smaller the extent to which local oscillations of parameters can be allowed. Does our brain have regulatory control over our body? Undoubtedly it does. Is every one of us in control of our own body? Only within a narrow range of parameters; the rest is "given" to us by Nature, in all its wisdom. Yet no one can give us a very complex social system, that is, undertake its regulation on our behalf. Our development will gradually lead us to a dangerous situation that Wiener spoke about, in which we will have to demand some "intelectronic reinforcements." The moment we start losing our overall grasp and thus also control, it will be impossible to stop a civilization, just like it is impossible to stop a watch: it will have to "go on."

But will it be going on "by itself," the way it has been so far? Not necessarily. These are, so to speak, negative aspects of progress, in a homeostatic sense. An amoeba is far less sensitive to a temporary loss of oxygen reaching its brain. A medieval city only needed water and food: a modern one turns into a nightmare when it runs out of electricity—the way it happened in Manhattan a few years ago, when the elevators in buildings and subway trains came to a halt. It is because homeostasis has two sides: it is an increase in insensitivity to an external perturbation, that is, one caused by a "natural" disturbance, and it is also an increase in sensitivity to an inner perturbation, that is, one caused by a disturbance within the system (organism) itself. The higher the artificiality of its environment, the more we are condemned to technology, to its working—and to its failure, if it should fail. And it may indeed fail. An individual's perturbation resistance may also be considered from two angles: as an isolated element and as an element of a social structure. All the "perturbation resistance" manifested by Robinson Crusoe was the result of his having been "preprogrammed" on the level of information by his civilization, before he became an "isolated element" on a desert island. Similarly, an injection a newly born baby receives, which provides it with some degree of resistance for its whole life, leads to an increase in the "perturbation resistance" of the baby as an isolated element, on a purely personal level. Yet in all those places where interventions have to be repeated, social feedback must function impeccably; that is, when a person suffering from a heart block is rescued from an imminent death by having a device implanted under his skin that serves as a prosthesis for his nerve stimuli, he has to receive regular energy supplies (batteries) for this device. Thus, on one hand, civilization saves man from death, yet on the other, it makes him extra-dependent on its own faultless functioning. On Earth, the human organism regulates the ratio of calcium in the bones to calcium in the blood by itself, but in space, when in zero gravity, calcium is being washed out from the bones into the blood, regulatory intervention falls down not on Nature but rather on *us*. In the systemic formations we know from history, there have been a number of sudden disturbances to homeostasis, caused both by external disruptions (epidemics, natural disasters) and by internal disruptions—a purely ideographic catalog of which is provided by historical chronicles. Systemic structures have manifested various kinds of resistance against such disruptions. On taking the whole of the system outside the realm of stability into the domain of irreversible transformations, some of those structures have brought about, by means of revolutions, a complete structural change. But humans have always entered into social relations with other humans; they have ruled over them or have been ruled and exploited by them. Anything that has occurred has thus been a consequence of human action. True, some of these actions have traversed an individual and a social group to become larger forces. Similar forms of material–informational feedback have acted in variable forms. At the same time, peripheral forms of support in systemic stabilization—such as, primarily, the family, which is one of the oldest such forms—have also been in operation. As technology develops, the complexity of regulatory processes grows so that it is necessary to use regulators that manifest a higher degree of variability than a human brain does. It is in fact a metasystemic problem because such a necessity is being experienced by countries characterized by diverse systems, as long as they find themselves on a high enough level of technoevolution. "Nonhuman" regulators, that is, those that are not human, will probably be capable of managing their tasks better than humans will—thus the improvement effect brought about by technological development will be significant also in this area. Yet the situation will change completely on a psychological level because there is a difference between knowing that the relations humans must enter into with one another generate statistical and dynamic regularities that can sometimes adversely affect the interests of individuals, groups, or whole classes and knowing that we are losing control over our fate and that it is being passed on to "electronic minders." We are then faced with a unique situation, which, on a biological level, would correspond to the situation of someone who knows that all his life processes are controlled not by him, not by his brain, not by all the internal systemic laws, but by some center outside him that prescribes the most optimal behavior for all the cells, enzymes, nerve fibers, and all the molecules of his body. Although such regulation could be even more successful than that carried out by the "somatic wisdom of the body," although it could potentially provide strength, health, and longevity, everyone is probably going to agree that we would experience it as something "unnatural"—in a sense that refers to our human nature. The same thing can probably be said when we apply this image to the relationship between "society" and "its intelelectronic coordinators." The more the complexity of the internal structure of civilizations grows, the higher the degree to which (i.e., the bigger the number of areas in which) we shall have to allow for thorough control and intervention to be enacted by such regulators to maintain homeostasis. Yet, subjectively, this process may seem to be a manifestation of "greed" on the part of those machines—conquering as they will be, one after another, various domains of what so far has been purely human existence. We are not thus faced with an "electronic God" or a godlike ruler but only with systems that, initially called on solely to watch selected processes, as well as processes of exceptional significance or complication, are slowly, in the course of their own evolution, taking control of practically the whole of social dynamics. Those systems will not be trying to "dominate over humanity" in any anthropomorphic sense because, not being human, they will not manifest any signs of egoism or a desire for power—which obviously can only be meaningfully ascribed to "persons." Yet humans could personify those machines by ascribing to them intentions and sensations that are not in them, on the basis of a new mythology of an intelectric age. I am not trying to demonize those impersonal regulators; I am only presenting a surprising situation when, like in the cave of Polyphemus, *no one* is making a move on us— but this time for our own good. Final decisions can remain in human hands forever, yet any attempts to exercise this freedom will show us that alternative decisions made by the machine (had they indeed been alternative) would have been more beneficial because they would have been taken from a more comprehensive perspective. After several painful lessons, humanity could turn into a well-behaved child, always ready to listen to (No One's) good advice. In this version, the Regulator is much weaker than in the Ruler version because it never imposes anything; it only provides advice—yet does its weakness become our strength?

5

PROLEGOMENA TO OMNIPOTENCE

BEFORE CHAOS

We discussed earlier the design factors that could result in the emergence of the "metaphysics of homeostats." In the process, we adopted a rather simplified classification of the sources of the "metaphysical attitude." This may have created an impression that, by referring to cybernetic analogies, we were attempting to solve on just a few pages problems as difficult and transhistorical as the meaning of existence, the finitude of an individual life, and the possibility of transcendence.

I want to defend myself against any such accusation of "shallowness." I do not want to withdraw anything I said; yet those earlier discussions, as well as any future ones—which will be even bolder—are primitive only if we see them as *initial approximations*.

If we are indeed situated at the pinnacle of all creation, if our existence is the result of a supernatural act, if therefore, as intelligent beings, we are a unique culmination of everything that can exist, then the future is surely going to enhance the power we have over matter. This will not change our attitude to the questions raised earlier because they can only be answered by metaphysics.

If we in turn consider ourselves to represent a very early developmental stage—which for us as a species began half a million years ago and as a civilization over twenty centuries ago—and if we assume that this development can go on for millions of years (although it does not have to), then our current ignorance in no way implies a similar ignorance in the future. This does not mean that we will be able to find answers to *all* questions of this kind. Instead, I believe we will grow out of questions to which there are no answers—not because such answers will be hidden from us but rather because these kinds of questions are badly posed. As long as we are only able to speculate on how we emerged and what shaped us the way we are now, and as long as Nature's actions in the animate and inanimate world fill us with wonder and offer an unmatched example for us, a realm of solutions that exceeds in its perfection and complexity everything we can achieve ourselves, the number of unknowns will be bigger than our knowledge. It is only when we are eventually able to compete with Nature on the level of creation, when we have learned to copy it so that we can discover all of its limitations as a Designer, that we shall enter the realm of freedom, of being able to work out a creative strategy subordinated to our goals. As I said earlier, the only way to deal with technology is by means of another technology. Let us expand on this point. Nature is inexhaustible in what it can do (a cyberneticist will say that the amount of information contained within it amounts to infinity). We cannot therefore catalog Nature, our finitude being one of the reasons for this. Yet we can turn Nature's infinity against it, so to speak, by working, as Technologists, on uncountable sets—in a similar way to what mathematicians do with set theory. We can remove the differences between "the artificial" and "the natural"—which will happen once the "artificial" first becomes indistinguishable from the natural and then exceeds it. We shall discuss later on how this is going to happen. But how should we understand this moment of "exceeding" the natural? It stands for carrying out with Nature's help what it cannot do itself.

OK, then, someone will say, all these lofty phrases were just intended to elevate human artifacts, all those various machines that Nature does not create.

Everything depends on what we understand by a "machine." It can, of course, refer just to what we have learned to construct so far. Yet if by "machine" we understand something that displays *regularity,* the situation will change. From such a broad perspective, it is not important anymore whether a "machine" has been constructed from existing matter, from those one hundred elements discovered by physics, or from air showers, or even from gravitational fields. It is also not important whether and how a "machine" uses, or even "creates," energy. It would, of course, be possible to construct a system consisting of intelligent beings and their environment in which our laws of thermodynamics would not apply. Someone will respond that this system would be "artificial" and that we would secretly have to provide it with energy from outside, in a way that would be unnoticeable to its inhabitants. Yet we do not know whether the Metagalaxy does not have any external energy sources that would be "connected" to this system from outside. Maybe it does; maybe its eternal supply of energy results from the infinity of the Universe. If that was the case, would this mean that the Metagalaxy was "artificial"? As we can see, everything depends on the scale of phenomena under discussion. A machine is thus a system that manifests some kind of regularity of behavior: statistical, probabilistic, or deterministic. From this point of view, an atom, an apple tree, a star system, or a supernatural world is a machine. Everything we construct, and that behaves in a certain way, is a machine: everything that has inner states and outer states, while the relations between sets of those states are subject to certain laws.

The question as to where the supernatural world can be found now is the same as the question as to where the sewing machine was before man came into being. It was nowhere—but it would have been possible to construct it.

It is no doubt easier to construct a sewing machine than such a world. Yet we shall attempt to demonstrate that there are actually no prohibitions that would disallow the construction of afterlife.

Following Ashby, we should note that there are two types of machines. A simple machine is a system whose inner and environmental states determine its future state. When the variables remain continuous, such a machine can be described as a system of regular differential equations, with time as an independent variable. Descriptions of this kind, which rely on the symbolic language of mathematics, are commonly used in physics, for example, in astronomy. With regard to systems (or "machines") such as a pendulum, a solid falling down in the gravitational field, or an orbiting planet, the system of such equations approximates the actual trajectory of the phenomenon with enough precision.[1]

In the case of a complex machine, such as a living organism, a brain, or society, such representation ("symbolic modeling") does not apply in practice. Of course, everything depends on how much we want to know about the system. The need for such knowledge is created by our goal and circumstances. If a hanging man is such a system, and if we are trying to determine his future states *as a pendulum,* then it is enough to take just two variables into account (angle of pendulum and its velocity). If this system is a living being whose behavior we want to predict, the number of significant variables we have to take into account becomes enormous, while our prediction will only be a designation of a future state whose probability will be higher the more variables we take into account. This probability will never equal 1 (it almost will, and in practice, the probability of 0.9999999 is more than enough for us). There are a number of mathematical methods for finding approximate solutions in situations when the number of variables makes the application of the standard analytical method impossible, for example, the so-called Monte Carlo method. But we shall not occupy ourselves with such matters because we are not practicing mathematics here. In any case, we can expect mathematical tools to give way to some others.

Phenomena that accompany the existence of "complex machines" are currently being examined by a number of new disciplines, such as information theory, operational analysis, theory of experiment planning, theory of linear programming, management theory, and dynamics of group processes. It seems that all of them, including several we have not mentioned, will be integrated by a universal systems theory. Such a general theory is likely to develop in two directions. On one hand, we can see it as a theory of physical systems, such as those provided by Nature. On the other, it will be a theory of mathematical systems, whose task will not be to study the actual existence of examined relations but only to ensure that such systems are free from internal contradictions. This bifurcation has not yet become apparent. We shall nevertheless dare to predict a moment when those two branches somehow become reintegrated again. This will mean the possibility of *constructing* systems with any characteristics—no matter whether they occur in the real world. We have to make a reservation here. In all its infinite relations, Nature is constrained by the existence of certain prohibitions (e.g., that one cannot draw energy "from nothing," overcome the speed of light, or simultaneously measure the moment and the momentum of an electron). As long as our world is to a large extent identical with Nature's world—which has only been slightly "altered" by us (as a result of technical activity)— and as long as we ourselves are a unique, or almost unique, outcome of natural processes (i.e., of bioevolution), Nature's limitations will also be ours. This is to say that it could be possible to replicate Napoleon one day but not in a way that would produce a faithful copy of the original, one that would also be able to fly merely by flapping his arms. This is not possible in the world we know. To enable Napoleon to fly, we would also have to create an environment for him in which "flight by wishful thinking" would be possible. In other words, we would have to create an artificial world, which would be isolated from the natural one. The higher the degree to which the world created by us has been successfully isolated from the natural one, the more the laws of that artificial world will differ from the natural ones. The opponent we have encountered will say that this is cheating because we would have to "embed" in a cunning way the fulfillment of wishes such as the desire to fly by means of flapping one's arms into our artificial world, a world that has been isolated from Nature. Yes, indeed. Yet because we consider Nature nothing more than a designer, in our opinion, it was Nature that embedded the spine, muscles, kidneys, heart, brain, and a whole series of other organs in our opponent. This would mean that he himself is a "fraud," even though he is quite an ordinary person, or maybe precisely because of this. We need to overcome the habit of considering outcomes of human activity as more imperfect than those of Nature's activity—understandable as such a habit may be at the current stage of development—if we are to talk about what is going to happen in a faraway future. We shall be competing with Nature on all levels: that of the infallibility and permanence of our creations, the universality of their actions, their regulatory potential, their homeostatic ranges, and many others. We shall discuss this matter separately.

For now, we shall continue with our introduction to "pantocreatics," which is a useful name for an ability to achieve all possible goals—also those that are not being realized by Nature. This ability will be anchored in the general theory of physical and mathematical systems.

CHAOS AND ORDER

As potential creators, we should first of all occupy ourselves with chaos. What is chaos? If, with event x occurring in A, any possible events can occur in B, and if such independence is universal, then we are faced with chaos. Yet if event x in A *limits* in some way what can occur in B, there is a relation between A and B. If x in A limits B unambiguously (we press a switch, then the light comes on), the relation between A and B is deterministic. If x in A limits B in such a way that, after event x has taken place in A, event y or z can occur in B, with y occurring forty times out of a hundred and z occurring sixty times, then the relation between A and B is probabilistic.

Let us now consider whether another "type" of chaos is possible, one that allows for the existing relations to be fully variable (i.e., neither deterministic nor probabilistic, since we already know that even this implies a form of order). Let us say that, after event x has taken place in A, there can follow y in B or u in B, or j in V, and so on. In such circumstances, the lack of any regularity does not allow us to detect the existence of any relations whatsoever. This is to say that variable relations are the same thing as no relations, which means that only one type of chaos is possible. Let us now consider how we can imitate chaos. If we have a machine with very many keys and lights, and if, on pressing a key, a light comes on, then even in the case of a strictly deterministic system, an observer of its behavior may arrive at a conclusion that he is facing chaos. This is because, if pressing the first key results in light T coming on, pressing this key for the second time results in light W coming on, pressing it for the third time results in light D coming on, for the fourth time in light Q coming on, and if this sequence is *very long,* so that only pressing key i for the millionth time will result in light T coming on again, after which the sequence will be repeated in exactly the same way, then the observer who will not be able to wait long enough to see the end of the first sequence will conclude that the machine is behaving in a chaotic way. Chaos can thus be imitated with a predetermined system, provided the duration of a sequence in which the same cause leads to seemingly random outcomes is longer than observation time. Fortunately, Nature does not work like this.

We have discussed the preceding not because we want to imitate chaos but rather to demonstrate that it is *not every* kind of order, that is, existence of relations, that the experimenter (science) is capable of detecting.

If event x occurring in A limits the possible events in B, we can say that there is a relation between A and B. Since event x in A to some extent determines what will happen in B, this relation can be used to transmit information. This also signifies the existence of *organization*: A and B constitute a "system."

There is an infinite number of relations in Nature. However, not all relations determine the behavior of a system or its parts to the same degree. Otherwise, we would have to deal with so many significant variables that science would become impossible. The nonuniform character of relations means that a system remains isolated to a greater or lesser degree from the rest of the Universe. In practice, we ignore as many relations (which we call "insignificant variables") as possible.

The relation between A and B, which limits B's possible states, is perceived as a restriction. A restriction on what? On "infinite possibilities"? No—their number is not unlimited. This is a restriction within the set of B's possible states. But how do we know which states are possible? On the basis of the knowledge acquired so far. But what is knowledge? Knowledge means expecting a particular event to occur after some other specific events have occurred. One who does not know anything can expect everything. One who knows something thinks that *not everything* can occur, but only certain things, while he considers other events impossible. Knowledge is thus a restriction placed on diversity; it is greater the lesser the uncertainty of the person expecting something to happen is.

Let us imagine that Mr. Smith, a bank clerk, is living with his puritanical aunt—who has a female lodger—in a multistorey house whose front wall is made of glass. As a result, the learned observer on the other side of the street is able to see everything that goes on inside. Let the interior of the house represent the "universe" we are supposed to examine. The number of "systems" that can be distinguished within this universe is practically infinite. We can approach it, for example, on an atomic level. We will then have groups of molecules from which chairs, tables, and the bodies of the three persons are made. The persons move; we want to be able to predict their future states. Since each body consists of around 10^{25} molecules, we would have to outline 3×10^{25} trajectories of those molecules, that is, their spatiotemporal paths. This is not the best approach, as before we manage to establish just the initial molecular states of Mr. Smith, the female lodger, and the aunt, around fifteen billion years will have passed, those people will have fallen into their graves, while we shall not have even managed to provide an analytic representation of their breakfast. The number of variables under consideration depends on what it is that we actually want to examine. When the aunt goes down to the cellar to fetch some vegetables, Mr. Smith kisses the lodger. In theory, we could arrive at who kissed whom just on the basis of the analysis of molecular behavior, but in practice, as we have demonstrated earlier, the Sun is likely to go out first. We would be unnecessarily diligent because it is enough to treat our Universe as a system that consists of three bodies. Conjugations of two bodies periodically occur within it when the third body goes down to the cellar. Ptolemy is the first one to appear in our Universe. He can see that the two bodies conjoin while the third one moves away. He thus develops a purely descriptive theory: he draws some cycles and epicycles, thanks to which one can know in advance which position will be taken by the two upper bodies when the lower one finds itself in the lowest position. Since in the very middle of his circles, there happens to be a kitchen sink, he declares it the center of the Universe, with all the significance this carries. Everything then revolves around the sink.

Astronomy is developing slowly. Copernicus arrives and invalidates the sinkocentric theory. After him, Keppler sketches out far simpler trajectories of the three bodies than the Ptolemaic ones. Then Newton comes. He declares that the bodies' behavior depends on their mutual attraction, that is, on the attractive force of gravity. Mr. Smith is attracted to the lodger, while she is attracted to him. When the aunt is nearby, they both revolve around her, because the aunt's force of gravity is correspondingly stronger. Now we are finally able to predict everything well. Yet suddenly the Einstein of our Universe arrives and subjects Newton's theory to a critique. He claims that it is completely unnecessary to postulate the existence of any forces. He creates a theory of relativity, in which the system's behavior is determined by the geometry of four-dimensional space. "Erotic attraction" disappears, just as attraction does in the theory of relativity itself. It is replaced by the curvature of the space around the gravitating masses (in our case—erotic masses). Then the coincidence of Mr. Smith's and the lodger's trajectories is designated by special curves, which are known as erotodesic. The aunt's presence causes a deformation of the erotodesic curves, as a result of which the coincidence between Mr. Smith and the lodger does not take place. The new theory is simpler, as it does not postulate the existence of any "forces." Everything is reduced to the geometry of space. Its general formula (that the energy of kissing equals erotic mass times the velocity of sound squared, since, as soon as the door closes loudly behind the aunt and this sound reaches Mr. Smith and the lodger, they throw themselves into each other's arms) is particularly beautiful.

Yet some new physicists then arrive—such as Heisenberg. They conclude that while Einstein has correctly predicted the system's dynamic states (the state of kissing, nonkissing, etc.), more precise observations involving large optical devices that allow us to see individual shadows of the arms, legs, and head demonstrate that it is possible to identify certain variables that the theory of erotic relativity overlooked. Those physicists do not question the existence of erotic gravitation, yet in observing the tiny elements from which cosmic bodies are constructed (i.e., those arms, legs, and heads), they notice the indeterminacy of their behavior. For instance, during the state of kissing, Mr. Smith's arms do not always occupy the same position. In this way, a new discipline starts to emerge, which is known as the micromechanics of Mr. Smith, the lodger, and the aunt. It is a statistical and probabilistic theory. Large parts of the system behave in a deterministic manner (immediately after the door has closed behind the aunt, Mr. Smith, and the lodger..., etc.), yet this is a consequence of *indeterministic* regularities coming together. At this point, real difficulties begin, because one cannot move from Heisenberg's micromechanics to Einstein's macromechanics. Bodies seen as fixed entities behave in a deterministic manner, yet erotic advances happen in a variety of ways. Erotic gravitation does not explain everything. Why does Smith sometimes hold the girl's chin and sometimes not? More and more statistics are produced. Suddenly a bombshell drops: arms and legs are not fixed entities; they can be divided into shoulders, forearms, thighs, calves, fingers, palms, and so on. The number of "elementary particles" is growing at an astounding rate. There is no unified theory of their behavior anymore. Between the general theory of erotic relativity and quantum micromechanics (the quantum of caressing has been discovered) lies an unbridgeable abyss.

Indeed, an attempt to reconcile gravitational theory with quantum theory (one that applies to the actual Universe, not to the one from our anecdote) has been impossible to achieve so far. Generally speaking, every system can be redefined so that it consists of any number of parts, which is in turn followed by looking for relations between those parts. If we only want to predict some general states, we can do so with a theory that contains a small number of variables. If we examine systems that are more and more subordinate to the original ones, the issue gets more complicated. Nature isolates stars from one another, but we have to isolate individual atomic particles ourselves: this is one of the many problems. We have to choose the kinds of representations that will reconcile the minimum number of variables considered with as high precision as possible for the prediction. Our anecdote was a joke, since the behavior of those three persons cannot be represented deterministically. They lack the necessary regularity of behavior for us to do this. This approach is nevertheless possible and actually imposes itself when a system shows great regularity and a great degree of isolation. Such conditions occur high in the skies, but not in an apartment. Yet when the number of variables increases, even astronomy has difficulties with using differential equations. These difficulties are already brought on by the attempt to outline the trajectory of the three gravitating bodies, while for six bodies such equations are impossible to solve.

Science exists owing to the fact that it creates simplified models of phenomena, ignoring less significant variables (e.g., by assuming that masses of smaller bodies within a system equal zero) and searching for *constants*. The speed of light is one such constant. It is easier to find constants in the actual Universe than in the aunt's apartment. If, rightly so, we are unwilling to consider kissing to be a phenomenon as universal as gravitation, but we want to find out why Smith engages in kissing, then we are at a loss. With all its limitations, mathematical mechanics is so universal that it allows us to calculate the position of cosmic bodies thousands and millions of years ahead. But how should we calculate the trajectories of Mr. Smith's brain impulses to predict his "oral coincidences"—or, to put it less academically, kisses—with the lodger? Even if that were possible, the symbolic representation of the subsequent brain states would turn out to be more complicated than the phenomenon itself (i.e., impulses traversing the neural network). In a situation like this, a neural equivalent of an act of sneezing would be a volume whose cover would have to be lifted with a crane. In practice, the mathematical apparatus will have got stuck in the ensuing complexity long before we actually get around to filling such volumes. What are we left with then? With considering *the phenomenon itself* its most perfect representation, and with replacing analytical with creative activity. In other words—with *imitological* practice.

SCYLLA AND CHARYBDIS: ON RESTRAINT

We now find ourselves in the most dangerous place in our delibera-tions.[2] We have multiplied questions while delaying answers; we have multiplied promises—which we have equipped with daring names, such as "pantocreatics"; we have said this and that about chaos; we have arrived at the origins of "imitology," and this whole train of thought is inevitably pushing us toward some new problems. They are the question of mathematics and its relation to the real world; the question of this world's linguistic and semantic problems; the question of various forms of "being." In other words, we are getting close to the realm of limitless philosophical entities, in which all our design-focused optimism can disap-pear without a trace. It is not that all these phenomena are enormously complex; that covering each one would require at least a volume, if not a whole library; or even that we lack comprehensive knowledge. It is rather that our knowledge will not be very useful, as the issues we are dealing with are contentious.

I should explain it further. Books that popularize the current state of knowledge—for example, in physics—and that do it well, nevertheless present a situation in which there exist two clearly separate areas: what science has determined once and for all and what has not yet been fully illuminated. We are thus treated with a visit to a magnificent building, which has been excellently decorated from top to bottom. We are stepping into various apartments and seeing some yet unsolved puzzles lying on tables here and there. We leave this building in the belief—strengthened by the greatness of the edifice as a whole—that these puzzles will be solved sooner or later. It does not even occur to us that solving these puzzles may lead to half of the edifice being destroyed. This is a similar situation to the one we experience when reading textbooks on mathematics, physics, or information theory. The amazing structure is what we see first. The fuzzy problems are hidden from our sight even more than in a popular lecture. It is because the popularizer (one who is also a scientist) realizes what an amazing effect bringing the Secret into light in a lecture will have. In turn, the author of a textbook (designed, e.g., for university students) is first of all concerned about the coherence of the outline presented and about its uniformity. He thus disregards all special effects and, not feeling obligated to translate the multistorey formulae into everyday language, finds it easier to avoid contentious interpretations. An expert will no doubt realize that the physical, material significance of the whole symbolism of quantum equations can be interpreted in many ways and that this or that formula hides layers of conflicting positions. He will also realize that another theorist would write a book that would differ in many ways from the one he has in front of him.

All of this is both understandable and necessary because it is impossible to popularize or educate while also at the same time introducing an audience into the very center of current debates. A reader of a popular book is not going to take part in resolving them anyway, while an expert in a given theoretical discipline must first of all get to know its weapons and the configuration of its battlefield before he can himself participate in scientific councils of war, once he has learned the basics of muster and tactics. Yet our goal is neither to popularize what has already been achieved nor to acquire a degree of expert knowledge but rather to look into the future.

It would be completely ridiculous if we were to expand our expectations to a significant degree and wanted to find ourselves in one go at the highest echelons of science, where it is not the authors of introductions or textbooks that are engaged in a discussion but the very creators of what is being taught and disseminated afterward, and if we dared to take part in their deliberations. It would certainly be a mistake. Ridiculousness aside, what exactly would we need to do? Let us say we understand everything that information theorists, mathematicians, and physicists say when they support this or that position. These are contradictory positions. A theory of the quantification of space cannot be reconciled with classical quantum mechanics. The "hidden parameters" of elementary particles either exist or they do not. Accepting that in the microuniverse processes are disseminated at infinite speed contradicts the finite speed of light. "Intelectronic engineers" say that it is possible to construct a model of the brain from binary (discrete) elements. "Fungoidists" claim that this is not possible. On both sides we have outstanding specialists, coauthors of subsequent scientific turns. Are we supposed to aim at an eclectic reconciliation of their premises? This would be futile: scientific progress is not born out of compromises. Are we to acknowledge the correctness of one side's arguments against the other? But how are we to settle on the selection criteria if Bohr is in argument with Einstein, or Brouwer with Hilbert? Should we perhaps turn to philosophers for a new criterion? But not only are there many philosophical schools; interpretations of the foundations of mathematics or physics are being disputed within every one of them!

Besides, all these are not merely academic problems; these are not arguments about the meaning of some details but rather about the most fundamental assumptions that shape knowledge: the problems of infinity, measurement, the relationship between atomic particles and the structure of the Universe, the reversibility or irreversibility of phenomena, the passage of time, not to mention issues of cosmology or cosmogony.

This is thus how our Scylla manifests itself: as an abyss, the limits of which we have been approaching recklessly, eyeing up the future that lies thousands of years ahead. Are elementary particles distinguishable? Can we really postulate the existence of an "antiworld"? Is there a limit point to systemic complexity? Is there a limit to a descent toward infinitely small values and to an ascent toward limitless numbers—or do they perhaps connect in a circle in some incomprehensible way? Can we assign infinitely large energy to particles? Of what interest is all this to us? What does it all mean? It does indeed encapsulate everything, if the so-called pantocreatics is not to be an empty word, a form of vacuous bragging worthy of a child or a fool. Even if we were to incarnate in some mysterious way the knowledge of the most distinguished specialists on Earth, this would not get us anywhere: it is not that today one cannot have universal knowledge but that such a person, were we to encounter him, would have to decide to which camp he would belong. The wave or corpuscular nature of matter comes to the fore depending on what we are examining. Is it the same case with length? Is length similar to color—is it something that emerges and not a feature of phenomena that occurs on all levels of reality? When asked this question, even the most outstanding specialist will say that the only answer he has is just his personal opinion—and even though it is based on a solid theoretical construction, it still remains unacceptable to other, equally distinguished specialists.

I would not want to give an impression through this that contemporary physics or cybernetics is an ocean of contradictions and questions. This is not the case. The achievements of science are enormous, but their glory does not obscure its dark moments. There have been periods in the history of science when it seemed that the edifice under construction was almost complete and that the only thing that future generations would have to do is work on finishing some tiny details. Such optimism reigned in the nineteenth century, during the period of the "indivisibility" of the atom. But there are also periods like the current one, when practically no inviolable scientific propositions exist, that is, propositions that would be univocally considered impossible to invalidate by all experts. This is a time when the famous physicist's witty remark that the new theory is not crazy enough to be true is actually being uttered in all seriousness, and when scientists are ready to sacrifice the most fundamental and sanctified truths on the altar of a newly anticipated theory—about the existence of a microparticle in a fixed place in space, about the emergence of matter from nothing (a hypothesis put forward by Hoyle), and finally, about how the notion of length does not apply anymore in inner-atomic phenomena.[3]

Yet the Charybdis of frivolous "shallowness" is not any less dangerous: playing with omnipotence, it is a whirlwind of cosmic garrulousness straight from science fiction, a space where anything can be said because no responsibility is claimed for anything. It is an area where everything is treated superficially, where gaps and fragments of logical thinking are hidden behind pseudocybernetic rhetoric, where banalities about "machines that compose poems the way Shakespeare did," and nonsensical pronouncements about extraterrestrial civilizations with which it is no more difficult to communicate than it is with a neighbor across the fence, keep flourishing.

It is truly difficult to navigate between these two whirlwinds. I doubt whether it is possible at all. Yet even if our passage were to end tragically, *navigare necesse est,* because if we do not set off, we will certainly not get anywhere. We thus need some restraint—but on what level? The level of design. It is because we only want to get to know the world to the extent that we can improve on it. And if we do not succeed, surely it is better to drown in Scylla than in Charybdis.

THE SILENCE OF THE DESIGNER

I have said that restraint on the level of design will serve as a compass in our navigation between the abyss of knowledge and the chasm of stupidity. Such restraint stands for belief in the possibility of acting effectively and in the need to give up on certain things. It means first of all giving up on asking "definitive" questions. This is not a silence of someone pretending to be deaf but rather active silence. We know far more about the fact that it is possible to act than about how it happens. The designer is not a narrow pragmatist, like a builder who is constructing his house from bricks, uninterested in where these bricks came from and what they are, as long as the house gets built. The designer knows everything about his bricks—except for what they look like when no one is looking at them. He knows that properties belong to situations, not things. There is a chemical substance that, according to some people, lacks flavor, whereas according to others, it tastes bitter. It is bitter to those who have inherited a particular gene from their ancestors. Not all people have it. According to the designer, the question as to whether this substance is "really" bitter makes no sense. If someone tastes the bitterness of this substance, then it is bitter to him. We can examine the difference between these two types of people; that is all. Some claim that apart from properties that are functions of a situation (such as bitterness or length) and are therefore variable, there are also some constants, and the task of science is to look for these very constants, such as the speed of light. This is also the Designer's position. He is absolutely certain that the world will continue to exist after he is gone; otherwise, he would be working for a future he is never going to see. He is told that the world will continue to exist after the last living being has died, but it will be a world of physics and not of sensory perceptions. Atoms and electrons will still exist in this world, but there will be no sounds, smells, or colors. The Designer then asks what kind of physics this world will actually represent: nineteenth-century physics, with its pelletlike atoms; a physics that is contemporary with its wave–corpuscular atomic duality; or a future physics, which will perhaps combine atomic properties with galactic ones? The reason he asks this question is not because he does not believe in the reality of the world: he accepts it as his foundational premise. Yet he can see that properties of the bodies discovered by physics are also functions of situations, that is, of the current state of knowledge in physics. One can say that an ocean exists when there is no one there, but one cannot ask what it looks like then. If it looks a certain way, it means that there is someone looking at it. If the Designer is in love with a capricious woman who sometimes seems to reciprocate his feelings and sometimes not, he can create contradictory opinions about her, but this does not in any way undermine that woman's objective existence. He can examine her behavior, write down her words, record the electric potentials of her brain; he can analyze her as a living organism, as a collection of molecules or atoms, and, last but not least, as a local space-time curvature, but this does not mean that the number of those women equals the number of possible examination methods. He is unsure whether it will ever be possible to reduce all these various methods to one so that atomic collisions can reveal love. Yet he acts as if it were possible. He thus practices a form of philosophy, although he resists getting drawn into its debates. He believes in the existence of one reality—which can be interpreted in an infinite number of ways. Some of these interpretations allow him to achieve the desired goals. He turns them into his tool. He is thus indeed a pragmatist, and "real" means the same as "useful" to him.

The Designer suggests in response that the person posing the question should join him in observing human activity. Whatever people do, they do it with a particular goal in mind. There surely exist hierarchies and complex structures of such goals. Some people act so that it looks like their actions have no goal. Yet the very structure of the preceding sentence ("act so that") demonstrates that they are also pursuing a particular goal: pretending that their actions are goalless. Some act from the conviction that they will only achieve their goal after their death. Many people seemingly aim at some other goals than those they have set their eyes on. Nevertheless, there is no goalless activity.

What is the goal of science? To get to know the "nature" of phenomena? How can one find out that one has already mastered it? That this is already its whole "nature" and not just part of it? So perhaps the goal is to explain phenomena? But what does explanation consist of? Comparing things? It is possible to compare the Earth with an apple, and bioevolution with technical evolution, yet with what can we compare Schrodinger's psi wave from the electron's equation? With what can we compare the "weirdness" of particles?

According to the Designer, science involves making predictions. Many philosophers share this opinion: neopositivists speak the most about it. They also claim that philosophy of science is, broadly speaking, a theory of science, and that they know how science creates and verifies or invalidates every new theory. A theory is a generalization of observed facts. We can predict future states on their basis. When these predictions begin to come true, and when they also begin to foresee the existence of phenomena unknown so far, the theory will be considered true. As a general rule, this is the case, yet in reality, matters are more complicated. The previously mentioned philosophers behave like an elderly lady who is working as an agony aunt at a newspaper. It is not that her advice is useless. Not at all. It can actually be very sensible—but it cannot be applied. The elderly lady has much life experience, and she advises the girl, on the basis of "erotic statistics," to dump the reckless boy. The philosopher knows the history of science and advises the physicists to give up on their theory because this theory is "betraying" them, since it is unable to predict many phenomena. It is not difficult to give such sensible advice. The girl believes that she will manage to change the boy for the better; the physicists think the same about their theory. In any case, the girl can have several boys she likes: it is the same with the physicists. They have to give up on such and such position for the sake of another one. If they give up on fixing a particle, they will gain one way of making predictions, but they will lose another. If they start quantifying space and introduce the idea that changes are distributed with infinite velocity, they will simultaneously be able to predict the existence of those subatomic particles that definitely exist, yet at the same time, this decision, made within the foundations of the edifice that is physics, will fundamentally shake all of its floors. There is no theory in science that would take "everything" into account and predict it. Yet in the majority of cases, it is possible to accept this state of things because what is being omitted is less relevant for the time being when it comes to scientific predictions. In physics, however, we have a dramatic situation: we do not actually know what is less relevant and what can thus be discarded. It is easy to decide when we find ourselves in a basket of a rapidly descending balloon and we can throw out either a sandbag or our companion. But imagine a situation in which you do not know what constitutes ballast and what counts as a value! The same equations of quantum mechanics can stand for either the "ballast," that is, emptiness, or a certain formal gesture, or they can have objective physical significance.

When analyzed post factum, after they have already become part of the personal history of two people or of the history of science, such issues allow the elderly lady and the philosopher to confirm that their convictions were right. Of course, a wonderful boy in love is better than a reckless good-for-nothing, and a theory that predicts everything without any mathematical override is better than a theory that is patched with some last-minute corrections—but where shall we find such a prince and such a theory?

The elderly lady and the philosopher are benevolent observers. The Designer has become involved in action—as have the physicists. This is why he realizes that usefulness can be understood in various ways: the way a morphine user understands it or the way Newton does. He does not thus allow himself to be drawn into disputes he considers pointless. If the brain is built from atoms, does this mean that atoms have "psychic" potential? If a wave throws up three sticks onto the shore, it is possible to make a triangle out of them, or pick them up and hit someone on the head. Are the hitting and the geometrical "potential" properties of those sticks? The Designer suggests that everything should be resolved by means of an experiment, and if an experiment is not possible, and never will be, the issue disappears for him. He leaves questions such as, How does mathematics exist? or Why does the world exist? without any answers—not because of his fondness for ignorance but because he knows the consequences of considering such questions. He is only interested in what he can do with mathematics and with the world, nothing else.

METHODOLOGICAL MADNESS

Let us imagine a mad tailor who makes all sorts of clothes. He does not know anything about people, birds, or plants. He is not interested in the world; he does not examine it. He makes clothes but does not know for whom. He does not think about it. Some of his clothes are spherical, without any openings for heads or feet; others have tubes sewn into them that he calls "sleeves" or "legs." Their number is random. The clothes consist of various numbers of elements. The tailor is only concerned about one thing: he wants to be consistent. His clothes are symmetrical and asymmetrical, large and small, stretchy and permanently fixed. When he starts producing a new item, he makes certain assumptions. They are not always the same. But he follows the assumptions he has made, and he expects them not to lead to a contradiction. If he sews on the legs, he does not cut them off later; he does not unstitch what he has sewn; they are always clothes and not bunches of randomly sewn rags. He takes the finished clothes to a massive warehouse. If we could enter it, we would discover that some of the clothes fit an octopus, others fit trees, butterflies, or people. We would find clothes for a centaur and for a unicorn as well as for creatures that have not even been imagined yet. The great majority of his clothes would not find any application. Everyone will admit that the tailor's Sisyphean labor is pure madness.

Mathematics works in the same way. It builds structures but it is not clear of what. These are perfect models (i.e., perfectly accurate), but a mathematician does not know what they are models *of*. He is not interested. He does what he does because such an action has turned out to be possible. Of course, a mathematician uses words we know from everyday language, especially when making initial assumptions. He talks, for example, about spheres and straight lines or about points. But he does not mean the same things we understand by them. The surface of his sphere has no thickness, while points have no size. The space of his constructions is not our space, as it can have an infinite number of dimensions. A mathematician knows not only infinities and transfinities but also negative probabilities. If something can undeniably happen, then its probability equals 1. If it absolutely cannot happen, then its probability equals 0. It turns out that something can less-than-not-happen.

Mathematicians know very well that they do not know what they are doing. A very competent person, Bertrand Russell himself, said that "mathematics may be defined as the subject in which we never know what we are talking about, nor whether what we are saying is true."

Mathematics is an example of "pantocreatics," performed on paper with a pencil. This is why we are talking about it: it will likely be mathematics that will start the "omnipotent generators" of other worlds in the future. We are no doubt far from this point. Part of mathematics will also probably always remain "pure," or, to put it differently, empty, just as the clothes in our mad tailor's warehouse are empty.

Language is a system of symbols that enable communication, as these symbols correspond to phenomena in the external world (thunderstorm, dog) or in the internal world (sad, sweet). If there were no real thunderstorms or no real sadness, there would be no such words. Everyday language is fuzzy, as the limits of its signification remain unclear. It is also evolving in its totality, alongside the ongoing social and cultural changes. It is because language is a nonautonomous structure, since linguistic creations refer to extralinguistic situations. In certain circumstances, it can become highly autonomous—as in poetry ("'Twas brillig, and the slithy toves / Did gyre and gimble in the wabe"),[4] thanks to creative word formation, or as a logician's language, which is exposed to some severe workout. It is nevertheless always possible to detect its genetic link with reality. The symbols of a mathematical language, in turn, do not refer to anything outside it. Chess is somewhat similar to a mathematical system: it is a closed system with its own laws and rules. We cannot ask about the truthfulness of chess, just as we cannot ask about the truthfulness of pure mathematics. We can only ask whether a given mathematical system, or a given game of chess, has been played correctly, that is, in accordance with the rules. Yet chess does not have any practical application, whereas mathematics does. There is a theory that explains this usefulness very simply. It says that Nature itself is supposed to be mathematical. Jeans and Eddington thought so, and I think that even Einstein was not entirely opposed to this idea. This can be seen in his statement, "God is sophisticated, but he is not malicious." I understand this sentence to mean that the complexity of nature can be grasped by enclosing it in (mathematical) formulas. If Nature were indeed malicious, a-mathematical, it would be like a malicious liar because it would be illogical, contradictory, or at least confused about events and hence unaccountable. As we know, until the end of his life, Einstein refused to accept quantum indeterminism and, in many of his thought experiments, tried to reduce its phenomena to deterministic laws.

Since the sixteenth century, physicists have been searching through the warehouses of "empty clothes" created by mathematics. Matrix calculus was an "empty structure" until Heisenberg found a "piece of the world" that matched that empty structure. Physics is full of such examples.

The steps of theoretical physics and hence also of applied mathematics are as follows: an empirical theorem is replaced by a mathematical one (i.e., certain mathematical symbols are given physical meanings such as "mass" and "energy"); a mathematical expression obtained in this way is transformed according to the rules of mathematics (this is a purely deductive and formal part of the proceedings), while the final result is transformed into an empirical theorem by once again inserting material meanings into it. This new theorem can be a prediction of a future state of phenomena, or it can express some general equations (e.g., that energy equals mass times the speed of light squared) or physics laws.

Physics is thus translated into mathematics; we treat mathematics in a mathematical way; we then translate the result back into the language of physics and obtain correspondence with reality (provided, naturally, that we have been proceeding on the basis of "good" physics and mathematics). This is, of course, a simplification, as contemporary physics is already so "overstretched" by mathematics that even initial theorems contain rather a lot of it.

It seems that, owing to the universality of relations in Nature, theoretical knowledge can only ever be "incomplete, imprecise, and uncertain," at least when compared with pure mathematics—which is "complete, precise, and certain." It is therefore not the case that mathematics, which is used by physics or chemistry to explain the world, says too little about this world, or that this world "escapes" through its equations—equations that are unable to capture it comprehensively enough. In fact, it is the other way around. Mathematics says (i.e., attempts to say) more about the world than it is possible to say about it—which apparently causes a lot of problems for science. These problems will no doubt be overcome. Perhaps one day matrix calculus will be replaced in quantum mechanics by another form of calculus, one that will enable us to make some more accurate predictions. But then it will only be contemporary quantum mechanics that will become outdated. Matrix calculus will not date. This is because chemical systems lose their validity, while mathematical ones never do. Their emptiness guarantees their immortality.

What does the "nonmathematical" character of Nature actually mean? The world can be approached in two ways. Either every element of reality has a strict equivalent (a mathematical "double") in physics theory, or it does not (i.e., it cannot have it). If it is possible for a given phenomenon to create a theory that not only predicts a certain final state of a phenomenon but also all the in-between states, while at every stage of a mathematical transformation, it is possible to indicate a material equivalent of a mathematical symbol, then we can speak about the isomorphism of theory and reality. A mathematical model is thus a double of reality. This postulate characterized classical physics and led to the belief in the "mathematical character" of Nature.[5]

There is also, however, a different possibility. If we shoot accurately at a flying bird and it drops dead, the final result of our actions will be the one we have hoped for all along. Yet the trajectory of a bullet and that of a bird are not isomorphic. They only converge at a certain point, which we will call "final." Similarly, a theory can predict a final stage of a phenomenon, even though there is no explicit mutual correspondence of real elements and mathematical symbols. Our example is primitive, but maybe it is better than none at all. There are few physicists today who would be convinced about the "doubling" relationship between mathematics and the world. This does not mean, however—as I have attempted to explain using the example of the shooter—that the chances of making successful predictions have diminished; it only emphasizes the nature of mathematics as a tool. In this viewpoint, mathematics stops being a faithful representation, a transient image of a phenomenon. It becomes more like a ladder that can be used to climb a mountain, even though it does not itself resemble a mountain in any way. Let us stay with the mountain example for a moment. By using an appropriate scale, we can read its height, the inclination of the slope, and so on, from the image. The ladder can tell us many things about the mountain against which it has been placed. Yet the question as to which parts of this mountain correspond to the ranks of the ladder makes no sense. The ranks are used for climbing to the top. In the same way, we must not ask whether the ladder is "true." It can only be better or worse—as a tool for achieving a goal.

The same can actually be said about the image. It seems to be a faithful representation of the mountain, yet if we continue examining it with ever stronger lenses, the details of the mountain slope will eventually fall apart into tiny black spots of photographic grain. This grain in turn consists of the molecules of silver bromide. Do the particular molecules explicitly correspond to something in the mountain slope? Not at all. The question as to where length "can be found" within an atom's nucleus is similar to the question as to where the mountain "can be found" when the image is being looked at through a microscope. The image is true as a whole, just as a theory (e.g., quantum theory) that will allow us to predict better the emergence of baryons and leptons, and that will tell us which particles are possible and which are not, will be true as a whole.

Similar theses may result in the bleak conclusion that Nature is unknowable. This is a terrible misunderstanding. The person who said this was secretly hoping that despite everything, mesons and neutrons would eventually turn out to be similar to really tiny droplets or to ping-pong balls. They would then be behaving like billiard balls, that is, in accordance with the principles of classical mechanics. I shall confess that the "ping-pong nature" of mesons would surprise me more than would the discovery that they are not similar to anything we know from everyday experience. Should the yet nonexistent theory of nucleons enable us, for example, to regulate stellar transformations, this would be a generous payback, I believe, for the "mysteriousness" of those nucleons—whereby "mysteriousness" simply means that we are unable to visualize them.

Hereby we close off our speculations about the mathematical or nonmathematical character of Nature to return to matters of the future. Pure mathematics has so far been a warehouse of "empty structures," in which the physicist has been looking for something that would "fit" Nature. Everything else has been cast aside. The situation may turn around, however. Mathematics is an obedient slave of physics, from which it receives appreciation as long as it is capable of imitating the world. Yet mathematics can start giving orders to physics—not to contemporary physics but to artificial physics, in a faraway future. As long as it exists only on paper and in mathematicians' minds, we call it empty. But what if we are able to actualize its constructs one day? What if we become capable of producing worlds "imagined in advance" by using mathematical systems as our building plans? Will they be machines? No—if we do not consider an atom a machine. Yes—if an atom is a machine for us. Mathematics will be a phantomological generator, a creator of worlds of a different reality from the one we know. How can we envisage this? Is it possible at all?

We are still inadequately prepared to discuss this latest technical revolution, which we are already capable of imagining today. So, yet again, we have yet got ahead of ourselves. We need to go back from pantocreatics to imitology. But first of all, we must say a few words about the systematics of those nonexistent objects.

A NEW LINNAEUS: ABOUT SYSTEMATICS

Let us start with an explanation. We would like to look into the future, hence we have to accept that contemporary knowledge is very limited when compared with the knowledge that will exist in future millennia. This position may sound like frivolous and nonchalant ignorance with regard to twentieth-century science. This is not the case. As civilization has been in existence for more than ten thousand years, and because we want to make an educated guess, at the risk of a complete failure, about what will happen in an era that is at least as remote from now, we shall not consider any of the current achievements as climactic. From the heights we have to climb, a cybernetic revolution will seem only one step away from the Neolithic one, just as the unknown, anonymous inventor of the zero will seem only one step away from Einstein. I am saying "we must" and "we want" to emphasize that it would be impossible to achieve anything if we were to use a different perspective in this "thought expedition." We can consider this perspective, elevated as it is above the future and the present, a groundless usurpation, and I completely understand someone who would adopt such a position. Yet if I shared it myself, I would have to remain silent.

There are also some practical difficulties involved in the argument I am developing here. I will present things that should be discussed simultaneously one by one instead. The reason for this is that it is not my intention to catalog and enumerate some "future inventions" but rather to show some general possibilities, without any technical descriptiveness (which in itself would be a truly empty claim). General yet not generalized ones—because they determine the future in a particular way. We shall never say that things will be a certain way, only that they *can be* a certain way—because this is not a science fiction book but rather a set of variously substantiated hypotheses. They constitute a unity that cannot be described together. A physiologist who is attempting to include knowledge about the working of an organism in a textbook is facing the same difficulty. He discusses, one by one, breathing, blood circulation, basal metabolism, and so on. The situation is advantageous given that people have been writing textbooks for a long time, and thus such a division of topics, although problematic, has been sanctioned by tradition. Yet I am not describing anything or almost anything that exists and cannot therefore refer either to general models (again, with some rare exceptions) or to textbooks about the future, because I am not familiar with them. Thus I am obliged to use an arbitrary classification; I reiterate certain issues and problems two or even three times as a result of those difficulties, and sometimes I discuss separately what should really be discussed alongside some other issues, but I am unable to do the latter.

After all these explanations, I shall introduce the "systematics of the subject." From now on, it will serve as a central theme for us. The names I shall be using are of a provisional nature; they are abbreviations that facilitate an overview of the disciplines in question and nothing else.

This is why I have placed the word "systematics" in quotation marks. "Pantocreatics" refers to everything man or another intelligent being can achieve. It refers to gathering information and using it with a particular goal in mind. Such separation also exists today to some extent; it can be seen in the differentiation between science and technology. In the future, the situation will change in that information gathering will have become automatized. Information gathering systems will not determine the direction of action. They are like a mill that delivers flour—but it is up to the baker (i.e., the technologist) what will be made from this flour. Yet what kind of grain should be thrown onto the millstone is determined not so much by the baker as by the mill manager: science will be such a manager. The grinding of the grain itself stands for information gathering. We shall discuss separately how this can be envisaged.

The part of pantocreatics that deals with information use, and that has emerged from the combination of the general theory of physical and mathematical systems, can be divided into two fields. For the sake of brevity, and also to provide a kind of overview, we shall call the first of these Imitology and the second Phantomology. They overlap with each other to some extent. We could, of course, try to be more precise and say, for example, that imitology is a design theory based on the mathematics and algorithms that can be identified in Nature, while phantomology stands for actualizing in the real world mathematical structures that have no equivalent in Nature. Yet this would already mean assuming that Nature basically has a mathematical character—while we do not want to make such an assumption. This would also involve accepting algorithmic universalism—which is highly dubious in itself. It is therefore more sensible to leave our definitions partially open.

Imitology is an earlier stage of pantocreatics. It is derived from the modeling of real phenomena in scientific theories, digital machines, and so on, which is already practiced today. Imitology involves launching probable material processes (a star, a volcanic explosion) as well as improbable ones (a microfusion cell, a civilization). A perfect imitologist is someone who is capable of repeating any natural phenomenon, or a phenomenon whose emergence is enabled by Nature, even though it itself has not been created spontaneously by Nature. We shall explain separately why I go so far as to call even the construction of a machine an imitative activity.

There is no sharp boundary between imitology and phantomology. As a later, more advanced phase of imitology, phantomology involves constructing ever less likely processes, all the way to entirely impossible ones, that is, processes that cannot take place under any circumstances because they contradict the laws of Nature. It seems to be an empty class because the unrealizable cannot be realized. We shall attempt to demonstrate, albeit only briefly and rather crudely, that such an "impossibility" does not have to be absolute. For now, we shall only signal briefly how we could envisage the first step toward phantomology. The model of an atom is supposed to help us get to know the original, that is, Nature. This is why we have constructed it. If it does not correspond to Nature, we consider it worthless. This is the situation we find ourselves in today, yet we could change our strategy. That model could be used for another purpose: to turn the model of an atom that is completely different from the actual atom into a building element of "some other matter" that differs from actual matter.

MODELS AND REALITY

Modeling is an imitation of Nature that takes into account few of its characteristics. Why only few? Is it because we cannot do better than that? No, it is mainly because we have to defend ourselves against the excess of information. Such an excess can actually signify inaccessibility. A painter paints pictures, yet, even though he has a mouth and we can talk to him, we are not going to find out how he does it. He does not know himself what is going on in his brain when he is painting. The information is contained in his head, but it is inaccessible. In modeling, one has to simplify: a machine that is capable of painting a very poor picture will tell us more about the material, that is, cerebral, foundations of painting than the "perfect model" of the artist—his twin brother— would. Modeling practice involves selecting certain variables and ignoring others. There would be an ideal correspondence between the model and the original if the processes of both were identical. This is not the case. The results of model development are different from those of any actual development. This difference can be caused by three factors: the simplification of the model in relation to the original, the model's own characteristics that are lacking in the original, and last but not least, the indeterminacy of the original itself. When we imitate a living brain with an electric one, we must consider a phenomenon such as memory as well as consider an electric network that represents the neural network. A living brain does not have a separate memory container. Actual neurons are universal—memory is "disseminated" all over the brain. Our electric network does not manifest any such characteristics. We thus have to connect special memory banks (e.g., of ferromagnetic kind) to the electric brain. Besides, an actual brain shows certain "randomness," an incalculability of actions, while an electric one does not. What does a cyberneticist do? He builds a "generator of accidentality" into the model—which, on being switched on, sends randomly selected signals into the net. Such randomness has been prepared in advance: this additional device uses random number tables, and so on.

We have thus arrived at what looks like an analogy of "incalculability" or "free will." After taking these steps, the similarity of output parameters in both systems, the neural and the electric, has increased. Yet this similarity has only increased with regard to the corresponding "inputs" and "outputs." The similarity does not increase—and does, in fact, decrease—if, alongside the dynamic "input–output" relation, we take into account the entire structure of both systems (i.e., if we take into account a higher number of variables). Even though the electric brain now has "volition" and "memory," the actual brain does not have either an accidentality generator or a separate memory bank. The closer this model moves toward the original one within a range of certain imitated variables, the further away it moves from that original model within a range of others. If we also wanted to take into account the changeable excitability of neurons, which is conditioned by the existence of its limit point, while every organism achieves this state through the very biochemistry of its transformations, we would have to equip each of the switch elements ("neuristors") with a separate electrical system, and so on.[6] However, we consider variables that do not manifest themselves in a modeled phenomenon as insignificant. This is a special case of the general mode of information gathering, one that assumes that an initial selection always takes place. For example, for an ordinary person speaking on the telephone, the crackling sound counts as "noise," whereas for a communications engineer who is examining the line, certain information can be conveyed precisely by such noise (this example is provided by Ashby).

If we thus wanted to model any phenomenon by taking into account *all* of its variables (assuming for a moment that this would be possible), we would have to construct a system that would be *more extensive* than the original one, as it would be equipped with additional variables that are characteristic of the modeling system itself but that the original one lacks. This is why, as long as the number of variables is small, digital modeling works well. On increasing their number, this method quickly reaches the limit of its applicability. The modeling approach therefore has to be replaced by a different one.

In theory, it is most efficient to model one phenomenon with another identical phenomenon. Yet is this possible? It seems that to model man, it is necessary to construct him; to model bioevolution, it is necessary to repeat it on a planet that is exactly like Earth. The most perfect model of an apple is offered by another apple, of the Universe by another Universe.

This may sound like a reductio ad absurdum of imitological practice, yet let us not be too quick in passing such a verdict.

They key question is as follows: is there something that, in not being a faithful (model) repetition of a phenomenon, contains more information than this phenomenon? Absolutely: a scientific theory. It covers a whole class of phenomena; it discusses every single one but, at the same time, all of them. Of course, a theory does not take into account many variables of a *given* phenomenon, yet, owing to the goal that has been set, these variables are not significant.

We are faced with a new difficulty here. We should ask whether a theory contains only as much information as we ourselves have introduced into it (having created it on the basis of observed facts as well as some other theories, e.g., measurement theory) or whether it can contain more information. The latter is impossible, you say? Yet it was on the basis of the theory of a physical vacuum that quantum field theory predicted a number of phenomena. Alongside the beta decay theory emerged the results of the theory of superfluidity (liquid helium) and also of the solid state theory. If a theory is largely supposed to predict phenomenon x, and then it turns out that some other phenomena that have been deduced from it—whose existence we did not know about before—also take place, where did this "additional" information actually come from?

It came from the fact that, generally speaking, there exists a continuity of transformations in the world. It came from their feedback. We have "guessed" one thing, and this one thing has subsequently "led" to the others.

This sounds convincing, but how does this information balance actually work? We have inserted x bits of information into the theory, and then we get $x + n$? Does this mean that if a system is complex enough (the way the brain is), it is capable of creating additional information, more extensive than the information it possessed in the preceding moment, without receiving any additional information from outside? This would be a true informational *perpetuum mobile*!

Unfortunately, this issue cannot be resolved on the basis of current information theory. The amount of information is greater the lower the probability of the arrival of a given signal. This means that if a message arrives that stars are made of Emmentaler cheese, the amount of information received will be truly enormous because the arrival of such a signal is extremely unlikely. Yet an expert will accuse us here, and justly so, of confusing two different types of information: selective information—that is, information that can be drawn from a set of possible signals (stars made of hydrogen, of entelechy, of borogoves, of cheese, etc.), which has nothing to do with the correctness, or appropriateness, of information about a certain phenomenon—and structural information, that is, information that is a representation of the situation. And thus the sensational news about the cheesing of stars contains a great amount of selective information and zero structural information because it is not true that stars are made of cheese. Perfect. Let us thus take a look at the theory of physical vacuum. It shows that beta decay happens in such and such a way (which is true) as well as that an electron's charge is infinitely great (which is not true). The first result, however, is so valuable to a physicist that he is prepared to make up for it with interest paid on the incorrectness of the second one. Yet information theory is not interested in the physicist's choice because this theory does not take into account the *value* of information, even in its structural state. Besides, no theory exists "on its own"; no theory is "sovereign": it is partly derived from other theories and partly combined with them. And thus the amount of information contained in it is very difficult to measure, since, for example, information contained in the famous $E = mc^2$ formula "gets into" this formula from a whole lot of other formulas and theories.

Yet maybe it is only today that we need theories and models of phenomena? Maybe, on being asked such a question, a wise man from another planet would silently hand out a piece of an old shoe sole picked up from the ground to us, communicating in this way that the whole truth of the Universe can be read from this piece of matter?

Let us stay for a moment with this old sole. This anecdote can have some amusing consequences. Please take a look at the following equation: $4 + x = 7$. An obtuse student does not know how to access the x value, although this result is already "entailed" in the equation, but it remains hidden from his misty eyes and will only "reveal itself" after a basic transformation has been performed. Let us thus ask, as righteous heresiarchs, whether it is not the same case with Nature. Does Matter by any chance not have all of its potential transformations "inscribed" in it (i.e., the possibility of constructing stars, quantoplanes, sewing machines, roses, silkworms, and comets)? Then, taking the basic building block of Nature, the hydrogen atom, we could "deduce" all those possibilities from it (modestly starting from the possibility of synthesizing a hundred elements all the way through to the possibility of constructing systems that are a trillion times more spiritual than man). We could also deduce all that is *unrealizable* from it (sweet kitchen salt NaCl, stars whose diameter equals a quadrillion of miles, etc.). From this perspective, matter already entails as its foundational assumptions all those possibilities as well as impossibilities (or prohibitions); we are just unable to crack its "code." Matter would thus be a kind of mathematical problem—with us, like that obtuse student mentioned earlier, being unable to get all the information out of it, even though it is already contained within it. What we have just said is nothing else than tautological ontology . . .

PLAGIARISM AND CREATION

What does this horror we have just dared to articulate mean? It means that it is in fact possible to "see" an atom's "cosmic potential," "evolutionary potential," "civilization potential," and, generally speaking, any other potential. We are not being entirely serious, of course. So far, we have been unable to infer the characteristics manifested by kitchen salt from the atoms of sodium and chlorine, respectively. We can infer some of those characteristics, but our very cleverly sounding "tautological ontology" is at most a project that can enable us to construct a different world than our own because it is impossible to infer "everything" from an elementary building block of matter in our world. The following approach seems a little more realistic: could we perhaps obtain the final result of natural processes not by plagiarizing things exactly the way Nature made them but by entering into those processes "sideways"? Then, by starting from entirely different positions than those that accompanied Nature's starting point, we could, after a certain number of stages, arrive at a result that would be identical with Nature's result.

Here is a rather crude example. A seismic shock on the surface of the Earth is needed. Rather than "construct" volcanoes, the way Nature does, we can cause a shock by means of a TNT explosion. We shall have obtained the shock we needed because the final results of a phenomenon, or of a series of phenomena, are not ultimately determined by the whole chain of causes and effects that have led to this final result.

And now for a rather less crude example. The *Penicillium notatum* mold creates penicillin. Instead of growing mold and extracting the necessary bodies from it, we can take some simple substances and synthesize penicillin from them.

Let us now look at an example that is quite close to being realized. The greatest amount of energy can be achieved via an annihilation process, that is, the process of combining matter and antimatter. As far as we know, antimatter does not exist in our Metagalaxy. We are already capable of artificially creating some of its parts. If we could do this on an industrial scale, then antimatter, stored, say, in "magnetic bottles" (so that an annihilation reaction does not ensue all of a sudden), would be the most efficient fuel for space vehicles. Interestingly, we would thus create the kind of matter that normally does not appear in Nature.

Finally, let us now consider an example that is absolutely unrealizable at the moment. In a section of a sperm's head—in the volume of three thousandths of a millimeter—we can find a design plan for a human brain, "coded" in the language of chemical molecules, which could develop from this sperm after it has been combined with an egg. This plan apparently consists of "production rules" and "directions for action." In that microscopic space, there is information about *what* has to be done, information about *how* it has to be done, and last but not least, a mechanism that will enact all this. Let us assume that we are capable of stimulating the sperm, or indeed the egg (from the point of view of the amount of information, this is irrelevant; fertilization is beneficial to maintaining the heterozygotic character of a population, which is why evolution has produced sexes, yet it is also possible to stimulate an egg so that it undergoes parthenogenesis by acting on it in a certain way), to undergo embryogenesis. Initially the whole fetus is developing, yet at a certain point of this development, we remove parts that are "unnecessary" with regard to our previously set goal and only concern ourselves with the development of the brain. We then transfer the "neuron preparation" obtained in this way into nutritional solution, where it will be combined with some other "preparations," or brain parts, as a result of which something like an "artificial brain," created from natural tissue, will emerge.

We can expect to encounter some ethical problems here. To avoid them, we do not start the development of a human egg but only copy all the information, that is, the entire genetic inscription, included in it. We know today, at least in principle, how to do it. To some extent, the process resembles photocopying or making photographic prints from a negative. The role of film or paper is played by a system of ribonucleic acids synthesized by us (i.e., not derived from an organism); an egg only provides "instructions" with regard to how these acid molecules must be combined. And thus we have made a "cast" of the egg's chromosomes, in the same way one makes a plaster cast of a sculpture. It is just our "artificial" chromosomes that become the starting point of the development process. Should there be some reservations about this too, we shall act in an even more roundabout way. We shall write down the chromosome information contained in an egg on a piece of paper, in the symbolic language of chemistry; on its basis we shall synthesize chromosomes; and the "laboratory egg" obtained in this way will go into embryogenetic "production." As we can see, our actions blur the boundary between what is "natural" and what is "artificial." Modeling thus allows us to cross the boundary between plagiarism and creation because our comprehensive knowledge of the genetic code obviously allows us to introduce all kinds of changes into it. Not only would we be able to program freely the color of the child's eyes, but also, on the basis of our comprehensive knowledge of the "genetic codes" that enact particular "talents" in the brain, we would even be able to mass-produce "talent matrices" and "embed" traits chosen by parents (musical talent, mathematical talent, etc.) into the genotype of any egg.

Evidently, we do not need to be familiar with the entire evolutionary road that Nature has traveled to construct a human being. We do not need millions of bits of information about particular stages of development, about Sinanthropus, about Mustier or Aurignacian civilizations. On having produced a "model" of a sperm or an egg that "matches" the original, we will have obtained a genotype that is more perfect than all the originals (owing to the accumulation of valuable genetic traits)— thanks to which we will have opened a "side entrance" into the process of creating a human organism for ourselves. Encouraged by this, we will then be producing ever better models, until we reach a fully balanced (on a corporeal and spiritual level) chromosomal design—a model devoid of genes that would carry a functional and organic propensity to disease. By causing controlled mutations (i.e., altering the genetic code provided by Nature; altering the chemical structure of particular genes), we will finally be able to succeed in producing traits that so far have remained absent from the *Homo* species (the development of gills that enable underwater existence, the enlargement of the brain, etc.).

It was not our intention to analyze man's "autoevolution" at this point. We shall present its future prospects, as well as outline a critique of evolutionary solutions, in the final section of the book. Here we only wanted to demonstrate the possible working of Imitology—which is Nature's competitor.

ON IMITOLOGY

Man has a long-standing tendency to come up with alternative theories that are mutually exclusive. In biology, preformism fought against epigenesis; in physics, determinism fought against indeterminism. Such theories rule each other out at a "low" level, that is, they silently assume that one of them is "final." But it usually turns out that one of the theories was closer to reality, yet it was merely a step on the correct path, nothing else.

In the period of advanced imitology, all this will count as the prehistory of science. A "better" theory will be the kind of theory thanks to which we will manage to control evolution, change the speed and range of an organism's regenerative potential, or orchestrate genetic traits in fetuses. All this will turn out to be possible long before, say, we gain the ability to create the chromosome apparatus of a nucleus in an artificial manner. All sciences construct theories, yet their attitude to those theories differs from branch to branch. The apparent perfection of astronomy theories results from the considerable isolation between systems examined by this discipline. Yet when this isolation diminishes, as was the case with the problem where several bodies were mutually affecting one another, the solution becomes more difficult to find. The "trial and error" nature of a theory is particularly visible when the scope of observed phenomena is negligible in comparison with the scope of the phenomenon itself (cosmogony, biogenesis, planetogenesis). Yet it seems that in thermodynamics or in the chromosome theory, more is at stake than a mere confrontation of our guesswork with Nature—and that those theories contain almost the "purest" truth.

I am unable to say whether imitology is going to annul such differences. At the end of the day, the current state of the Universe could have arrived "from any direction"; that is, what we are observing could have emerged in a variety of different ways. Yet there is still a lot to discover, so one should not take on an additional risk by predicting any future development of individual sciences.

As we know, imitology is not supposed to stand for "perfect imitation"—unless someone demands such a thing from it. We know that the number of variables it will be inserting into models "wound up" by it will vary, depending on the goal that this modeling production will have to serve. Given the existence of a predecided set goal, there is an *optimum* amount of information that needs to be gained to achieve this goal—which is by no means the same thing as the *maximum* amount of information.

According to imitology, everything man does is a form of modeling. This seems nonsensical. It is fine when we are talking about modeling phenomena that occur in stars or in living organisms, but how about "modeling" a microfusion cell? An electric cooker? A rocket?

Let us try to present a rather simplified classification of "modeling activities."

1. *Models of existing phenomena.* We want it to rain. We model climactic, atmospheric, and other such phenomena. We find out what the "starting point" of rain is. When we actualize it (in nature), it will start pouring. Sometimes, but very rarely, the rain happens to be colored: for example, a volcanic explosion throws multicolored mineral dust, which colors the water drops, into the atmosphere. We can also create this kind of rain as long as we "weave" a system—which will introduce an appropriate dye into the clouds or into the condensing water—into the "knot" of the interwoven causal chains that start the rain. In this way, we will have increased the probability of a given natural phenomenon— a phenomenon that is normally rare. But it rains quite often, which is why our contribution to increasing the probability of rain will not have been too significant. Colored rain, however, is a rarity. In this case, our action as an "amplifier of the states of low probability" has reached a rather high level.

2. *Models of "nonexistent" phenomena.* Nature does not actualize all possible courses of action, but it actualizes more of them than it is normally assumed. Not every engineer knows that some sea creatures have developed sails; that the principle of recoil, echolocation, is used in evolution; that fish are equipped with a "manometer" that tells them what depth they find themselves at, and so on. And, speaking more generally, Nature "arrived at" the "idea" of coupling more likely processes (an increase in entropy or disorganization) with some less likely ones (the

emergence of living organisms)—an idea that has resulted in an increase in organization and a decrease in entropy—billions of years ago. In a similar way, it created levers, chemodynamic and chemoelectrical machines, and transformers of solar energy into chemical energy (skeletons of vertebrates, their cells, photosynthesizing plants); it also created pumps (hearts)—regular and osmotic ones (kidneys), "photographic" cameras (organs of sight), and so on. Within the realm of bioevolution, it did not touch nuclear energy because radiation destroys genetic information and life processes. But it "applied" such energy in stars.

Thus, generally speaking, Nature couples various processes with itself. We are able to *imitate* it in this regard, and we do indeed do so. We couple processes always and everywhere: turning mills with water, melting ore, casting iron, constructing machine tools, planting cotton, and weaving clothes from it. Consequently, an increase in entropy always takes place somewhere, which results in its decrease locally (motor, cooker, microfusion cell, civilization).

Electrons behave in a particular way in an electric field. On coupling this process with some others, we get television, or ferromagnetic memory, or processes of quantum enhancement (masers, lasers).

Yet we are always imitating Nature. However, this needs to be understood correctly. A herd of running elephants and giraffes would be able to tread on clay so hard that a "car negative" would emerge in it, while a nearby volcano could throw out molten magnetite ore. This ore would pour itself into the "pattern," and in this way, a car, or something resembling it, would emerge.

This is, of course, highly unlikely. Yet, from a thermodynamic point of view, it is not impossible. The consequences of imitology can be reduced to increasing the probability of phenomena that are highly unlikely to occur "naturally" yet that are possible. Theoretically, a "spontaneous" emergence of a wooden wheel, a bowl, a door handle, or a car is possible. Besides, the probability of such a "synthesis" via a sudden convergence of atoms of iron, copper, aluminum, and so on, is incomparably higher than a spontaneous creation of a living organism via a simultaneous convergence of atoms and their "shifting" to correct places to produce an amoeba or our friend Mr. Smith. A car is constructed from, at most, a dozen or so thousand parts, an amoeba from millions of parts. Also, the location, moment, and crystallization of particular atoms and solid bodies within the body of a car or its motor have no relevance as far as its function is concerned. Yet the location and characteristics of the molecules from which an amoeba is "made" significantly determine its existence. Why then did amoebas emerge but not cars? Because only a system that has been equipped with a capacity for self-organization from the very beginning is highly likely to emerge spontaneously, and also because the "starting conditions" of the Earth allowed for it.

We would like to propose a general principle. The distribution of the probability of Nature's design activity is entirely different from the distribution of man's design activity—nonetheless, the latter has, of course, to be embedded in the former. The normal distribution (bell curve) of probability that is typical of Nature makes the emergence of pots or counting machines across the whole Universe by means of spontaneous generation extremely improbable. Having plundered all the dead planets and burned out stellar dwarves, maybe we would find a couple of "accidental spoons," maybe even a spontaneously crystallized tin can, but we would need to wait all eternity for it to contain, by accident, some beef or something else that would be at least partially edible. Yet such phenomena are not "impossible" in the sense of being prevented by nature's prohibitions (or laws, since—while also serving as imperatives for things to be a certain way—such laws prohibit an alternative set of occurrences). Our design activity is thus a special case within the realm of probability of Nature's design activity, and in addition, it is located where the probabilistic values suddenly start to decrease and become incomparably microscopic. In this way, we arrive at some highly improbable thermodynamic states, such as a rocket or a TV set. Yet where Nature is "in its element" as a designer, we are at our weakest: we are *(still)* unable to initiate the processes of self-organization on a scale that Nature can, nor can we match its skill. Also, if nature were incapable of doing this, neither the reader of this book nor its author would exist. So far, man has been interested in an extremely narrow section of "Nature's productive spectrum," among all the things that are possible on the level of design. We have not attempted to construct meteorites or comets or Supernovae (although we are on the right track here, thanks to the hydrogen bomb). Yet cannot we go beyond the limits designated by Nature in any way? It is, of course, possible to invent Universes and Natures that are different from ours. Yet how can they be actualized?

We are now leaving this topic—but not for long.

6

PHANTOMOLOGY

THE FUNDAMENTALS OF PHANTOMATICS

We are faced with the following problem: how do we create realities for the intelligent beings that exist in them, realities that are absolutely indistinguishable from the standard reality but that are subject to different laws? By way of introduction, we shall start with a more modest task. We shall ask, Is it possible to create an artificial reality that is very similar to the actual one yet that cannot be distinguished from it in any way? The first topic focuses on the creation of worlds, the second on the creation of illusions. But we are talking about perfect illusions. I do not even know whether they can be called *just* illusions. Please judge for yourselves.

The branch under examination will be called phantomatics: it provides a stepping-stone to creative engineering proper. We shall start with an experiment that—let us make it clear right from the start—does not belong to phantomatics itself.

There is a man sitting on a veranda, looking at a garden and simultaneously smelling a rose he is holding in his hand. We are recording (e.g., on a cassette or reel tape) a series of impulses that are traveling along all his nerves.

Several hundred thousand of such recordings need to be made at the same time because we have to record all the changes that are taking place in his sensory nerves (of shallow and deep sensation) and in brain nerves (i.e., signals traveling from the sensory receptors of the skin and from the muscle proprioceptors as well as from organs of taste, smell, hearing, sight, and balance).

Once we have recorded all these signals, we place our man in complete isolation, for example, in a bath with tepid water in a dark room;

we then place electrodes on his eyeballs, insert them into his ears, attach them to his skin, and so on. In sum, we are connecting all this individual's nerves to our tape recorder, which we then start—and, in this way, we introduce the recordings made earlier into his nerves.

This is not as simple as it sounds. Depending on the significance of the topological location of impulses within the nerve trunk, it is easier to do this with some types of nerves than with others. The optic nerve will pose particular difficulties. The main olfactory area, at least in man, is almost dimensionless: if we detect three smells at the same time, it is very difficult to determine which one is coming from where. But the location within the field of vision is of high quality: the initial organization of impulses takes place already in the retina, and the optical nerve is like a multivein cable, with each vein conducting a bunch of impulses aimed at the primal visual cortex. The difficulty of "locating" the recorded impulses inside this nerve is significant (and so is the process of recording itself). Similar but lesser difficulties are caused by the auditory nerve. We can think of several technical ways of overcoming this problem. The simplest one seems to be through introducing impulses cortically, from the back of the head and thus directly into the visual cortex. Given that we cannot envisage exposing the cortex surgically, though it is impossible to stimulate it with exact localized precision through the skin and bones, it would be necessary to transform electric impulses into some other ones (e.g., bunches of radiation produced by the maser, with ultrashort waves whose diameter does not exceed one neuron). Provided they are focused and weak enough, such waves are capable of stimulating the brain tissue without harming it in any way. Yet this is a little desperate, while the results are not fully guaranteed.

It would also be possible to construct a special "eyeball attachment" so that it constitutes a kind of "anti-eye," an optically balanced system that is "linked up" with the natural eye through the pupil (not directly, of course: in front of the pupil are the eye's anterior chamber and the retina—but both of these are transparent). The eye and the "anti-eye" together constitute a uniform system, in which the "anti-eye" is a transmitter and the eye is a receiver. Now, when looking (in normal situations)—not directly, with his own eyes, but rather with his "anti-eyes"—our man can see everything normally, the only difference being that he is wearing these (rather elaborate) "spectacles." The "spectacles" are not only an insert between his eye and the world that transmits light but also a "pointing" device that breaks the perceived image into elements whose number equals the number of rods and cones in the retina. The elements of the anti-eye's field of vision are connected (e.g., via a cable) with a recording device. In this way we can cunningly gather the same information that is collected by the retina—yet not by inserting ourselves behind it, that is, behind the eyeball, directly into the optic nerve, but *in front* of it, into the "information gathering attachment." If we then want to reverse the reaction, we place those "spectacles" on the man once again but do it in the dark this time. We send the recording from the device via the *device–"anti-eye"–eye–optic nerve* route all the way to his brain. This solution is not the best one but at least it is technically feasible. We should add that this solution has nothing to do with screening a movie or some kind of microfilm into the eye, with the camera placed against the pupil. A movie or any other recording of this kind has a *predetermined* focus, which is why we cannot shift our gaze from the foreground, which is in focus, to the background, which remains out of focus. A movie thus determines in advance what should be seen clearly, and what with less detail, even if it is a 3-D (stereoscopic) movie. But the power of muscle contraction that causes the flattening and expansion of the lens provides another source of information for the brain, facilitating, for example, the evaluation of distance—although to a lesser degree than binocular vision does. This is why, when aiming at perfect imitation, we must provide the eye with freedom also with regard to accommodation, not to mention the fact that the cinematic image is not optically perfect "from the point of view of the human eye." This substantial aside was not so much supposed to offer a concrete solution, because our ideas are very primitive, but rather to emphasize, on one hand, the difficulties involved in actualizing the problem and, on the other, the ultimate possibility of doing so.

And thus when our man is resting in the dark, while a series of impulses are traveling along all of his nerves into his brain—impulses that are exactly the same as those that traveled along his nerves when he was sitting on a veranda with a rose in his hand—he will subjectively find himself in that situation again. He will see the sky, a rose in his hand, a garden in the background behind the veranda, grass, children playing, and so on. A somewhat similar experiment has already been conducted on a dog. First, the impulses traveling along the dog's motor nerves while it was running were recorded, after which the dog's spinal cord was cut. Its hind legs thus became paralyzed. When the electric recording was inserted into the nerves of the paralyzed limbs, the hind part of the dog that had been paralyzed "came back to life," performing the same movements that are performed by a normal dog while running. The change to transfer speed will change the speed of movement. The difference between our thought experiment and the other, actual one lies in the fact that the dog had its impulses introduced centrifugally (into its motor nerves), while we have introduced such impulses into the centripetal nerves. What would happen, however, if the man on whom the experiment was being performed would want, for example, to get up from his armchair and go out into the garden? He would not be able to do so, of course. It is because the impulses we have introduced into this man's nerves are permanently recorded and fixed. If he was trying to get up, a strange confusion would arise; in trying to grab the railing of a staircase one meter away, he would grab air. A bifurcation would occur within his sensations—between what he is feeling and observing and what he is doing. This bifurcation would be a consequence of the separation of his current motor activity from his earlier sentient activity recorded by us.

Do similar situations take place in real life? It sometimes happens that someone, on coming to the theater for the first time, addresses the actors loudly, offering advice (e.g., to Romeo not to commit suicide), and is very surprised that the actors are not hearing his advice. They are not reacting because any kind of art—theater, film, literature—is "preprogrammed in advance," determined once and for all, and no intervention into the plot is going to alter the course of events. Art is a unidirectional information transfer. We are just its addressees, mere recipients of a cinematic projection or a theatrical performance. We are passive recipients, not participants in the action. A book does not offer the same illusion theater does because it is always possible to take a look at the epilogue right away and find out that it has already been predetermined. The future course of action in a play, however, is only recorded in the memory of the actors (at least for a spectator who has not read the script). We read sometimes in science fiction literature about future entertainment, which is to consist of activities that will be similar to those described in our experiment. A character places the required electrodes on his head, thanks to which he suddenly finds himself in the middle of the Sahara or on the surface of Mars. The authors of such descriptions do not realize that the only difference between this "new" art form and the currently available one is the way in which the two are "linked up"—which itself is of little significance—to the fixed and predetermined content, and that even without the electrodes, we would obtain a similar illusion in a stereoscopic "circarama," perhaps one equipped with an "additional olfactory channel" as well as stereo sound. The field of vision is the same as it is in nature, that is, potentially 360 degrees; everything one sees is three-dimensional; the colors are natural; the olfactory apparatus simultaneously creates "desert" and "Martian" breezes. We do not thus need to project into the year 2000 if we can actualize all this today, as long as we have proper investment. It is of little relevance where one decides to place the electrodes—unless those electrodes themselves are supposed to carry the burden of the thirty centuries of civilization.

And thus, while in "traditional" art, between the content of the message and the brain of the receiver, we find the latter's sensory organs, in this "new" art straight from science fiction, those organs disappear because information content is inserted directly into the nerves. The unidirectional character of the linkup is the same in both cases. This is why neither the experiment demonstrated by us as an illustration nor the "new" art belong to phantomatics. Phantomatics stands for creating *bidirectional* links between the "artificial reality" and its recipient. In other words, phantomatics is a feedback art. Someone could, of course, just hire actors, dress them as seventeenth-century courtiers while himself donning a French king's costume, and, together with the other fancy-dress wearers, perform his "reign on the throne of Louis" in an appropriate setting (e.g., that of an old hired castle). Such activities do not count even as basic phantomatics—one of the reasons for this being that it is possible to leave that realm.

Phantomatics stands for creating situations in which there are no "exits" from the worlds of created fiction into the real world. Let us now explore, one by one, the various means by which phantomatics can be actualized and also consider the interesting question as to whether there exists a conceivable way to enable the phantomatized person to become convinced that his experiences are only an illusion, separating him from temporarily lost reality.

THE PHANTOMATIC MACHINE

What can a person connected to a phantomatic generator experience? Everything. He can climb the Alps, wander around the Moon without a spacesuit or an oxygen mask, conquer medieval towns or the North Pole while heading a committed team and wearing shining armor. He can be cheered by crowds as a marathon winner or the greatest poet of all time and receive a Nobel Prize from the hands of the Swedish king; he can love Madame de Pompadour and be loved back by her; he can enter into a duel with Iago to avenge Othello or get stabbed himself by Mafia hitmen. He can also experience enormous eagle's wings growing on his back; he can fly and then become a fish again and spend his life on a coral reef; as a shark he can career with his mouth wide open toward his victims; he can even capture the swimmers, eat them with great relish, and digest them with gusto in a quiet corner of his underwater cave. He can be a six-foot-four Negro, pharaoh Amenhotep, Attila, or—contrarily, so to speak—a saint; he can be a prophet, with an additional guarantee that all his prophecies will be fulfilled 100 percent; he can die and be resurrected many times over.

How can we actualize such experiences? This is not quite straightforward, of course. This man's brain will have to be connected to a machine that will send streams of olfactory, optical, sensory, or other stimuli into it. As a result, he will be standing on tops of pyramids, or lying in the arms of Miss World 2500, or bringing death with his sword to armor-clad enemies. The machine will have to send the impulses produced by his brain in response to the impulses it receives to its own subsystems, and it will have to do it instantly, in a split second. Thanks to the corrective game of feedback, as well as the organization of the streams of impulses by appropriately designed self-organizing systems, in those subsystems, Miss World will be responding to his words and kisses; the stems of flowers he will aim to pick will bend flexibly, while the chest of the enemy he will take delight in piercing will spurt blood. Please excuse such a melodramatic tone of this lecture. Yet, without wasting too much space or time, I would like to outline the workings of phantomatics as a "feedback art." It is a domain in which the recipient becomes an active participant and a hero, finding himself in the center of preprogrammed events. It is probably better to resort to this somewhat operatic language than to use technical jargon—which would not only make the discourse extremely clunky but which would also be rather futile because, for the time being, neither a phantomatic machine nor any programs for it exist.

The machine cannot run a program that would predict all possible actions of its recipient and protagonist in advance. That would be impossible. Yet at the same time the machine does not have to represent the level of complexity that would match the complexity of all the persons appearing in the vision (enemies, courtiers, Miss World, etc.). As we know, in a dream we find ourselves in various unusual environments; we meet many people, sometimes weird ones: people behaving in an eccentric way, surprising us with what they say; we can even find ourselves in a conversation with a whole crowd. All of this—that is, various environments and our dream partners—is a product of the working of only one entity: the dreaming brain. The program for a phantomatic vision can thus only be of general nature, something like "Egypt in the days of the eleventh dynasty" or "the marine life of the Mediterranean," and the machine's memory banks must contain the full load of facts referring to a given theme—facts that will become transformed and activated from the lifeless recordings whenever a need arises. This need is, of course, dictated by the very "behavior" of the phantomatized person—for example, when he turns his head to look at the section of the pharaohs' throne chamber that is situated "behind his back." The impulses sent to his neck muscles by his brain must be immediately "countered" so that the afferent projection of the optical image changes in such a way that "the back section of the chamber" actually enters his field of vision. The phantomatic machine must react instantly to even the tiniest change to the stream of impulses sent by the human brain, in a way that is demanded by this change. Of course, these are just the first steps. The laws of physiological optics, gravity, and so on, must be faithfully reproduced (unless the subject of the selected vision contravenes this: somebody will want to fly by just "flapping his arms," i.e., against gravitation). Yet apart from the strict deterministic chains of causes and effects mentioned earlier, the vision must also contain a mobile group of processors with a certain degree of freedom: this simply means that the characters appearing in it, our protagonist's phantomatic partners, need to exhibit human characteristics—a (relative) autonomy of speech and action from the actions and words of the protagonist. They must not be mere puppets—unless, again, this is demanded by the phantomatology enthusiast before the "show." Of course, the complexity of the equipment used will vary; it is easier to imitate a Miss World than Einstein. In the latter case, the machine would have to have the complexity, or intelligence, that would match the brain of a genius at its disposal. We can only hope that there will be many more of those wishing to have a chat with such Miss Worlds than those desiring a conversation with the creator of relativity theory. Let us also add, to make our account complete, that this "insert," those "anti-eyes" we talked about in our introductory example, would not be of much use to a phantomat that would run at full power and offer a total freedom of illusion; some other, more perfect solutions are needed. Yet the principle remains the same: we have a person connected to the environment, imitated by the phantomatic machine via two information channels, an afferent and efferent one. In this case, the machine can do everything except one thing: it does not directly command the brain processes of the receiver; it only commands the facts that enter the brain, and thus we cannot demand of the phantomat to create the experience of an identity split or a severe attack of schizophrenia. But this remark comes a little too soon. So far we have discussed "peripheral phantomatics," which is derived from the body's "peripherals," because the game and the antigame of the impulses take place in the nerves, without intervening directly in brain processes.

The question as to how it is possible to learn about the fictitious nature of the phantomatic vision is therefore prima facie analogical to the question that is sometimes posed by a dreaming person. There are dreams in which the sense of the real with regard to what is going on is extremely strong. Yet it should be remembered that a dreaming brain never manifests the full command, awareness, and intelligence that it exhibits when a person is awake. In normal circumstances, a dream can be mistaken for reality, but not the other way round (i.e., reality for a dream)—unless we find ourselves in a very special situation (just after waking up, during an illness, or while experiencing increasing mental fatigue). Yet in those latter cases, we are always faced with "gullible" consciousness because it has been "dulled."

Unlike a dream vision, a phantomatic one takes place when one is awake. It is not the phantomatized person's brain that creates other people and other worlds: they are produced by a machine. A phantomatized person is a slave of the machine when it comes to the amount and content of afferent information: he does not receive any other external information. However, he can freely use that information, that is, interpret it and analyze it in any way he wants, as long as he is insightful and clever enough, of course. Does this mean that a person in full possession of his mental faculties is capable of detecting phantomatic "deception"?

We can say in reply to this that if phantomatics becomes a form of contemporary cinema, the very fact of going to the phantomatics theater, purchasing a ticket, and other preparatory activities—the memory of which the phantomatized person will also retain during his session—and, last but not least, knowing who he is in real life, will enable him to maintain a correct, and not entirely serious, relation to his experience.

Yet there is a double aspect to this state of events: on one hand, while knowing about the constructed nature of the experienced action, that person would be able, just like in a dream, to allow himself to do more than he does in real life (and thus his martial, social, or erotic courage would challenge his behavioral norm). This aspect, rather pleasant on a subjective level, as it allows for the freedom of behavior, would be matched by another, somewhat contradictory one: the awareness that neither the performed actions nor the persons appearing in the vision are material and hence real. The craving for authenticity would thus remain unsatisfied even by the most perfect vision.

Of course, it can happen, and it *will* happen if phantomatics does become a form of entertainment or art. The management of the hypothetical Phantomat will not be interested in covering up too well the fictitious nature of those experiences—if such experiences were to lead, for example, to a patron's nervous breakdown. Also, certain wishes—that is, those of a sadistic nature—would probably not be able to be realized owing to appropriate legislation.

Yet we are not interested in user or administrative problems but rather in an entirely different issue: one that belongs to gnoseology. There is no doubt that "entering" the vision can be covered up perfectly. Someone goes to the phantomat and orders a trip to the Rockies. The trip is very pleasant and beautiful, after which the person "wakes up," that is, the vision comes to an end; the assistant removes the patron's electrodes[1] and politely bids him good-bye. On being walked to the door, the patron goes out into the street and suddenly finds himself in the midst of a terrible cataclysm: the houses are collapsing, the earth is trembling, while a massive "saucer" full of Martians is falling from the sky. What happened? The "waking up," the removal of the electrodes, the leaving of the Phantomat were *also* part of the vision—which started with an innocent sightseeing trip.

Even if no one were to play such "pranks," psychiatrists would still see various neurotics in their waiting rooms, haunted by obsessions of a new type—the fear that what they are experiencing is not true at all and that they have become "trapped" in a "phantomatic world." I mention this point because it clearly indicates how technology not only shapes normal consciousness but also makes its way onto the list of diseases and disorders whose emergence it initiates.

We have only listed one of the many possible ways of masking the "phantomatic nature" of experience. Many other, equally effective ones could be outlined here, not to mention the fact that a vision can contain any number of "levels"—the way it happens in a dream, when we are dreaming that we have already woken up, while in fact we are dreaming another dream, which is sort of embedded in the previous one.

"The earthquake" is suddenly over; the saucer disappears; the patron discovers that he is still sitting in an armchair full of wires connecting his head to the equipment. A polite technician explains with a smile that it was a higher-level program; the patron leaves, returns home, and goes to sleep. Next day he goes to work—only to find out that his office does not exist anymore: it has been destroyed in an explosion by an old shell, left from the last war.

Of course, this too can be a continuation of the vision. Yet how is one supposed to know?

First of all, there is one very simple method. As we have said before, the machine is the only source of information about the external world; there is no doubt about it. But it is not an exclusive source of information about the state of the organism itself. It only partly serves this role by replacing the body's *neural* mechanisms, which provide information about the location of one's arms, legs, and head, about the eyeball movement, and so on. Yet *biochemical* information produced by the organism is not subject to such control—at least in the phantomats discussed so far. It is therefore enough to do about a hundred squats: if we get sweaty, if we feel a little out of breath, if our heart is pounding and our muscles feel tired, we are in the real world and not in a vision because our muscle fatigue has been caused by the accumulation of lactic acid in the muscles. The machine cannot influence the level of sugar in the blood, the amount of carbon dioxide in it, or the concentration of lactic acid in the muscles. In a phantomatic vision, it would be possible to do even a thousand squats without any sign of fatigue. Yet this could also be circumvented if someone wanted to develop phantomatics even further. To begin with, a phantomatized person could be allowed some freedom of movement; he would just need to be positioned in a particular way to exercise this freedom (i.e., to use his muscles). Of course, if he was picking up a sword, the movement would only be real from an external observer's point of view: his hand would not be clasped on the hilt of the sword but would grab emptiness instead. This rather primitive method could be replaced with a more sophisticated one. The organism's chemical information is transmitted to the brain in various ways—either via the nerves (a tired muscle "refuses to play" and thus the transmitted impulses do not animate it, or we feel muscle pain—which also results from the irritation of the nerve ends, which, of course, can be imitated in a phantomatic way) or directly: the excess of carbon dioxide in the blood irritates the medulla of the respiratory center, leading to the deepening and acceleration of breathing. Yet a machine can simply increase the amount of carbon dioxide *in the air* we breathe; when the amount of oxygen drops to a certain level, the ratio of the first gas to the second in the blood will shift the way it does during hard physical labor. An advanced machine thus thwarts "the biochemical and physiological method."

"Playing intellectual games with the machine" is the only thing left. The possibility of distinguishing between vision and reality depends on the "phantomatic potential" of the device. Let us imagine we have found ourselves in the situation described and are trying to find out whether this is actual reality. Let us say we know a certain renowned philosopher or psychologist; we go to see him and engage in a conversation with him. This could be an illusion, yet a machine that imitates an interlocutor endowed with intelligence must be much more complex than a machine that creates "soap opera" kinds of scenes such as the landing of a saucer full of Martians. As a matter of fact, a "travel" phantomat and a "human-producing" one are two different kinds of devices. It is incomparably more difficult to construct the second one than the first.

Truth can also be discovered in another way. Like everyone else, we have our secrets. They can be insignificant, but they are our own. A machine is unable to "read our thoughts" (it is not possible for it to do it: memory's neural "code" is an individual property of every person, and so "cracking" one individual's code does not tell us anything about other people's codes). And thus neither the machine nor anyone else knows that a particular drawer in our desk gets stuck. We rush back home to check whether that is the case. The drawer that gets stuck makes the reality of the world in which we find ourselves highly probable. The creator of the vision would have really needed to have followed us very carefully if he was to be able to detect and record even such a triviality as the warped drawer on his tape, just before our trip to the phantomat! The easiest way to debunk a vision is through similar details. Yet a machine can always resort to a tactical maneuver. The drawer is not stuck. We grasp the fact that we are still inside the "vision." Our wife appears; we tell her that she is only an "illusion." We prove this by waving the drawer we have just pulled out at her. The wife laughs pityingly and explains that in the morning the drawer had been fixed by the carpenter, whom she had called in. And thus, once again, we do not know anything at all. Either this is the real world or the machine has performed a cunning maneuver, matching it with our move. A "strategic game" played against a machine no doubt assumes a comprehensive familiarity with our everyday life. We must not exaggerate though: in a phantomatic world, every unusual phenomenon is going to awake suspicion, suggesting that it is a fabricated vision—while old shells do explode and wives do call in carpenters sometimes in real life too. We shall thus determine as much: the statement that person x finds himself in the real world rather than in a phantomatic one can always be only probable, sometimes highly probable, but never absolutely certain. Playing against a machine is like playing chess: a contemporary electronic machine will lose to an excellent player, but it will beat a mediocre one; in the future, it will beat every human being. The same can be said about phantomats. The principal weakness of all efforts focused on discovering the true state of events lies in the fact that a person who has suspicions as to the nonreality of the world in which he lives must act on his own. It is because any act of turning to other people for help is, or rather can actually be, *an act of feeding the machine with strategically valuable information.* If this is indeed a vision, by letting our old friend in on the secret about the issues concerning existential uncertainty, we pass on additional information to the machine—which it is going to exploit to enhance our belief in the reality of our experience. This is why the person undergoing the experience cannot trust *anyone* but himself—which severely narrows down his options. He acts defensively to an extent, as he is surrounded from all sides. This also means that a phantomatic world is a world of total solitude. There cannot be more than one person in it at any one time, just as it is impossible for two real persons to find themselves in the same dream.

No civilization can become "fully phantomatized." If all its members were to start experiencing phantomatic visions from a certain point on, the real world of this civilization would come to a halt and die out. Yet since even the most delicious phantomatic dishes do not sustain life functions (although the sense of gratification can be generated through introducing certain impulses into the nerves!), the person who is undergoing phantomatization for a prolonged period of time will have to receive real food. Of course, it is possible to envisage some kind of omniplanetary "Superphantomat," to which the inhabitants of a given planet have been connected "for ever," that is, for as long as they have been alive, while their bodies' vegetative processes are being supported by automatic devices (e.g., those introducing supplements into their blood). Naturally, a civilization of this kind looks like a nightmare. Yet similar criteria cannot determine its probability: something else determines it. This civilization would only exist for the duration of one generation—the one that remains connected to the "Superphantomat." This would thus be a peculiar form of euthanasia, a kind of pleasant suicide of a civilization. For this reason, we consider its implementation to be impossible.

PERIPHERAL AND CENTRAL PHANTOMATICS

Phantomatics can be positioned in a sequence that would consist of some more or less direct ways of influencing the brain we know from history, by means of using peripheral stimuli ("peripheral prephantomatics") or acting centrally ("central prephantomatics").

The first set includes the rituals, mainly developed in ancient civilizations, of inducing people into a particular state of ecstasy by means of motor stimuli (e.g., dance rituals), aural stimuli (influencing emotional processes in a "rocking" manner by means of rhythmic impulses—since melody is younger than rhythm in evolutionary terms), visual stimuli, and so on. They facilitate the induction of a group of people into a trance so that their individual consciousness becomes blurred or, rather, constricted—something that always accompanies very strong emotions. Such climactic collective excitement is nowadays associated with "unbridled group behavior," with orgies, but in ancient communities, it used to be a half-mystical, half-demonic amalgamation of individual experience in a state of general excitement—in which the sexual element was not dominant at all. Instead, such practices were considered appealing owing to their mysteriousness and the fact that they released powers in people that were unknown to them from their everyday experience.

The second set includes the ingestion of substances such as mescaline, psilocybin, hashish, alcohol, fly agaric extract, and so on. Through influencing brain chemistry, they evoke subjectively elevating and blissful experiences—which can appeal to aesthetic or emotional sides of the soul. These two kinds of practices have actually often been combined in an attempt to achieve a culmination of sensations. What links those activities with phantomatics is the active influencing of the information introduced to the brain to evoke a desired state in it—not because such a state can function as an appropriate regulator for the environment but rather because it offers delight or shock (catharsis), that is, simply a powerful and deep experience. Were those ancient practices a collective manifestation of sadism or masochism? Or were they a manifestation of a religious life? Or perhaps they constituted the early beginnings of "mass art" that does not separate creators from the audience but rather turns everyone into a coauthor of the "work"? Why is this of interest to us? This issue has some connections with the classification of phantomatics itself.

Psychoanalytic approaches tend to reduce all human activity to elementary drives. They tend to label both puritan asceticism and the most explicit debauchery as "masochism" or "sadism." The problem is not so much that those statements are untrue as that this truth is too trivial to be of any use to science. Discussions about pansexuality, and so on, are as pointless as debates as to whether a sexual act is a manifestation of solar activity. The latter statement is probably indeed true: as life owes its emergence to solar radiation, and thus, by presenting long cause-and-effect chains that lead all the way from our star to the Earth's crust, and then further through evolution's developmental cycles, it is possible to demonstrate that the energetic deterioration of radiating quanta in plants, which in turn serve as nutrition for animals (among which the human is one), eventually leads, at a certain point very far from the energy source, to sexual acts—thanks to which this whole process can actually be continued (because without procreation, all those organisms would die out). Similarly, we could say that sexual drive is sublimated into artistic work. Whoever says this is speaking metaphorically rather than conveying the truth—or at least this is not a scientific truth. Not everything constitutes scientific truth: an ocean of insignificant variables is larger than an ocean of stupidity—and this is already saying something.

When cause-and-effect chains become long enough, any attempt to connect very remote stages becomes a metaphor rather than a scientific proposition. This applies in particular to complex systems such as the neural network, where—owing to the numerous inner connections and feedback loops—it is difficult to determine what constitutes an effect and what constitutes a cause. Looking for "first causes" in a network as complex as the human brain is pure apriorism. Even though a psychiatrist–psychoanalyst will deny this, his theses show that there is as much difference between a strict tutor and Jack the Ripper as there is between two cars, whereby the first one has better brakes than the second, which is why it does not cause crashes. Hundreds of years ago, artistic, magical, religious, and entertainment activities did not use to be as separated as they are now. We call phantomatics "an entertainment activity" owing to its genetic connections with similar contemporary technologies, but this does not determine its future or perhaps even universal aspirations.

In our classification system, peripheral phantomatics is a method of directly acting on the brain, in the sense that phantomatizing stimuli only provide information about *facts*. Reality functions in the same way. It always determines external states but not internal ones, because the same sensory observations (that there is a storm, that we are sitting atop a pyramid), no matter if generated artificially or naturally, evoke different feelings, emotions, and reactions in different people.

"Central phantomatics," that is, direct stimulation of selected brain centers that enable pleasant sensations or the feeling of ecstasy, would also be possible. Those centers are situated in the midbrain and the brainstem. Anger and anxiety centers (i.e., centers of aggressive and defensive reactions) are located very close to them. Olds and Milner's article is well known by now.[2] An animal (a rat) was placed in a cage; it had an electrode chronically (i.e., permanently) implanted into its brain (into the diencephalon) and was thus able to stimulate this place electrically by pressing a pedal which closed the circuit with its paw. Some animals shocked themselves permanently over the period of twenty-four hours, with the frequency reaching eight thousand times per hour, that is, over twice per second. If the electrode was inserted a little further, then, on suffering a shock, they would not do it again. As H. W. Magoun puts it, we can imagine that there are two opposite nerve centers located in this part of the brain: one linked with "reward," the other with "punishment." "In these studies," he asks, "have Heaven and Hell been located in the animal brain?"[3]

Jasper and Jacobsen have identified similar relations in the human brain, in which, depending on the place being stimulated, the examined person would experience either anxiety and fear, just like before an epileptic attack, or some pleasant sensations. On the basis of such anatomical and physiological data, "central phantomatics" would thus be a form of "central masturbation," even though the sensations experienced during the stimulation of the hippocampus are not equivalent to sexual discharge (orgasm). Naturally, we are inclined to denounce such electrically generated "happiness attacks," just as we would denounce any simple act of self-pleasuring. Cyberneticists, such as the already mentioned Stafford

Beer, are actually aware of the need to introduce punishment and reward mechanisms into a complex homeostat. A simple homeostat (such as the one constructed from four elements by Ashby) does not have the need for such a subsystem; this kind of "algedonic control" is only required by very complex systems, with many states of equilibrium and many self-programming actions and goals.

Given that up to this day, people have not stopped using substances, including poisonous ones (alkaloids, alcohols, etc.), which precipitate "pleasant states," we cannot rule out the emergence of a future "central phantomatics" just because it causes moral outrage as a "technology of facilitated ecstasy." This form of phantomatics cannot in any case be considered an "art," just as drug taking or drinking alcohol cannot. It is another matter with peripheral phantomatics, which, under some circumstances, could become an art—but also a realm of all possible excess.

THE LIMITS OF PHANTOMATICS

Peripheral phantomatics is a way of introducing the human into the world of experiences whose lack of authenticity cannot be detected. We have said that no civilization can become "totally phantomatized" because this would mean its suicide. Yet a similar *reductio ad absurdum* can be also applied to television. A civilization that would divide itself into two parts—people who transmit programs and people who receive them on their TVs—would not be able to exist either. Phantomatics is thus only possible, or even probable, as a technology of entertainment but not as a path that—when followed—would lead to a society cutting itself off from the real world and becoming "encysted" in the way discussed earlier.

Phantomatics seems to be a sort of pinnacle around which various contemporary technologies of entertainment converge. They include "amusement parks," "cinema," "ghost castles," and, last but not least, Disneyland—which in fact is one big primitive pseudophantomat. Apart from such officially recognized technologies, there exist some illicit ones (activities such as those described, e.g., by Jean Genet in *The Balcony*, where "pseudophantomatization" takes place in a brothel). Phantomatics has a certain potential to become an art. At least, this seems to be the case to begin with. It could thus experience a bifurcation of a kind that happens in cinema as well as other domains of art into artistically valuable production and worthless trash.

The menace of phantomatics is nevertheless significantly greater than that posed by cinema—in its degenerate and at times even socially transgressive form (such as pornography). The reason for this is that, in its uniqueness, phantomatics offers a kind of experience whose "privacy" can only be matched by a dream. It is a technique of superficial wish fulfillment that can be easily abused through actions that violate social conventions. Someone could say that such possible "phantomatic debauchery" cannot pose a social threat but is rather a form of "release." It is just that "doing wrong to one's neighbor" in a phantomatic vision does not harm anyone. Has anyone ever been held accountable for the most horrible content of his dreams? Would it not be better if someone assaulted, or even killed, his enemy at the phantomat, instead of doing it in real life? Or if he "coveted his neighbor's wife," which otherwise could bring disaster on some happy couple? In other words, cannot phantomatics absorb the dark forces lurking in man, without doing any harm to anyone?

This attitude could be countered with an opposite one. A critic will say that criminal acts committed in a vision can actually encourage their reenactment in real life. As we know, man desires the most what he cannot have. Such "perversity" is ubiquitous. There is no rational basis for it. What drives an art lover when he is ready to give up anything for an authentic van Gogh—which can only be distinguished from a perfect copy by an army of experts? A quest for "authenticity." Thus the inauthenticity of phantomatic experiences would take away their "buffer" value; they would instead become a school or a training ground for socially banned activities rather than their "absorber." Making a phantomatic vision indistinguishable from reality will lead to some unpredictable consequences. A murder will be committed, after which the killer will defend himself by saying that he was deeply convinced that it was "just a phantomatic vision." Many people will get so entangled in the indistinguishable truths and fictions of life, in the subjectively indivisible world of authenticity and illusion, that they will not be able to find a way out of this labyrinth. Such situations could become powerful "generators" of frustrations and nervous breakdowns.

There are thus good reasons why phantomatics should not become a universe of complete behavioral freedom such as dreaming, a realm in which the madness of nihilistic debauchery would only be constrained by one's imagination and not conscience. Illegal phantomats can also, of course, be set up. Yet this is more a policing problem than a cybernetic one. Cyberneticists might be expected to build a kind of "censorship" into the device (akin to the Freudian "dream censor"), which would stop the vision as soon as the phantomatized person starts showing some aggressive or sadistic tendencies.

This seems to be a purely technical problem. For someone who is capable of building a phantomat, it will probably not be too difficult to introduce such controls into it. Yet here we come across two entirely unexpected consequences of such postulated controls. Let us present the simpler one first. The phantomatization of the great majority of artworks would be impossible: they would have to find themselves beyond the limits of permissibility. If the protagonist in our vision expresses even so noble a desire as to become Podbipięta,[4] we shall not avoid harm, because, as Podbipięta, he will decapitate three Turks in one go, while as Hamlet, he will stab Polonius, thinking the latter is a rat. And—please excuse this particular example—if he wanted to experience religious martyrdom, this would also turn out to be somewhat tricky. It is not that there are hardly any works in which no one gets killed or comes to some harm (even children's stories are not exempt—just think how grue-some the Grimms' fairy tales are!). It is that the scope of the regulation of stimuli, that is, the phantomat's censorship, does not in fact extend to the phantomatized person's actual domain of experience. Maybe he wants to be flagellated as a result of his need for religious martyrdom, or maybe he is just an ordinary masochist? It is only possible to control the stimuli introduced into his brain but not what is actually going on in his brain and what he is experiencing. The context of the experience remains outside our control (in this case, it seems to be a disadvantage, but as a general rule, we could say that this is in fact rather fortuitous). This limited experimental material, obtained by stimulating various sections of the human brain (during surgery), already demonstrates that the same, or similar, input is recorded differently in every brain. The language neurons use to communicate with our brains is practically identical in all humans, while the language, or rather the coding method, used to form memories and association maps is highly individual. This is easy to verify, since memories are organized in a particular way only for single individuals. And thus, for example, pain can be associated by someone with elevating suffering and a punishment for misdemeanors, while for someone else, it can be a source of perverse pleasure. We have thus reached the limits of phantomatics because it cannot be used to determine attitudes, opinions, beliefs, or feelings directly. It is possible to shape the pseudo-material context of the experience but not the opinions, thoughts, sensations, or associations that accompany it. This is why we have called phantomatics a "peripheral" technology. Just like in real life, two people can draw entirely different and contradictory conclusions from two identical experiences (emotionally and ideologically, not scientifically). Because, as we know, *nihil est in intellectu, quod non fuerit prius in sensu*[5] (or rather, in the case of phantomatics, *in nervo*), yet neural stimulation does not explicitly determine emotional and intellectual states. A cyberneticist will say that "input" and "output" states do not explicitly determine the state of the network between them.

Someone will ask, How come they do not determine it, since we have said that phantomatics enables one to experience "everything," even, say, being a crocodile or a fish?

A crocodile or a shark—for sure, but in a "pretend" way, and this in a double sense. First, it is a pretend way because we are only dealing with an illusory vision, as we already know. Second, to be really a crocodile, one must have a crocodile, not a human, brain. Ultimately, a person can only be himself. But this should be understood properly. If a National Bank clerk is dreaming about becoming an Investment Bank clerk, his wish can be fulfilled perfectly. Yet if someone desires to become Napoleon Bonaparte for a couple of hours, he will be that (in a vision) only superficially: when he looks in the mirror, he will see Bonaparte's face, he will have his "old guard" around him—his faithful marshals—yet he will not be able to converse in French with them if he did not know this language beforehand. He will also, in this "Bonapartian" situation, manifest his own personality traits and not those of the Napoleon we know from history. At best he will attempt to *perform* Napoleon, that is, pretend to be him, more or less successfully. The same applies to the crocodile...

Phantomatics can lead to a literary wannabe receiving, say, the Nobel Prize; it can place the whole world at his feet. Everyone will be adoring him for his great poems—yet he will not be able to create those poems in his vision, unless he agrees to have them delivered to his desk.

Put briefly, the more the character one wishes to impersonate differs in personality traits and historical period from his own, the more fictitious, naïve, or even primitive his behavior and the whole vision will be. Because, to be crowned a king or receive the Pope's emissaries, one has to be familiar with the whole court protocol. The persons created by the phantomat can pretend that they cannot see the idiotic behavior of the ermine-clad National Bank clerk, and thus his own pleasure will perhaps not diminish as a result of his mistakes, but we can clearly see that this whole situation is steeped in triviality and buffoonery. This is why it will be very hard for phantomatics to develop into a mature dramatic form. First, it is impossible to write scripts for it, only some rough situational outlines. Second, drama needs personalities: characters in a play have them assigned in advance, while the phantomat's patron has his own personality and will not be able to perform the role outlined in the script because he is not a professional actor. This is why phantomatics can mainly be a form of entertainment. It can become a kind of "Thomas Cook" for possible and impossible space travel, alongside a wide range of its very useful applications—which nevertheless do not have anything to do either with art or with entertainment.

Phantomatics can help us create training and educational situations of the highest caliber. It can be used to train people for any profession: medics, pilots, engineers. There is no threat of a plane crash, a surgical accident, or a catastrophe brought on by a miscalculated building project. Second, it allows us to examine psychological reactions: this will be particularly useful for selecting astronautics trainees, and so on. The possibility of disguising a phantomatic vision will enable the creation of an environment in which the examined person will not know whether he is really flying to the Moon or whether this is only an illusion. Such a disguise is necessary because there is a need to get to know his *authentic* reactions in the face of a real breakdown and not a fictitious one, when anyone can easily show some "personal courage."

"Phantomatic tests" will allow psychologists to get to know a wide range of people's reactions better; to learn how panic develops, and so on. They will facilitate a fast selection of candidates for various university courses and professions. Phantomatics may prove irreplaceable for all those whose location (a scientific base in the Arctic, a space flight, a stay at a space base, or even star exploration) makes them spend long periods of time in solitude, in a relatively narrow and enclosed space. Thanks to phantomatics, years of travel needed to reach a particular star may become filled with everyday activities such as those that would have been taken up by the crew members on Earth—that is, years of traveling across terrestrial lands and oceans or even years of study (since, in a vision, one can also listen to lectures by renowned professors). Phantomatics will be a true blessing for the blind (except for those who suffer from total blindness, i.e., whose primary visual cortex has been damaged)—it will open a whole massive world of visual experience to them as well as to those who are ill or convalescent, to the elderly who want to experience a second youth—simply, for millions of people. In other words, we can see from the preceding that its entertainment role may turn out to be rather marginal.

It will no doubt also cause some negative reactions. Groups of its stringent opponents will appear—such as authenticity lovers who will despise the immediacy of wish fulfillment allowed by phantomatics. However, I think some sensible compromises will be reached because every civilization is actually a facilitated life, while development is to a large extent reducible to expanding the range of such facilitations. Phantomatics can also, of course, become a real menace, a social plague, but this possibility applies to all products of technology, although not to the same degree. We know that consequences of the misuse of steam and electrical technology are far less dangerous than consequences of the misuse of atomic technology. Yet this is a problem concerning social systems and the existing political relations, and it does not have anything to do with phantomatics or with any other branch of technology.

CEREBROMATICS

Is it possible to influence brain processes, and thus states of consciousness, while avoiding the normal—that is, biologically created—access routes? No doubt it is: pharmaceutical chemistry has at its disposal a great number of substances that cause various forms of excitation or that impede brain activity, or even substances that can channel brain activity in a particular way. Many hallucinogenic substances have their own unique effects: some cause only "visions," others only vague states of confusion or happiness. But would it be possible to *form* and *shape* those brain processes according to our wishes? In other words, can we "transform" Mr. Smith's brain so that he becomes, albeit temporarily, the "real" Napoleon Bonaparte, or so that he displays an amazing musical talent, or finally, so that he becomes a fire worshipper, believing in the absolute truth of this cult?

Some clear differentiations should be made at this point. First, the preceding "transformations" can mean many different things. They are all changes to the dynamic structure of the brain's neural network, which is why we use the collective term *cerebromatics* to describe them. Phantomatics provides the brain with "false" information; cerebromatics falsifies, that is, "transforms," the brain itself. Second, there is a difference between inserting one trait, for example, musical talent, into a *given* personality (this will no doubt change that personality, but we can concede it will still be the same personality, only slightly altered) and turning Mr. Smith into Napoleon.

One must cut one's coat according to one's cloth. This is to say, cutting off the activity of certain brain sections (e.g., the frontal lobes) can make an adult infantlike, with his reactions, intellect, and emotional stability resembling those of a child. It is also possible to remove the constraining activity of the parietal lobes, which will release personal aggression. (Alcohol does this, especially in people with aggressive tendencies.) In other words, the activity of the neural network as a whole in an individual can be shifted or narrowed down, within certain limits. However, it is impossible to add, in the proper sense of this term, absent traits to one's psyche. An adult used to be a child; at that time, his frontal lobes contained unmyelinated fibers—this is actually what makes a child resemble, to some extent, a person suffering from frontotemporal dementia. It is thus possible to put an adult "back" to a childlike stage— even if there are limitations to this process, as the remaining parts of his brain are not "childish": a child does not have the number of memories and experiences that an adult has. It is possible to "remove the breaks" from a particular driving function and thus turn a normal person into a glutton, a sex maniac, and so on. Personality can therefore become diverted from its normal balanced state, that is, from its original course: but that is all. We will not turn Mr. Smith into Napoleon through such maneuvers.

A parenthetical remark is needed here. Even though we have said that input and output states do not explicitly determine states of consciousness—which can be evidenced by the fact that in an analogous environment, various ideological stances develop because the same information can be interpreted in various ways—this does not mean that consciousness remains completely independent from the content inserted into it. To give a simplified example, when someone believes that "people are good," while we constantly expose him, for a prolonged period of time, to human wickedness and meanness—be it through phantomatic visions or through appropriately staged events—this person may abandon his conviction about the nobility of our species. And thus peripheral phantomatics can also, through appropriate maneuvers, lead to a change of opinion, even of a deep-rooted one. The more life experience a person has accumulated, the more difficult such a change is to accomplish. It is particularly difficult to undermine metaphysical beliefs due to the preceding tendency, as already mentioned earlier, to block any information that contradicts the structure of those beliefs.

It is a different matter when it comes to the cerebromatic "molding of the soul" directly, that is, influencing mental processes while bypassing afferent neural pathways—by modeling their neural basis in a different way.

The brain is not a unified or indivisible entity. It, too, has various interconnected "subsystems." Those connections can be physiologically variable, which means that it is not always the same parts of the brain that serve as "input" points for the stimuli that come from its other parts—and the other way round. The universal plasticity and modeling dynamics of the neural network means precisely this: that the network is potentially capable of being connected and disconnected, as a result of which such combinations lead to the emergence of various subsystems. If someone can ride a bicycle, he possesses a given set of such "prearranged" connections, which remain on the alert, ready to switch on together automatically as soon as he mounts a bicycle. To teach someone how to ride a bicycle by just introducing the appropriate information into his brain, while bypassing the normal way, that is, the necessary practice, is not simple even on a theoretical level.

Two approaches are possible here. The first one is "genetic": being able to ride a bicycle (or knowing the Koran, or trampolining) must become a "genetic" trait; that is, it must already be programmed into the genotype of the egg from which a given individual and his brain are going to develop. In this way, we could reach a situation when it is actually unnecessary to learn anything because all theoretical and practical knowledge is "implemented" into the chromosomes before fetal development takes place, as a result of which such knowledge becomes hereditary. This would entail having to increase substantially the amount of genotypic information, to complexify the structure of the nucleus, and so on. Perhaps the genotype would not be able to contain such an excess of information beyond a certain limit. We do not know, yet we should also consider this possibility. Then we would have to limit ourselves to perfecting the traits in the genome that would at least *facilitate* learning, if not replace it entirely. It would no doubt be rather strange if we managed to make all human knowledge hereditary so that every newly born baby would come into this world with the knowledge of more than ten languages and of quantum theory. This would not necessarily mean that he would immediately "speak with the tongues of men and angels" or lecture from his cradle about nuclear spins and quadrupole moments. Specific knowledge would develop in his brain in the course of time, as his organism developed and grew on the way to maturity.

This in turn conjures a vision of a world (somewhat resembling that of Huxley) in which children are "preprogrammed" so that their skills and hereditary knowledge (or rather the knowledge that has been inserted into and preserved in the egg's chromosomes) are accompanied by an affinity for doing what this hereditary knowledge and those skills allow them to do. Various abuses and attempts to "produce human types of different qualities," that is, with "higher" and "lower" minds, are also obviously possible. They are possible, but poisoning all of the Earth's atmosphere so that its biosphere burns out in a few hours is also possible. As we know, many things are possible that are not being actualized. During an early stage of a new technological turn, or while just "sensing" the imminent change, the tendency to universalize its novelty and assume that from now on it will completely dominate all human activity is common. It was the same in previous centuries, or more recently with atomistics, when it was believed that electric plants and chimney stacks would give way to microfusion cells almost everywhere within a few years. Such exaggerated and simplified predictions usually do not come true. The programming of heredity can also be practiced in a way that is thus both rational and modest: the innate knowledge of higher mathematics surely does not contradict human dignity.

The second, cerebromatic approach means transforming the mature brain. We spoke earlier about programming scientific information rather than shaping one's personality. It goes without saying that it is much easier to model genetically (chromosomally) a given personality type than a given form of knowledge. It is because the amount of genotypic information does not actually change that much, no matter if we are "designing" a future Mr. Smith to be a choleric or a phlegmatic. When it comes to cerebromatics, it is very difficult both to transform a mature personality into a new one and to insert the missing knowledge into it by conducting a procedure on the neural network. Contrary to what it may seem, this approach is much more difficult than the "genetic–embryonic" one. It is easier to plan development in advance than to transform the dynamics of a fully formed system in any significant way.

This difficulty has two aspects: technical and ontological. It is difficult to insert the knowledge about how to ride a bicycle into a neural network. It is very difficult to "append" a "sudden" mathematical talent to a forty-year-old Mr. Smith. This would need to involve some surgical and cybernetic procedures, the opening of the neural circles (circuits), and the insertion of some biological, electronic, or other "appendages" into them. Technically, it would be a most thankless task. It would consist of having to rearrange billions or at least tens of millions of connections. And even though, according to Lorente de Nó, there exist no more than ten thousand main (large) neural circuits along which impulses circulate in the cortex, we can expect each neuronal circuit *as a whole* to have a unique significance (both as a basis for thinking and as a functional element). Opening it and inserting an "appendage" would thus amount to destroying completely the primary subjective and objective signification— it would not just be an "organizational and informational appendix."

But enough of the details for now because they really fade into the background in the face of the ontological problems raised by such procedures. If we want to transform a dynamo machine into a centrifugal pump, we must reject so many of its parts and add so many new ones and rebuild the structure as a whole to such an extent that the newly constructed pump will not be an "ex-dynamo machine" anymore but simply a pump—and nothing else. Similarly, the "transformations" that are supposed to turn Mr. Smith into Napoleon or Newton can in the end produce a completely new personality, one that will have so little in common with the old one that our procedure will practically have to be described as murder. It is because we will have annihilated one person and created another one in his skin. Besides, the differences are always fluid, and it is impossible to draw a strict line between "murderous" cerebromatics and a cerebromatics that "transforms certain traits of the continued personality." A procedure as brutal as the cutting off of the frontal lobes (lobotomy) causes some significant changes to one's character, personality, drives, and emotional life. Consequently, lobotomy has been banned in many countries (including ours). Such procedures are especially dangerous given that a person undergoing surgery does not usually realize on a subjective level what kinds of changes have taken place inside him. However, if it is of any consolation, all our knowledge is based on injurious procedures.

Yet is it possible to create an "appendix" that, as a carrier of "musical talent" "connected" to Mr. Smith's brain, will enhance his personality without destroying it? We are not going to solve this problem arbitrarily, once and for all. The criteria for action cause most difficulties here because a cerebromaticist who promises to proceed "carefully" is like someone who is removing a few blades of grass from a haystack. Every time the difference is microscopic, yet after a while, the haystack will cease to exist, without anyone being able to say when it happened. This is why a cerebromaticist who wants to "remake" Smith into Beethoven in small steps is as dangerous as the one who is planning to carry out such a transformation in one go.

We have simplified the technical side of the problem in our preceding discussion, since different parts of the brain make different contributions to the development of personality. The influence of firmly located centers (those responsible for cortex analysis), such as the visual and aural fields, on its constitution is minimal. Tiny ganglia and thalamic nodes, in turn, play a much more important role than the other areas of the brain. But this has no particular relevance for the outcome of our deliberations. It is ethics rather than "problems of material" that make us reject any suggestions to "transform the soul," whereby a given personality, even if it was somewhat stupid, was to be transformed into a terribly nice and talented one—which would nevertheless be *different*. "Soul technology," both in its current and future forms, encounters here the problem of the subjective uniqueness of individual existence. By the latter I do not mean a mysterious phenomenon that cannot be explained but rather a dynamic path of a system. The decision as to which diversions from this path would have to be described as a complete personality change and which would only function as its "correction" that would not disturb its continued existence can only be made in a purely arbitrary manner. In other words, "cerebromatics" can kill people imperceptibly, since, instead of presenting us with a corpse—which would serve as undeniable evidence of the crime committed—a different person appears. The "killing" itself can be divided into any number of stages, which makes the detection and assessment of such procedures even more difficult.

We have thus explained that Mr. Smith would be wise not to demand being "transformed" into Casanova or a famous explorer, since, in the end, the world would perhaps gain a great man, but Mr. Smith would lose what should be most precious to him—that is, himself.[6]

It could be said that human life, from birth through to maturity, involves a continuous "dying" of subsequent personalities—that of a two-year old toddler, six-year old mischief, twelve-year old youngster, and so on, all the way to the adult age, in which our personality is very different from those earlier ones. And thus, if someone wishes for a spiritual makeover, which will present society with a more worthy person than he currently is, why should we actually refuse him?

It will certainly be easy to imagine a civilization in which cerebromatic procedures are allowed, just as it will be easy to imagine a civilization in which, say, criminals are subjected to compulsory personoclastic cerebromatization. Yet we have to state very clearly that these are destructive processes: "switching" from one personality to another is not possible either as a reversible or as an irreversible process, since such metamorphoses are divided by the zone of psychic annihilation—which is equivalent to bringing individual existence to a halt. Thus one can only be oneself or nobody—with two caveats, which will be dealt with separately.[7]

TELETAXY AND PHANTOPLICATION

The categorical statement with which we closed the previous section—that it is only possible to be either oneself or nobody—does not contravene phantomatics' potential. We already know that Mr. Smith, who is "living" Nelson's life at the phantomat, is only playing at, that is, pretending to be, the famous seaman. Only great naïveté could lead him to believe that he is indeed a well-known historical figure. If he was to live long enough in the phantomatic world, the fact that the orders he would give as an admiral would be obeyed without a murmur would no doubt affect his mind after some time. This could then raise fears that, on returning to his office, he would order—out of pure absent-mindedness, if nothing else—for the attorney general to be hanged in the foreyard. And if he entered the phantomatic world as a child or a teenage boy, he would be able to identify with a scene to such an extent that a return to everyday reality would cause him enormous difficulties. It could even prove impossible. An infant who has been phantomatized in a "cave vision" from the very first weeks of his life can without a doubt develop into a savage—at which point it will become impossible even to think about trying to civilize him. I am saying this not to entertain myself with paradoxes or to be witty but rather to point out that personality is not something that is pregiven, while phantomatics is not an equivalent of daydreaming, only slightly more vivid and colorful. The only way a phantomatized person can assess the superficiality of phantomatics is by comparing it with real life. Nonstop phantomatization will, of course, prevent one from conducting such an assessment and will lead to permanent changes—of the kind that would never occur in an individual's actual life. This is really a special case of adaptation to a given environment and temporality.

We mentioned earlier that this lack of authenticity in a phantomatic vision, the fact that it represents a biotechnologically executed escapism, is a big problem. Cybernetics offers two ways of overcoming such inauthenticity of experience. Since they have to be called something, we shall call them teletaxy and phantoplication.

Teletaxy does not mean performing a "short circuit," that is, connecting someone to a reality-faking machine while separating him from the real world. Instead, it involves connecting someone to a machine that only functions as a link between him and the real world. An astronomical telescope or a TV set count as prototypes of the "teletactor." These prototypes are nevertheless highly imperfect. Teletaxy connects a person to a randomly chosen fragment of reality, as a result of which he experiences this reality as if he were really placed inside it. The technical aspect of the problem can be solved in various ways. It is, for example, possible to construct exact models of the human—models whose receptors (those of sight, hearing, smell, balance, sensation, etc.) will be respectively connected to his sensory pathways. The same applies to all his motor nerves. His double, or "remote-I," who has been "directly linked up with his brain," can find himself inside the crater of a volcano, atop Mount Everest, in circumterrestrial space, or chatting to someone in London, while the person commanding him will be in Warsaw during all this time. The limited speed of communication, in this case involving radio signals, will no doubt prevent the "alter ego" from straying too far away from the person commanding him. Movement on the surface of the Moon will produce significantly delayed reactions, since a signal needs around one second to reach our satellite and the same amount of time to get back. Thus, in practice, the person controlling the "remote-I" must be located no farther than a dozen or so thousand kilometers away. The illusion of being present on the Moon or inside a volcano will be perfect, but it will not pose any danger, since the annihilation of the "remote-I" as a result of some kind of disaster, such as a rock slide, will just result in the sudden termination of the vision for the connected person, but it will not endanger his health. This communication system will be no doubt particularly useful in the exploration of celestial bodies, but it may also come in handy in various situations that have nothing to do with entertainment. The remote-I's external similarity to the person commanding him is unnecessary, and actually superfluous, during space exploration; it can only be desired in special cases of "teletaxic tourism" if the illusion is to be complete. Otherwise, the person will indeed see the white sunlit rocks of the Moon and feel its stones under his feet, but, on raising his hand to his eyes, he will of course just see the remote-I's limb, while in the mirror, he will not see himself, a human being, but rather his other—a device, a machine. This might shock many people, since they will seem not only to have been transported into a different situation but also to have lost their body, together with the previous location.

It is a short way from teletaxy to phantoplication—which simply means connecting one person's neural pathways to those of another person. Thanks to this procedure, in a "phantoplicator" designed to facilitate this, a thousand people can simultaneously "take part" in a marathon, see through the runner's eyes, experience his movements as their own—in other words, deeply identify with him in their sensations. The name comes from the fact that any number of people can take part in such a show at the same time (phanto-*plication*). However, this method only involves a *unidirectional* information transfer, since those "connected" to the runner are not all able to command his movements. The principle behind this procedure is already familiar to us. This is how individual microtransmitters, placed in various parts of astronomers' bodies, send information about what is going on in their hearts, blood, and so on, to scientists on Earth. A new branch of science called bionics deals with similar problems (i.e., imitating, by technical means, the working of selected receptors in living organisms, or directly connecting the brain or nerves to performing devices while bypassing certain natural connectors, such as the hand). We have said before that it is not possible to switch over from one personality to another—with two caveats. Of course, neither teletaxy nor phantoplication has anything to do with it, since they are just different ways of "connecting the brain" to specific "information containers." What we are most interested in, however, is the possibility of connecting one brain to another and the potential consequences of this maneuver, that is, of "switching" from one consciousness to another, "merging" two or more consciousnesses, or a metamorphosis of an individual consciousness that would not result in the annihilation of individual existence. If we consider that the National Bank clerk Mr. Smith (whom we have known since he was little, and who has some unique personality traits—which correspond to the particular dynamic traits of his brain's neural network) and the person who is very different from him (who has a different personality, different interests and talents, yet who says that he is Mr. Smith but has undergone surgery to be "inserted into the brain" of an "amplifier" of some undeveloped mental characteristics) are two different people, then the whole problem collapses. Reincarnations or "spiritual switchovers" are impossible, while the new Mr. Smith only *thinks* that he is the old Mr. Smith, the bank clerk—yet this is just an illusion.

If, in turn, on hearing him out and becoming convinced that he remembers his earlier life really well, all the way back to his childhood, as well remembering his original decision to undergo surgery, and also that he has the ability to compare his previous (by now lost) personality traits with the new ones, we will conclude that it is one and the same person. In that case, the problem will turn out to be fully actualizable. This is our first caveat: depending on the originally adopted criteria, we shall accept the equivalence between the two Mr. Smiths (i.e., the Mr. Smith from T1, or time before the surgery, and the Mr. Smith from T2, or time after the surgery) or we shall not.

Unfortunately, cybernetics has at its disposal truly unlimited capabilities. A person shows up whom we identify as our friend Mr. Smith. We are talking to him for a long time and become convinced that this is our old friend, completely unchanged, and that he perfectly remembers both us and his life. He is exactly the same he has always been. Then a devious cyberneticist turns up and informs us that this alleged Mr. Smith is "in fact" an entirely different person, whom he has "converted" into Smith, having transformed his body and brain and endowed the latter with the sum total of Mr. Smith's memory. In the course of that procedure (which involved the preparation of an inventory of memories), Mr. Smith, sadly, died. The cyberneticist is even inclined to give us access to the dead man's body, for research purposes. The criminal aspect of the matter does not interest us as much as the ontological one here. In the first case, the same person has been "transformed" into another one—but has retained the memory of his original past. In the second case, a completely new person "imitates" Mr. Smith in every respect "without being him," since Mr. Smith is buried deep in the ground.

If we take the *continuity* of individual existence as a criterion for continued existence, no matter what changes have taken place (we can mention here, e.g., "an infant's physiological transformation into Einstein"), then the first Mr. Smith (from our first example) will be *real*.

If we take the *immutability* of personality as a criterion, then the second Mr. Smith will be "real." This is because the first one already has "an entirely different personality": he is fond of mountaineering, grows cacti, has enrolled in a conservatoire, and lectures in natural evolution at Oxford, while the second one is still a bank clerk and "has not changed in any way at all."

In other words, the problem of the equivalence or nonequivalence of an individual turns out to be *relative* and depends on the accepted criteria for differentiation. A civilization that remains primitive on a cybernetic level fortunately does not have to occupy itself with such paradoxes. A civilization that has fully mastered imitology and phantomatology (which contains, as we are able to declare today, peripheral and central phantomatics, phantoplication, and cerebromatics), and that already practices pantocreatics with enthusiasm, must address problems concerning "the relativity theory of personality." Such solutions must not be absolute, since there are no absolute fixed criteria. When the transformation of personality can be actualized, individual identity stops being *a phenomenon to be investigated* and becomes *a phenomenon to be defined.*

PERSONALITY AND INFORMATION

Norbert Wiener was probably the first to formulate the idea about the theoretical possibility of "telegraphing" man as an unusual form of communication. This was seen as one of the potential applications of cybernetic technology. Indeed, what is man, or any material object, but a sum of information that, when encoded in the language of radio or telegraphic signals, can be sent over any distance? We would not be incorrect in saying that everything is information—a book, a clay jug, a picture, as well as mental phenomena—since memory, which is the basis of ongoing subjective existence, constitutes an information record in the brain. The erasure of this record as a result of illness or accident can annihilate all memories. Imitology stands for imitating phenomena by drawing on a key set of information. We are not saying by any means that information is *all* that exists. We can identify a clay jug if we have a full protocol of information that pertains to it (i.e., to its chemical composition, its topology, its dimensions). This protocol, or "description," is only equivalent to the jug in the sense that we can replicate this jug on its basis, whereby—provided we have an accurate enough device at our disposal (e.g., an atomic synthesizer)—it will not be possible to distinguish the "copy" we will produce from the original with any kind of test. If we do the same thing with, say, a Rembrandt painting, then the usual distinction between a "copy" and an "original" will become blurred, since the two will be indistinguishable. The procedure of this kind involves encoding the information provided by the jug, picture, or any other object and then decoding it again in an atomic synthesizer. Its middle part, that is, the stage when the original jug does not exist anymore (e.g., because it broke), but only its "atomic description" does, is, of course, not equivalent on a material level with the original. The protocol can be written down on paper, it can be turned into a series of impulses recorded in a digital machine, and so on. Naturally, there is no material similarity between such a system of signs and the actual jug or picture. At the same time, there exists a mutually equivalent correspondence of all the signs in this set with the original object—which is what enables its detailed reconstruction.

If we synthesize Napoleon out of atoms (assuming we have his "atomic description" at our disposal), Napoleon will come alive. And if we put together a similar description of any person, and then send it by wire to a machine in which the body and brain of this person will be constructed on the basis of the information received, we shall then see him emerge, alive and healthy, from this machine.

The question of the technical feasibility of undertaking such a procedure recedes into the background in the face of its unexpected consequences. What will happen if we send an "atomic description" not once but twice? Two identical persons will come out of the receiving device. And if we do not just wire this information in one direction but emit it as a radio wave, while the receivers will be located in thousands of places all over the globe and on the surface of numerous planets and moons, the "transmitted" person will appear in all these places. Even though we have transmitted Mr. Smith's description only once, he is now emerging from the cabins of the receiving devices a million times over—on Earth and in the sky, in cities, on mountaintops, in jungles and moon craters.

This only seems strange until we ask where exactly Mr. Smith finds himself at this point. Where has the telegraphic journey taken him? Since the persons who are leaving the receiving devices are by definition absolutely identical and are all called Mr. Smith, it is clear that even the most thorough examination or interrogation of them is not going to clarify anything. From a logical point of view, one of the two following possibilities is taking place: either all of these persons are Mr. Smith simultaneously or none of them is. Yet how is it possible for Mr. Smith to exist in a hundred million places at the same time? Has his personality been "copied"? How can we understand this? A person can visit different places, he can experience a certain reality, but only one at a time. If Mr. Smith is sitting at his desk, he cannot at the same time be in Eratosthenes Crater, on Venus, at the bottom of an ocean, or facing the jaws of a crocodile in the Nile. Telegraphed persons are normal, ordinary people. They cannot therefore be joined as one by some kind of psychic bond, which would make them all simultaneously experience the kinds of things mentioned previously.[8]

Let us imagine a crocodile has eaten one of the Smiths—the one who had reached the Nile. Who died? Smith did. Yet at the same time he is still alive, in countless places at the same time, right? What connects all the Smiths is a unique similarity, but it does not constitute any kind of physical or psychological bond. Identical twins, for example, are similar, yet they are emotionally independent. Each one is an autonomous and integral personality, and each one is living his own singular life. The same applies to the million telegraphed Smiths. They are a million different psychic subjects who are entirely independent from one another.[9]

This paradox seems impossible to solve. There is no experiment that would allow us to establish *where* exactly the continuation of the telegraphed Smith is located. Let us look at this problem from a different angle. There is this phenomenon, recognized by psychiatry, called "split personality." However, the split is never as absolute as it might seem from its various descriptions in literature. But it is possible to enact a separation by means of brain surgery that will result in the coexistence of two practically independent nervous systems in one skull. We know that one body can have two heads, because monstrosities of this kind are capable able of surviving some time after birth (it has even happened in humans), and also this phenomenon has already been brought about via artificial methods (e.g., in the USSR in dogs).

The division of the brain into two autonomous and separately functioning parts has been carried out with the help of neurosurgical procedures, for example, in monkeys. This involves cutting across the corpus callosum, which connects the two hemispheres of the brain, as deeply as possible. Let us imagine this procedure has been performed on Mr. Smith. The separation of the two hemispheres has occurred slowly and gradually so as not to cause a sudden disturbance to his brain functions and to allow for a full reconstitution of each hemisphere in the process of becoming independent from the other one—after what was no doubt a shock caused by such a cruel intervention. After some time, in Mr. Smith's head, there will exist two independently functioning brains. This will lead to the familiar paradox. The monkeys on which similar operations have been conducted behave during an examination as if they possess two relatively autonomous brains, one of which constantly dominates and controls the subsystems of the afferent neural pathways, and thus the body as a whole, or, alternatively, they both "get connected" to those pathways and take turns in controlling the body. It is, of course, impossible to ask the monkeys about their subjective states. But it is a different matter with Smith. Let us assume (in violation of the anatomical truth but for the sake of our argument) that the two hemispheres of the divided brain are completely equivalent (in reality, the left hemisphere is usually dominant). Each one contains the same memory record and the same personality outline that the brain as a whole contained previously. The question as to which hemisphere is a continuation of Smith, or which of the two brains is the "real Smith," turns out to be meaningless. We have here two analogical Smiths within one body. The dynamic path of consciousness, now bifurcated as a result of the material intervention, produces two independent personalities, each of which has equal right to consider itself a continuation of the original personality. The act of copying has thus indeed taken place. Conflicts may of course arise between these two systems, since they have just one shared organism, one sensory system, and one executive (muscle) system. Yet if we now transfer, by means of another operation, these hemispheres (which are now functioning as fully autonomous brains) into two specially prepared bodies, we will obtain two Smiths—who will also be physically separated. And thus, even though we cannot imagine or visualize it, the possibility of copying personality is real. From the point of view of the person who exits the receiving device, it is him and no one else who is the most exact, normal, and healthy continuation of the "telegraphed" person—and there are no grounds for questioning this statement.

Thus one person can be simultaneously transmitted in many directions. This does not mean that he will then exist as one in all those persons. There will be as many "hims" as there are atomic copies of him. The multiple continuation of an individual turns out to be a fact.

Yet this is only the first and, we should add, also the most primitive of the paradoxes.

As it turns out, we are faced here with a unique case of "existential relativity," one that is somewhat similar to the relativity of measurement in Einstein's theory, when the measurement result depends on the accepted frame of reference. As we already know, from the point of view of the Smiths who exit the receiving devices, *every* one of them is a continuation of the telegraphed person. However, from the point of view of the telegraphed Smith, none of them is.

How does the actual process of "telegraphing" happen? Mr. Smith enters the cabin of the device where his "atomic description" is prepared, say, by exposing him to some hard radiation. The "atomic outline" obtained in this way is then transmitted telegraphically. In a short while, numerous Smiths will be leaving the receivers in towns and villages all over.

But what about the original? If he leaves the cabin in which we undertook our atomic "stocktaking," he clearly has not gone anywhere: he has remained in his original place. Besides, even if one million copies have emerged within the receiving devices, the situation of the original Smith remains the same: unless we tell him about any of this, he will just go home, without any idea about what happened. It therefore seems that the "original" has to be annihilated immediately after the "atomic stocktaking" has taken place. Thus, having put ourselves in Mr. Smith's shoes, we can see that the prospects of his telegraphic journey are not too rosy. Indeed, he knows that he is going to die in the cabin, that is, he will be killed once and for all, while people who will emerge from the receivers will not just be similar to him—they will actually be *he himself.* There exists an explicit causal link between each state a person finds himself in and his previous state. I am experiencing sweetness at the moment of T_1, because at T_0 someone placed a cube of sugar on my tongue. There is also a causal link between Mr. Smith and his atomic description: the description looks the way it does because we have acted on Smith's body in a certain way, as a result of which full information transfer about Mr. Smith's constitution has taken place. Similarly, there exists an informational and causal link between the atomic description and the "copies" that come out of the receivers, since those copies have been constructed in accordance with the protocol of the "description." Yet what is the link between those transformations (Smith as a living organism, Smith as transmitted information, multiple Smiths—who have replicated on the basis of this information) and Mr. Smith's death—which we brought on right after preparing his atomic description?

Let us be clear about it: there is no link between the two. If we prepare an atomic copy of the original Rembrandt hanging on the wall, someone may say, I can recognize the original on the basis of its positioning. It is hanging on the wall, hence the other painting, the one on the easel, is a copy. If we burn the original, no one will be able to find it anymore. We will have destroyed the only object that allows us to doubt the originality of the atomic copy. However, the copy has not become the original as a result of this act, in the sense of turning into the object made of wood and canvas that the renowned Dutch painter had covered with paint several hundred years ago. It is empirically indistinguishable from the original, but it is not *the* original, as it has a different history.

If we kill Smith while assuring him that he is going to wake up soon in a million locations at the same time, it will certainly be a wicked act— a murder whose traces will be "cybernetically" erased, with a surplus. Instead of one murdered person, many such persons will appear, and they will all be identical.

The murderous aspect of this act seems self-evident, since, if we want to telegraph a person, it is clearly not enough to transmit his atomic description, as this person also has to be killed. To make things more explicit, let us imagine that we are transmitting Smith's description; copies of his person are already starting to appear in the receiver's exit, yet the original is still alive and does not know about what happened. Can we expect him to stay in our company until we approach him with a hammer and, the moment we have smashed his skull, suddenly to "become," in a way unbeknownst to us, either one of those other tele-graphed persons or all of them at the same time? What is it exactly that is supposed to transport him to the other end of the telegraphic wire if the signal transmission itself did not manage to do it? A blow with a hammer to the back of his head? As we can see, a suspicion of this kind does not signal a paradox but rather pure absurdity. Smith is going to be dead, forever and ever, and thus telegraphing a person remains out of the question.

This difficulty does not just concern the telegraphic transfer of infor-mation about a person. For example, in the future, every human being could have an "atomic matrix" of his body stored in a "personality bank." This matrix would provide a perfect record of his atomic structure—a record whose link to that person would correspond to the relation be-tween an architectural plan and the actual house. If that person were to die—say, in an ill-fated accident—his family could go to such a bank and have the matrix inserted into an atomic synthesizer, after which, to everyone's amazement, the tragically killed person would come out of the device and fall into the arms of his mourning relatives. This is indeed possible—but, as we know by now, such a joyous scene does not in any way cancel the death of the "original." Yet, since in this case no murder has been committed, but rather only a victim of an accident or illness has been successfully replaced by his "atomic double," there are no moral compunctions that would make such an undertaking unacceptable, at least within a given civilization.

However, it is impossible to use a similar method to create an "existential reserve" for oneself to guarantee one's continued individual existence. No matter if I merely have my "atomic description" stored in a drawer or in a bank, which will only become transformed into my living double after being inserted into the synthesizer (incidentally, the description is just a piece of software), or if I already have a living double while I am still alive—the course of my own life is not going to be affected in any way. When I fall down a ravine or die in some other way, my double is no doubt going to replace me, but I will not be alive then. The temporal coexistence of the original and the copy proves this. The relationship between them is the same as the one between twins—yet no one sane is going to claim that one twin is an "existential reserve" of the other.

We have so far established that it is not the very act of telegraphing information that irreversibly kills a human being but rather the subsequent annihilation of that person that does it. The latter is supposed to create an illusion that this person has really reached the other end of the wire. It therefore seems that the irreversibility of individual death is caused by an *interruption* to the continuity of existence.

It is here that we enter a true nightmare of a paradox. As we know, contemporary medicine has high hopes for hibernation—a field that is constantly evolving. Such a state of suspended and slowed existence, which occurs in the physiology of certain mammals (the bat, the bear), can actually also be induced in humans—who normally do not hibernate. (This would happen with the help of some pharmacological substances or through cooling down the body.) Hibernation can also be deepened so that it resembles authentic death. Such a state of reversible death—which involves not just slowing down but actually stopping all the life processes—is induced by significantly cooling down the organism as a whole. So far, it has been successfully carried out in several laboratory animals. After freezing, protozoa (to which sperm, including human sperm, also belongs in a sense) can be maintained in this state for a very long time, perhaps infinitely long. The possibility of fertilizing a woman with the sperm of a man who died many years before thus becomes entirely feasible.

Cooling down complex organisms such as man (or mammals in general) below the freezing point is posing great difficulties, since water tissue has a tendency to crystallize as ice, a reaction that leads to destroying vital protoplasmic structures. Yet these difficulties are not insurmountable. We can expect that such a freezing technology, promising as it does 100 percent chance of a future revival at any given moment, will be accomplished. Great hopes are pinned on it—with regard to space travel, among other things. Yet, in the light of the thought experiments discussed so far, this technology can evoke some doubts. Are we really dealing with irreversible death here? Is it not possible that the frozen person dies once and for all, while the person we revive will be his mere copy? He seems to be the same person. His life processes have just been stopped, in the way one stops a watch. Restarting it amounts to a revival. In any case, those processes never come to a complete standstill. We know that the way such phenomena work is not dissimilar from what happens to a dial that is made up of seven rainbow-colored sections. As long as it remains still or revolves slowly, we can see the individual colors. Increasing the rotation leads to flickering, while, as soon as the correct speed has been achieved, all the colors blur into uniform whiteness. Something similar happens with consciousness. The processes that are foundational to it must maintain a certain speed below which consciousness becomes blurry and then falls apart, long before the biochemical reactions in the brain have actually stopped. Consciousness thus fades away earlier than metabolic processes do. The latter effectively stop too—but some of them can actually continue, even if slowly. Of course, when approaching the temperature of absolute zero, their functioning basically comes to a halt, while the organism stops aging. Whichever way it goes, all the structures within the living tissue are preserved. We can thus say we have acquitted the freezing process of murder.

Let us perform yet another thought experiment. Imagine we have frozen our Mr. Smith at the exact temperature of absolute zero. His brain and the other organs of his body all take the form of crystalloid structures. Apart from the tiny oscillations that atoms manifest even at the lowest energetic level, we shall not be able to see any motion through an electronic microscope. Trapped by the frost, the atoms of Mr. Smith's brain, immobilized and thus more accessible, can be picked out by us from his skull one by one and placed in appropriate containers. For the sake of good order, we place the atoms of each particular element separately. Just to be sure, we store them in the extreme cold of liquid helium until we piece them back together when the time comes, fitting each one back where it belongs. Then we successfully put the reassembled but still frozen brain, together with its body, through a revival procedure. Defrosted, Mr. Smith gets up, puts on his clothes, and goes home. We have no doubt that it was genuinely him. Suddenly it turns out that our assistant broke every single test tube that contained, in the form of fine powder, the atoms of carbon, sulfur, phosphorus, and all the other elements that had constituted Mr. Smith's brain. We had placed those test tubes in the freezer on the table. The assistant tipped over the table and, on seeing the disaster he had caused, quickly removed all of its traces. He collected whatever was left from the spillage of the elements into some new test tubes and replaced the missing bits by referring to the notes in the lab diary—in which we had entered with utmost precision, down to a single atom, what each test tube contained. We have barely collected ourselves on hearing this piece of news; we can still see Mr. Smith walking away and waving his stick in the courtyard, when suddenly the door opens and another Smith enters. What happened? Falling down from the table, the test tubes broke. The assistant was in a hurry and only collected half of the spillage, yet his colleague, trying to be helpful, subsequently collected with great care the remains of the elements spilled on the floor. He also replaced what was missing by referring to the notes in the lab diary, after which he diligently started the defroster and revived Mr. Smith no. 2.

Which one of the two Mr. Smiths is thus a true continuation of the frozen one: the first or the second? Each carries approximately a half of the "original" atoms—which is actually not that relevant, since atoms lack individuality and get exchanged by an organism in metabolic processes. We seem to have experienced the doubling of Mr. Smith. But what about the original? Is he alive in both bodies, or maybe in neither of them? In contrast with the experiment that involved slicing across the corpus callosum of the two brain hemispheres, this question cannot be answered, since there are no empirical criteria on which we could draw. Naturally this dilemma could be solved in an arbitrary manner, by deciding, for example, that both Mr. Smiths are a continuation of our friend, whom we are subjecting to such arduous trials. This is convenient, and perhaps even necessary in this case, yet such a solution no doubt raises some moral concerns. Having placed his trust in us, Mr. Smith entered the hibernator as calmly as if he had been entering a telegraph cabin, from which we dragged him by his feet after pounding him with a hammer—reassured as we were by his multiple appearance in the planets of the solar system. We have shown that in that previous case, we were dealing with murder. But what about the present case? The lack of a corpse seems to work to our advantage, yet even in this case we could have dispersed Smith into an atomic cloud—while it is not so much committing murder in an invisible and highly aesthetic manner that is at stake here but rather not committing it at all.

Things are beginning to get muddy. Does this mean that some kind of immaterial soul exists that remains trapped in the brain's structure like a bird in a cage but that escapes the bodily fetters as soon as the cage bars, that is, the structure's atoms, are broken down and separated? Only despair drives us toward metaphysical hypotheses—yet they do not solve anything either. What happened after the corpus callosum had been cut? Did we then manage to split the immaterial soul into two parts? Also, did we not have whole hordes of Smiths, manifesting a regular level of spirituality, leave the receiving devices? The latter fact leads to an obvious conclusion that if a soul does exist, then every atomic synthesizer is capable of constructing it easily. In any case, the point is not whether Mr. Smith has an immaterial soul. Let us say he does. The point is that each new Smith was in every respect absolutely identical with the original Smith, but it was not him, since, as well as using an atomic description and a telegraph, we also had to use a hammer! The explanation presented in this paragraph is thus of no use to us.

Perhaps the source of the paradox lies in the fact that our thought experiments are as removed from the capabilities of the real world as is, for example, an imaginary journey at infinite speed or a perpetual motion machine. Yet even this is not going to help us much. Does Nature not offer extremely accurate copies of the human organism in identical twins? Agreed, such twins are not absolutely identical on the level of their atomic structure. But this also results from the fact that the technology of evolution, that is, selection, has never aimed to create such an absolute structure of similarity, since the latter is biologically irrelevant or redundant. And since such a degree of similarity between systems with the same degree of complexity has been achieved somewhat accidentally and randomly (since random elements play a significant role in the creation of twins, during the initial division of the fertilized egg), future biotechnology, in conjunction with cybernetics, will probably be tempted to beat this achievement—an achievement that has only been an accidental contribution from Nature.

To make our argument complete, we should also consider a situation in which the very act of constructing an atomic description destroys a living organism. Such a situation would eliminate certain paradoxes (e.g., the paradox concerning the possible coexistence of the original and its "continuation"). It could become the basis for a claim that things have to be this way, that is, that such coexistence is only thinkable but cannot be actualized. This is why we are spending a little bit more time on this issue. Let us imagine we have two devices for telegraphing people: device O and device N. Device O saves the person to be telegraphed. In other words, after the information about his atomic structure has been collected, the person remains in good health. Device N, in turn, destroys the atomic structure of the examined person during the process of collecting information, which is why at the end of this process we have one dead person, or perhaps his dispersed atomic remnants, as well as a full record of his structural information. Also, the amount of information obtained is the same in both cases, that is, complete and adequate enough to allow the reconstruction of this person, after he has been telegraphed to the receiving device.

Device O as a saving device is both more subtle and more complex. Historically, it will no doubt appear later, as a product of a more advanced technology than the technology that produced device N, with its destructive orientation. But we shall consider device O first. Its working is based on the position vector similar to that in the picture tube of a TV. The ray from the device runs along the examined person's body. Every time it touches an atom or an electron, it makes a recording in the memory of this device, as a result of the ray "tripping" over every particle of matter. Having had their locations recorded, atoms in the outer layers of the body will become transparent, as it were, for that particular ray. Naturally, for this to happen, the ray cannot be material (corpuscular). Let us say it is not a ray but only a center of pressure in electromagnetic fields. We are able to control those fields so that they overlap with one another. Consequently, when they encounter just vacuum, the device's needles remain still. Yet, depending on the mass of an atom that will find itself on the path of those fields, the resulting interaction will lead to a change in the value of those fields and will make the needles move. This will then be recorded by the "memory" system. The device simultaneously records the temporal and spatial locations of the readings, their order, and so on. Having undertaken 10^{20} individual readings, millions of which occur naturally every second, we already have a complete information record about the location of all atoms within a body, that is, about this body's material configuration. The device is sensitive enough to react differently to an ionized atom and to a nonionized one; it also reacts differently to an atom located in a given position in the protein chain, since this reaction depends on the thickness of the electronic layer of a molecule. These propagating electromagnetic fields, used for recording purposes, must lead, through their interaction, to slight declinations of the body atoms from their previous state. Yet those declinations are so tiny that the organism is going to cope with them without any harm to itself. Once we have a complete record, we send it by wire. On receiving the information, the receiver switches on, and then a copy of the person is produced at the other end of the wire. This individual is extremely similar to the original one, but the original does not have to know about it. He can leave the cabin and go back home, having no idea that, in the meantime, a copy of him has been created somewhere—or even a whole legion of them. This was the first experiment.

Now we are starting the second device. Its operating principle is much more brutal, since the position vector is material, which means that the expelled particles hit the body's atoms one by one, starting from its outer layer and going deeper and deeper. A crash, that is, an accident, happens every time. From the declination of the expelled particle, whose momentum we know, we then read out the initial position and mass of the particle that has been hit (i.e., of the body's atom). We obtain a second reading, as exact as the first one, yet through this procedure, we disperse the organism itself, turning it into an invisible cloud through our actions.

Please note that in both cases, we obtain exactly the same amount of information, except, in the second case, we destroy the original organism when taking the reading. Since the destruction has been caused solely by the brutality of the device—which in itself *has not in any way increased* the amount of information obtained—the destruction event is incidental to the act of information transfer itself; it is by no means inherent to it or to the atomic synthesis of the copy that emerges at the other end of the wire.

The information transfer and the synthesis enabled by it occur in exactly the same way in both cases. It therefore becomes clear that the original's fate is of no consequence for what happens at the other end of the wire. In other words, at the other end of the wire, inside the receiver, identical individuals emerge in both cases. But since we have shown that in the first case, the person who emerges cannot be a continuation of the original, the same must apply to the second case. And thus we have demonstrated that the person created in the synthesizer is always an imitation, a copy, and not "an original sent by wire"—which in turn shows that the "insert" into the cause-and-effect chains of the organism's existence, created in the process of recording and transferring information, is in fact more than just an insert, a caesura between two parts of an uninterrupted life line of a given person: it is an act of creating an "imitator"—who is like a twin to the original individual, while the latter either remains alive or dies. The original's fate is of no relevance for the copy, because such a copy is never a continuation of the original. When it remains alive in the first case, the original, in turn, through his very presence, disproves the belief that he has supposedly just been "telegraphed" somewhere. As a consequence of his annihilation in the second case, the original creates an impression (a false one, as we have demonstrated) that he has indeed taken a "trip by wire."

To close our deliberations, we shall now present a variation on the preceding experiment, one that occurs without having to create an atomic matrix or use an atomic synthesizer. It still cannot be performed today, but significant progress has already been made in this respect. We are referring here to the possibility of growing a fertilized human egg outside an organism. This egg needs to be cut in half. We freeze one half, while allowing the other half to develop in a normal way. Let us imagine that a person develops from it and then dies at the age of twenty. We then defrost the other half of the egg and, after twenty years, we have "the other twin"—who can be described as a continuation of the dead person, using exactly the same explanation we did with regard to the atomic copy created in the synthesizer. The fact that we had to wait twenty years for the "continuation" to emerge does not mean anything, because it is quite possible that the synthesizer will also have to run for twenty years before it can produce an atomic copy. If we then consider this "other twin" as a continuation of the dead person and not as his very similar-looking double, the same will apply to the question of creating an atomic copy. In the latter case, however, any ordinary twin whose development is delayed by hibernation is a "continuation" of his brother. And because hibernation time can be reduced at random, ultimately, every twin will turn out to be a continuation of the other twin—which is, of course, pure absurdity. A twin is indeed a perfect molecular copy of the "original." Yet the similarity between two states of the same person—when he is eight and eighty years old, respectively—is no doubt even smaller than the mutual similarity between the twins. Despite this, anyone would admit that the child and the old man are the same person—something we cannot say about the two brothers. It is therefore not the amount of analogous genetic information that determines the continuation of existence but rather the fact of being genetically identical, even if the dynamic structure of the brain undergoes significant changes during one's lifetime.

7

THE CREATION OF WORLDS

We seem to be at the end of an era. I am not referring here to the age of steam and electricity, which then mutates into the age of cybernetics and space science. Such terminology indicates yielding to various technologies—which will become too powerful for us to be able to cope with their autonomy. Human civilization is like a ship that has been built without any design plans. The construction process was extremely successful. It led to the creation of enormous propeller machinery and resulted in an uneven development of the inside of the ship—but this is something that can be remedied. Yet this ship is still rudderless. Civilization lacks knowledge that would allow it to choose a path knowingly from the many possible ones, instead of drifting in random tides of discoveries. The discoveries that contributed to its construction are still partly accidental. This situation is not altered by the fact that we continue to move toward the edge of the Galaxy without knowing what lies ahead. One thing is sure: we actualize what is already possible. Science is playing a game with Nature, and even though it wins every time, it allows itself to be drawn into the consequences of this victory and exploit it, as a result of which, instead of developing a strategy, it ends up just practicing tactics. Thus, paradoxically, the more such successes, or victories, there will be in the future, the more difficult the situation will become, since—as we have already demonstrated—it is not always possible to exploit everything we have conquered. The embarrassment of riches, the deluge of information that engulfs man as a result of his cognitive greed, need to be tamed. We have to learn how to regulate scientific progress too; otherwise, the random character of any future developments will only increase. Victories, that is, suddenly appearing domains of some new wonderful activity, will engulf us with their sheer size, thus preventing us from noticing some other opportunities—which may turn out to be even more valuable in the long run.

It is important for a developing civilization to gain the freedom of strategic maneuvering to be able to control its path. The world today has other concerns. It is divided; it does not satisfy the needs of millions—but what if those needs are eventually satisfied? What if the automatic production of goods takes off? Will the West survive this? This is a grotesque vision: of humanless factories producing billions of objects, machines, nutritional elements, through the energy of a star to which our civilization is "connected." Will some kind of General Apocalyptics become the owner of that star?

Never mind property rights. If I say that one era is coming to an end, I am not even thinking about the demise of the old systems. Satisfying the basic needs of humanity is a necessary task, a preparation for a final exam; it is the beginning of a mature age rather than its end.

Science develops from technology. Having become established, it subsequently takes control. To speak about the future, especially a remote future, involves having to speak about the transformation of science. What we shall be discussing here may never be actualized. The things that are absolutely certain to come about are those that are already taking place, not those we are capable of imagining. I do not know whether Thales and Democritus were more audacious in their thinking than we are today. Perhaps not, since they were unable to grasp the labyrinth of facts, the entangled jungle of hypotheses through which we have had to make our way over the centuries. The whole history of science is actually a harsh terrain marked with traces of defeat—whereby failures have been much more numerous than successes. It is full of systems that have been abandoned like old wrecks, theories that have become as outdated as old flints, and broken truths that used to command general respect. We can see today that the centuries-long conflicts fought within science were ostensibly futile since their arguments focused on words and concepts that actually lost their meaning over time. This happened with Aristotle's legacy and, hundreds of years after him, with the fight between epigeneticists and preformists in biology. I say "ostensibly" because we might as well say that, similarly, all those extinct organisms, those fossils of animals that preceded the existence of man, were ostensible or superfluous. The statement that those animals paved the way for the arrival of man does not seem very fortuitous to me: it is a sign of an excessively selfish anthropocentrism. Perhaps we could just say that those mines of creation, just like those old theories, formed a chain made up of links that have not always been necessary or inevitable, that have sometimes been excessively costly and sometimes misguided, yet that together have arranged themselves into an upward path. Anyway, it is not a question of acknowledging their individual value.

It is very easy to describe the extinct forms of organisms as primitive and the creators of false theories as fools. When I write this, there is an issue of a scientific journal on my desk that reports an experiment, the results of which challenge one of the basic truths of physics: Einstein's thesis about the speed of light. Perhaps this law will still defend itself. Yet something else is important: the fact that there are no authorities impervious to challenge in science. Its errors and mistakes are not ridiculous because they are the result of a conscious risk. Such consciousness allows us to formulate hypotheses because, even if they are to collapse soon, the defeat will take place on the right path. From his early days, man has always embarked on this journey—even when he still did not realize it.

INFORMATION FARMING

Many cyberneticists are currently working on the possibility of automating hypothesis formation. A "theory" produced in a machine is an information structure that successfully encodes a limited amount of information that is relevant with regard to a certain class of phenomena in the environment. This information can then be successfully applied in formulating infallible predictions for that class. A machinic theory of class represents a certain *constant* value in machinic language that is shared by all the elements of that class. The machine receives information from the environment and creates particular "constructs," or hypotheses, which compete with one another until they become redundant, or fixed during the "evolution" that this "cognitive process" represents.[1]

The biggest difficulties are posed by the question of the original emergence of constants in the machine—which determines any further processes of hypothesis formation; the question of the capacity of machinic memory and of the speed of access to the information contained within it; as well as the question of how to regulate the growth of *association trees*—a term that refers to the rapidly growing alternative working approaches. Besides, a small increase in the number of initially considered variables (if we take the pendulum as our phenomenon, the question arises as to how many variables we need to take into account to be able to make predictions about its future states) leads to the collapse of this whole program. With five variables involved, a large digital machine, working at a speed of one million operations per second, can go through all of their values within two hours. With six variables, the same process requires thirty thousand such machines, working at maximum speed over a period of several dozen years. This means that if these are random variables (at least for us, i.e., as long as we cannot see any link between them), no system whatsoever, no matter if artificial or natural, will be able to deal with more than several dozen variables, even if it was as large as the Metagalaxy.

If, for example, someone wanted to build a machine that would model sociogenesis, while every person who has existed since *Australopithecus* would need to have a number of variables assigned to him, the task would be impossible to perform, now and in the future. Fortunately, this is not necessary. Indeed, if Nature had had to consider on a regulatory level the momentum, spin, and moment of every separate electron, it would have never constructed any living organisms. It did not do this on an atomic level either (there are no organisms that would consist of just several million atoms) because it was unable to regulate quantum fluctuations and Brownian motions. The number of independent variables turned out to be too high at that level. The cellular structure of organisms did not so much result from the emergence of protozoa as primary systems as from a requirement whose roots go much deeper into the fundamental characteristics of matter. A hierarchical design stands for a relative autonomy of its levels, which remain dependent on the main regulator, but it also entails *inevitably* giving up control over all the changes taking place in the system.

The design of our hypothetical fruits borne by the Imitological Tree must also be hierarchical. We shall look at this problem more closely in a short while. For now, we shall concern ourselves with the scope of imitological activity.

Let us recap what we have concluded so far.

Up to a certain level of complexity, it is cost-effective to construct models that involve a dynamic coupling of ostensibly significant variables. It is very important to know the scope of the model's applicability, that is, the degree to which it represents the behavior of actual phenomena. The selection of significant variables does not mean having to give up on accuracy. On the contrary, in protecting us against the barrage of irrelevant information, it enables a faster detection of the whole class of phenomena that are similar to the phenomenon under examination—or, in other words, the construction of a theory. Specific circumstances can determine what counts as a model and what counts as an "original" phenomenon. If neurons in a chain reaction multiply at the same speed that bacteria in food do, then, with regard to the parameters of exponential growth, one phenomenon can serve as a model of the other. As it is more convenient to examine, say, bacteria, we shall consider bacteria culture as our model. However, when models become too complicated, we either search for models of another type or draw on an "equivalent" model (we "model" a human being with another human being, through a "side entrance" into the process of embryogenesis, as discussed earlier).

The amount of initial knowledge must be greater the more precise a model we want. The educational value of the model is of no relevance. It only matters that we should be able to "ask questions of it" and then receive answers. We should notice the difference in approaches to modeling between a scientist and a technologist. Faced with the possibility of carrying out "synthesis in a living organism," should that be his goal, a technologist will be satisfied with just obtaining "the final product." A scientist, in turn, at least in the conventional understanding of the term, will wish to learn in detail about "the theory of synthesis in an organism." A scientist wants an algorithm, whereas a technologist is more like a gardener who plants a tree, picks apples, and is not bothered about "how the apple tree did it." A scientist considers such a narrow, utilitarian, and pragmatic approach a sin against the Laws of Full Knowledge. It seems that those attitudes will change in the future.

Models are close to theories in that they ignore a number of variables that are considered insignificant in a phenomenon. The more variables a model takes into account, however, the more it becomes a reiteration of the phenomenon itself and not its "theoretical" representation. A model of the human brain is a dynamic structure that takes into account the significant variables of all human brains, yet the lesser the "significance" of the model of Mr. Smith's brain for other brains, the more "the surface of its dynamic contact" with all the processes of Mr. Smith's brain increases. And thus, eventually, such a model takes into account both the fact that Mr. Smith is lousy at math and that he met his aunt yesterday. Of course, we do not need such a faithful model—which is to some extent a "literal" reiteration of the phenomenon (of the star Capella, of the little pug called Rex, of Mr. Smith). As we can see, a machine that would copy, with utmost exactitude, every material phenomenon would be a *universal plagiarizer*. Its full consideration of all the variables of phenomena would in a sense automatically cut it off from any creative activity—an activity that must involve *selection*, that is, choosing certain variables while rejecting others to find a *class* of phenomena that share dynamic trajectories of such variables. The properties of behavior of this class are what we call theories.

Theories are possible because the number of variables in a single phenomenon is incomparably higher than the number of variables this phenomenon shares with a whole lot of other phenomena, whereby it is fine to ignore the former on scientific grounds. This is why it is possible to ignore the history of individual molecules, the fact that Mr. Smith met his aunt the day before, or a million other variables.

In fact, the ways in which physics and biology approach phenomena differ considerably. Atoms are interchangeable, whereas organisms are not. In contemporary physics, the individual history of an atom is irrelevant, except for a certain hypothesis concerning the reddening of photons sent by an atom. Such an atom might have come all the way from the Sun, or from a piece of coal in the cellar, but this does not change its properties in any way. But if the aunt told Mr. Smith that she was going to disinherit him, after which Mr. Smith became crazy with despair, the variable would become highly significant. As a matter of fact, we can understand Mr. Smith to some extent because we are a lot like him. It is a different matter with atoms. If we create a theory of nuclear forces and then ask what exactly those pseudoscalar couplings "really" are, then this question does not make any sense. Having assigned certain names to the operations of our algorithm, we cannot demand that they have a meaning other than in relation to those algorithmic transformations. At best, we could say, "If you perform such and such transformations on paper, and then replace this with that, then you obtain two and a half as a result, while, if you do this and that in a lab, and look at the needle of this device, it will stop between two and three." The experiment has confirmed the theoretical result, which is why we shall be using the notion of pseudoscalar couplings and all the rest of the nomenclature.

And thus the opposing photons and all the rest are rungs of a ladder that we are climbing to get to the attic. There may be something precious in the attic, like a new source of atomic energy, yet we must not ask about the "meaning" of the ladder "itself." The ladder is just a fragment of an artificial environment that we have created to get to the top, while the opposing photons are fragments of an operation performed on paper that allows us to predict future states—and nothing else. I am saying this to avoid creating an impression that Imitology is meant to "explain everything" to us. An explanation consists in reducing unknown features and behaviors to known ones by finding a similarity with familiar things. If the unknown does not resemble a pinball, a sphere, a chair, or cheese, then, instead of feeling helpless, we have to turn to mathematics.

The science and technology expert's approach to the world is likely to change. He will be connected to this world via Imitology. Imitology does not by itself outline any goals for action; they are provided by a civilization at a given stage of its development. Imitology is like a telescope: it shows at what it has been pointed. If we notice something interesting, we can increase magnification (i.e., point information-gathering devices in that direction). Thanks to countless processes that model various aspects of reality, imitology will present us with various "theories," various links and characteristics of phenomena. There is nothing that is completely isolated, but the world is favorably inclined to us: relative isolations do exist (between different levels of reality: atomic, molecular, etc.).

There are theories of systems (in mechanics). The theory of bioevolution would be a systems theory of systems, while the theory of civilization would be a systems theory of the systems of systems. Fortunately, quantum processes hardly ever manifest themselves at the scale of protozoa. Otherwise, we would drown in the ocean of diversity, without any hope for regulation. The latter is initially based on biological homeostasis (thanks to the existence of what are most certainly nonintelligent plants, the amount of oxygen in the atmosphere remains constant, and they regulate that amount), and then, once intelligence has developed, on the homeostasis that makes use of the products of theoretical knowledge.

"Ultimate modeling" is thus not only impossible but also unnecessary. "Fuzzy" representation that results from ignoring a number of variables makes a theory universal. Similarly, a blurry photograph does not allow us to determine whether this is Mr. Smith or Mr. Brown, but it still allows us to say this is a human being. For a Martian who would like to know what a human being looks like, a blurry photograph is more valuable than Mr. Smith's portrait; otherwise, he is going to think that all humans have such bulbous noses, gapped teeth, and bruises under their left eyes. In conclusion, all information assumes the existence of an addressee. There is no "information in general." A civilization and its scientists are an addressee of the "imitological machine." Today they have to enrich information "ore" themselves, through the process of elimination. In the future, they will obtain its very essence and build their theories not from facts but from other theories. (This is already happening to some extent: there are no theories that are completely isolated from others.)

The reader is no doubt expecting a discussion of "information farming" we promised a while ago. Instead, we are dealing with the nature of scientific theories. It seems like I am doing everything to put him off from reading further. But let us consider briefly what we actually want to achieve. We are supposed to go all the way toward automatizing Science. This is a terrifying task. Before we embark on it, we have to understand fully what it is exactly that Science does. What we have just said is only an early metaphorical approximation. However, metaphors need to be translated into a scientific language. I am really sorry about this, but this is the truth. And thus...

We are to invent a device that will gather information, generalize it in the same way the scientist does, and present the results of this inquiry to experts. The device collects facts, generalizes them, and checks the validity of this generalization by applying it to a new set of facts. On having undergone "quality control," the "final product" leaves the "factory."

Our device thus produces theories. In philosophy of science, a theory is a system that consists of symbols. It is a structural equivalent of an actual phenomenon, which can be transformed by means of rules that have nothing to do with this phenomenon so that subsequent sections of the phenomenon's trajectory (its successive states in time) correspond on the level of variables considered by this theory to the values of variables that are deducible from the theory.[2]

A theory does not apply to an individual phenomenon but to a class of phenomena. The elements of a class can coexist in space (billiard balls on the table) or follow one another in time (subsequent positions of a billiard ball in time). The more numerous the class, the "better," that is, more universally applicable, the theory.

A theory does not have to have any verifiable consequences (Einstein's unified field theory). Until we manage to derive such consequences from it, it remains ineffectual—not only as a tool for real action but also as a tool for cognition. To be useful, a theory must have an "input" and an "output": an "input" for the facts it generalizes and an "output" for the facts it predicts (thanks to which it can be verified). If it only has an "input," it remains as metaphysical as it would be if it did not have either an input or an output. The reality is less beautiful, that is, less simple. The "inputs" of some theories are at the same time "outputs" of other theories. There are theories that are more or less general, yet, from a developmental point of view, they should all form a hierarchical entity such as, for example, an organism. The theory of bioevolution is "connected" with theories derived from chemistry, geology, zoology, and botany, which are all subordinate to it, while it itself remains subordinate to the theory of self-organizing systems—of which it is a special case.

There are currently two approaches to theories: complementary and reductionist ones. A complementary approach means that the same phenomenon or class of phenomena can be "explained" with two different theories, while the decision as to which theory should be used is pragmatic. Complementary approaches are used, for example, in microphysics (electron as a wave and a particle). Yet some people think that this state of events is temporary and that we should always aim at reductionist approaches. Instead of complementing one theory with another, it is necessary to construct a theory that will unite the other two, reduce the first one to the second, or reduce both of them to yet another, more general one (this is what a "reduction" involves). And thus it is thought, for example, that phenomena of life can be reduced to physical and chemical processes. Yet this point of view is subject to debate.

A theory is more credible the more of its consequences come true. A theory can be entirely credible yet almost worthless (i.e., trivial, e.g., "all people are mortal").

No theory takes into account all the variables of a phenomenon. This does not mean that we are able to list any number of such variables in every case but rather that we do not know all the states of a phenomenon.

Yet a theory can predict the existence of some new values of the already established variables. But it would not do it in a way that would always specify what those newly discovered variables actually are and where they can be found. The "indication" of those new variables can be "hidden" in its algorithm; one has to be a real expert to know where the treasure is hidden. We are approaching some foggy terrains and mysterious concepts here, such as "intuition." It is because a theory is a structural piece of information, which could have been selected from among a gigantic number of thinkable structures that have no correspondence in Nature, whereby the selection process would take place after its countless competitors ("Bodies attract one another according to the diameter cubed; or according to the distance times mass quotient squared," etc., etc.) have been defeated. Yet in actual fact, nothing like this happens. Scientists do not work blind, using trial and error, but use guesswork and intuition.

This problem belongs to so-called Gestalt psychology. I am not able to describe a friend of mine so that you can immediately recognize him on the basis of my description. Yet I am myself able to recognize him right away. From the perspective of the psychology of sensory perceptions, his face thus represents a certain "Gestalt form." Sometimes a person resembles somebody else in our view, but we cannot always say how exactly: not individually through any of his body parts or face, but rather in his entirety, through the system, the harmony of all his features and movements, and thus, again, through his "form." This kind of generalizing perception does not just apply to the visual sphere. It can also work with regard to all the senses. A melody retains its "form" no matter if it has been whistled through one's fingers, played by a brass band, or tapped out with one finger on the piano's keyboard. Everyone can experience such ways of recognizing the "form" of shapes, sounds, and so on. If he is any good, a theoretical scientist well versed in the abstract formalism and symbolism of the theories among which he spends his life starts to perceive those theories as certain "forms." Naturally, such forms are devoid of any faces, features, or sounds; they are just abstract constructs in his mind. Yet he can manage to discover a similarity between the "forms" of two thus far disconnected theories, or, by aligning them together, he can grasp that they are special cases of a yet nonexistent generalization that needs to be constructed.

What we have just said is, of course, very crude. We shall come back to this issue, or rather, it will come back to us when we are ready to start "information farming."

Let us play a game now. We have two mathematicians, one of whom is a Scientist, the other one Nature. From prior assumptions, Nature derives a complex mathematical system that the Scientist is to deduce, that is, re-create. The way it happens is that Nature is sitting in one room and, from time to time, shows the Scientist, through the window, a piece of paper with a few numbers on it. Those numbers correspond to the transformations that occur at a given stage of the system being constructed by Nature. We can imagine Nature is a starry sky, whereas the Scientist is the first astronomer in the whole world. At first, the Scientist does not know anything, that is, he does not notice any link between the numbers ("between the movements of celestial bodies"), yet after a certain period of time, something begins to dawn on him. He decides to give it a try at last, which means that he himself constructs some kind of mathematical system. He then waits to see whether the numbers that Nature is going to show him through the window will match his expectations. It turns out that the numbers displayed by Nature are different. The Scientist tries again, and, if he is a good mathematician, after some time, he will hit on the right path and construct exactly the same mathematical system that is used by Nature.

In this case, we are allowed to say that these are really two identical systems, that is, that the mathematics practiced by Nature is analogous to the Scientist's math. We then repeat the game but change some of its rules. Nature keeps showing the numbers (say, in pairs) to the Scientist, yet they are not derived from its mathematical system. Instead, each time they are created by means of an operation selected from the list of fifty possible ones with which we have provided Nature. Nature can change the first two numbers randomly but not any of the numbers that follow. It takes, at random, one of the transformation rules from the list, then performs the requisite division, multiplication, exponentiation, and so on; it shows the result to the Scientist, selects another rule, and then again transforms the (previous) results; it again shows him the result, and so on. There are operations that require any kinds of transformation to be abandoned. There are also operations saying that, if Nature's left ear is itching, then something must be subtracted, and if there is no itching, then the root of something must be extracted. Above all this, there are two operations that always apply. Each time Nature must order the two results obtained so that the first number shown by it is smaller than the second one; also, in at least one of the numbers next to an odd digit, there must always be a zero.

Even though this may seem strange, the sequence of numbers generated in this way will manifest its own regularities. The Scientist will be able to discover those regularities; that is, after some time, he will be able to predict, but only roughly, which numbers will appear in the sequence. Yet, since the probability of determining exactly the value of each subsequent pair decreases dramatically the more his predictions attempt to apply not only to the nearest stage but to a whole sequence of stages, the Scientist will have to create several systems of prediction. His prediction that a zero will appear next to an odd number will definitely come true: they appear in every pair of numbers, although in different places. It is also certain that the first number will always be smaller than the second one. All other changes are already subject to various distributions of probability. Nature thus manifests some kind of "order," but it is not the "order" of a singular type. It is possible to detect various kinds of regularities in it, which, to a large extent, depend on how long the game lasts. Nature seems to display the existence of some "constants" that are not subject to any transformations. Its future states that are not too remote in time can be predicted with a certain degree of probability, yet it is impossible to predict far-off states.

In this case, the Scientist could assume that Nature does indeed use one system, but it has so many variable operators that he is unable to re-create it. However, he is more likely to assume that Nature's actions are probabilistic. He will thus use appropriate methods, so-called Monte Carlo methods, to find some approximate solutions. The most interesting thing is that the Scientist may be suspecting that there exists "a hierarchy of levels in nature" (numbers, then operations above them, then superoperations of sequencing and "zeroing" above them). We thus have both different levels and "prohibitions" (the first number can never be higher than the second one), that is, "the laws of Nature," but this whole evolving numerical system is not a uniform mathematical system as a formal structure. However, this is only part of the problem. If the game lasts a very long time, the Scientist is eventually going to realize that Nature performs certain operations more frequently than others (the reason for this being that "Nature" is human and thus has to show predilection for certain operations; indeed, a human being cannot behave utterly chaotically or "randomly"). Following the rules of the game, the Scientist is observing only numbers and does not know whether they are being produced by a natural process, a machine, or a human being. However, he begins to suspect the existence of an even higher factor beyond the operations of transformation—a factor that determines what operations are to be performed. This factor (a human being pretending to be Nature) possesses a limited scope of action, yet gradually, the system of his predilections will become revealed in the numerical sequences he chooses (e.g., he uses operation no. 4 more often than operation no. 17, etc.). In other words, he will reveal the dynamic traits that are characteristic of his personality. But there is also another factor, a relatively independent one, because no matter which operations Nature prefers, from time to time, when coming across an operation whose result depends on the itching of the ear, he behaves in a particular way. This itching is not coupled with the dynamics of his consciousness anymore but rather with the peripheral molecular processes of his skin receptors. In the last instance, the Scientist is thus studying not only brain processes but also what is occurring in a particular section of the skin of the person pretending to be "Nature"!

Of course, he could ascribe to "Nature" traits that it does not have. He could think, for example, that "Nature" likes to have a zero next to an odd number, whereas in fact it is obligated to create such a result since it has been ordered to do so. This is a very crude example, but it shows that the Scientist can interpret the "numerical reality" he observes in various ways. He can see it *as a particular number of feedback systems*. Whatever mathematical model of the phenomenon he constructs, in no way will every element of his "theory of Nature," or its every symbol, have an exact equivalent on the other side of the wall. Even if he has learned all the rules of transformation after one year, he will never manage to create "an algorithm of the itching ear." Yet it is only in that latter case that we could speak about the equivalence, or isomorphism, of Nature and Mathematics.

The possibility of the mathematical representation of Nature does not therefore by any means imply its "mathematical character." It does not even matter whether this is a true hypothesis because it is entirely superfluous.

Having discussed both sides of the cognitive process ("ours," i.e., one that belongs to theory, and "the other one," or one that belongs to Nature), we can finally start on automatizing cognitive processes. It seems that the simplest thing would be to create an "artificial scientist," in the form of some kind of "electric superbrain," which would be connected through its senses, or "perceptrons," to the external world. This idea seems obvious because there is so much talk about the electronic imitation of thought processes and about the excellence and speed of the activities performed by digital machines today. Yet I do not think we should try to get there via an attempt to construct "an electronic superman." We are all fascinated by the complexity and power of the human brain, which is why we are unable to envisage an information machine that would not be analogous to the nervous system. The brain is without a doubt a great product of Nature. Yet now that I have paid it its due respect by saying this, I would like to add that it is a system that solves various tasks with extremely varied efficacy.

The amount of information that a skier's brain can "process" when he is slaloming downhill is much greater that the amount "processed" during that time by an excellent mathematician. By the "amount of information," I understand here in particular the number of variables that are regulated, or "controlled," by the slalom skier's brain. The number of variables controlled by the skier is incomparably higher than the number of variables to be found in the "field of selection" in the mathematician's brain. The reason for this is that the great majority of regulatory interventions performed by the slalom skier's brain have been automatized; they occur outside his field of consciousness, while the mathematician is not able to reach such a degree of automatization in his formal thinking (even though a good mathematician can achieve *some* degree of it). The whole mathematical formalism is a kind of fence that a blind person can walk along with confidence while holding on to it firmly. Why do we need this "fence" in the deductive method? The brain, as a regulator, manifests limited "logical depth." The "logical depth" (the number of subsequent stages of successive operations) of a mathematical deduction is incomparably greater than the "logical depth" of a brain that does not think in an abstract way but that, in accordance with its biological destiny, performs the role of a body-controlling device (a slalom skier on a downward slope).

This first type of "depth" is by no means something to be proud of—quite the contrary. It stems from the fact that the human brain is unable to regulate successfully any genuinely complex phenomena, *unless those phenomena are processes of its body*. The reason for this is that, as a body regulator, the brain controls a great number of variables: hundreds or maybe even thousands of them. Yet, someone will say, every animal has a brain that successfully controls its body. Alongside this task, the human brain is also capable of solving a countless number of others. In any case, it is enough to compare the size of a monkey's brain with that of the human's to see, at least roughly, how much bigger a section of the human brain's mass "is directed" toward solving intellectual tasks!

In fact, there is no point in discussing the intellectual superiority of the human over a monkey. The human brain is without doubt more complex, yet a considerable part of this complexity "is not any good" at solving theoretical problems because it controls bodily processes: this is what it is designed for. The problem thus looks as follows: the less complex part (*the part of the brain's neural system* that constitutes the foundation of intellectual processes) tries to obtain information about its more complex part (*the whole brain*). This is not impossible, but it is very difficult. In any case, it is not impossible indirectly (a single person would not even be able to formulate a problem). The cognitive process is social: it involves a kind of "adding up" of the "intellectual" complexity of many human brains that are all investigating the same thing. Yet since this "adding up" needs to be placed in quotation marks—since all those individual brains do not become conjoined to form one system—we have not yet solved our problem.

Why do individual brains not become conjoined to form one system? Is science not such a superior system? It is, but only metaphorically. If I understand anything at all, I understand this "something" as a whole, from beginning to end. It is impossible for brains of individual people, in becoming conjoined, to create a kind of "superior intellectual field," in which a form of truth will be formulated *that none of the individual brains on its own will be able to contain*. Scientists, of course, collaborate, but in the last instance, one person must formulate a solution to the set problem because no "choir of scientists" is going to do it.

But is it really true? Was it not rather the case that, first, Galileo formulated something, then Newton took it up and developed it, then several others added some things, then Lorentz performed his transformation, and then, having considered all of this together, Einstein unified all the data and created relativity theory? Of course it was this way, yet this is irrelevant. All theories use a small number of variables. Theories that are more universal do not contain a great number of variables: they are just applicable in a large number of cases. Relativity theory is a good example of this.

Yet we are talking about something else. The brain is perfectly capable of regulating a great number of variables in the body to which it is "connected." This happens automatically or semiautomatically (when we want to stand up and are not thinking about the rest, i.e., about the whole kinematic complex that has been started by this "order"). However, on the level of thought, that is, as a machine for regulating phenomena outside this domain, the brain is not a very efficient device, and more important, it is incapable of coping with situations in which a large number of variables have to be taken into account at the same time. This is why it cannot properly regulate, for example, biological or social phenomena (on the basis of their algorithmization). In fact, even far less complex processes (such as climactic or atmospheric ones) still escape its regulatory capacity (which is understood here only as the ability to predict accurately future states on the basis of earlier ones).[3]

Last, but not least, in its most "abstract" realm of activity, the brain remains under a much greater influence of the body (for which it functions as both master and servant thanks to the two-way feedback) than we normally realize. Being connected, in turn, to the surrounding world "through this body," the brain begins to express all the properties of this world through phenomena related to bodily experience (hence the search for the person who is carrying the Earth on his shoulders, who "attracts" stones to the Earth, etc.).

The traffic capacity of the brain as an information channel remains at maximum level in bodily phenomena. Yet the excess of information coming to it from outside, for example, in the form of a text being read, is blocked once it has reached a certain number of bits per second.

Astronomy, one of the first disciplines practiced by man, still has not provided a solution to "the problem of many bodies" (i.e., gravitating masses that mutually affect one another). But there is someone who would be able to solve this problem. Nature does this "without math," through the sheer activity of those bodies. This raises the question of whether it would be possible to address "the information crisis" in a similar way. This is, of course, impossible, you say; it is a pointless idea. The mathematization of all sciences is increasing, instead of being reduced. We cannot do anything without math.

Agreed, but let us establish first of all what kind of "math" we are talking about—the kind that uses the formal language of equations and inequalities, written on a piece of paper or encoded in the binary elements of large electron machines, or the kind that actualizes the fertilized egg without any formalism whatsoever? If we are to stay with the first one, we will be facing an information crisis. However, if we initiate the second one for our purposes, then matters can take a very different course.

Fetal development is a "chemical symphony," which starts the moment the sperm's nucleus becomes conjoined with the egg's nucleus. Imagine we have managed to track this development, from fertilization through to the emergence of a mature organism, on a molecular level and that we now want to represent it in the formal language of chemistry—the same language we use to represent simple reactions such as $2H + O = H_2O$. What would such a "score for embryogenesis" look like? First, we would have to write down, one by one, formulae for all the compounds that are there at the "start." Then we would have to list all the relevant transformations. Given that a mature organism contains at a molecular level around 10^{25} bits of information, the number of formulae we would have to write down would have to be in the realm of quadrillions. There would not be enough surface across all the oceans and the entire mainland for it. It is a completely hopeless task.

Let us not worry for now about how chemical embryology is going to cope with similar problems. I think that the language of biochemistry will have to undergo a very radical transformation. Perhaps we will end up with some kind of physical–chemical–mathematical formalism. Yet this is not our concern because, should someone "need" a particular living organism, none of this writing activity would actually be necessary. It would be enough to take some sperm and fertilize an egg with it—which, after a certain time, would have transformed "itself" into "the required solution."

It is worth pondering whether we shall be able to do something similar in the area of scientific information. Could we not take up "information farming," cross-breed bits of information with one another, and initiate their "growth" so that we eventually obtain a "mature organism" in the form of *scientific theory*?

Rather than the human brain, we would like to offer another product of evolution—the human genotype—as a model for our trials. The amount of information per cubic volume of the brain is incomparably smaller than the amount contained in the same volume of sperm. (I am talking about sperm rather than an egg as the former's information "density" is higher.) Of course, we need different sperm and different laws of genotypic development from those created by evolution. This is only a starting point and, at the same time, the only material system on which we can rely.

Information should develop from information, just as organisms develop from organisms. Its bits should fertilize one another; they should cross-breed, undergo "mutations," that is, small changes, as well as radical transformations of the kind that do not occur in genetics. Perhaps this will take place in some kind of container, where "information-carrying molecules"—in which particular bits of information will be encoded in the same way an organism's traits are—will react with one another in the protoplasm? Perhaps it will start a unique "fermentation of information leaven"?

Yet our enthusiasm is premature; we have raced too far ahead. Since we are supposed to learn about evolution, we should investigate how it gathers information.

Such information needs to be stabilized but also malleable. To achieve stabilization, that is, optimal information transfer, the following conditions need to be fulfilled: the absence of disturbances in the transmitter, low level of noise in the channel, the permanence of signs (signals), the possibility of connecting information into coherent and uniform units, and, last but not least, information surplus (excess). Connecting allows us to detect errors and minimize their effects—which destroys information transfer. Information excess performs the same role. A genotype uses these methods the same way a communications engineer does. A piece of information sent in the form of printed or handwritten text behaves in a similar way. It should be legible (the absence of disturbances) and indestructible (e.g., even when print fades); particular letters need to be connected into blocs (words), which in turn need to be connected into higher units (sentences). Information contained in writing is also excessive, which is evident in the fact that a partially destroyed text can be deciphered.

An organism enacts the protection of stored information against disturbances by isolating reproductive cells well; it undertakes information transfer by enacting the exact mechanics of chromosomal division. Information is then blocked in genes, which are in turn blocked in their superior units—chromosomes ("the sentences of the heredity script"). Finally, every genotype contains some surplus information, which we can recognize through the fact that a damaged egg can still produce a non-damaged organism—of course, within certain limits.[4] During its course of development, genotypic information is transformed into phenotypic information. A phenotype is the final shape of a system (i.e., its morphological traits, together with physiological ones, and thus its functions) that emerges as a result of the activity of hereditary (genotypic) factors and the influences of an external environment.

Using a visual representation, we could describe the genotype as a kind of empty and shrunk rubber balloon. If we place it in an angular dish, the balloon—whose "genotypic tendency" was to become a sphere—will adapt its shape to the shape of the dish. It is because plasticity—which results from the working of "regulatory buffers" that form an "amortization insert" between genotypic instructions and environmental demands—is an important characteristic of organic development. To simplify, we can say that an organism can survive even under very inauspicious conditions, that is, conditions that go beyond average genotypic programming. A lowlands plant can also develop in the mountains, yet it will start to resemble mountain plants in its shape. This is to say that its phenotype will change but not its genotype because, on being transferred to the lowlands, its seeds will once again produce plants in the original shape.

How does the evolutionary flow of information happen?

It happens in a circular manner. Its system consists of two channels. The source of the information transmitted in the first channel is to be found in mature organisms, during their reproductive acts. Yet, because not all of them can reproduce in the same way, but mainly those that have become privileged as a result of their most successful adaptation, their adaptive traits, including phenotypic ones, take part in the "selection of transmitters." This is why it is not the reproducing organisms themselves that are seen as a source of such information but rather their whole biogeocenosis, or biotope, that is, organisms together with their environment (and with other organisms living in it because it is also to their presence that our organisms have to become adapted). Ultimately, information thus travels from biogeocenosis through fetal development to the next generation of mature organisms. This is what happens in the embryogenetic channel, which transmits genotypic information. The second, reverse channel transmits information from mature organisms to biogeocenosis, but such information is already phenotypic because it is being transferred "on the level of" whole individuals and not "on the level of" regenerative cells. Phenotypic information simply stands for the entire life activity of organisms (what they eat; how they eat; how they become adapted to biogeocenosis; how they change it through their existence; how natural selection occurs, etc.).[5]

And thus information that has been encoded in the chromosomes, on the molecular level, travels through the first channel, while macroscopic, phenotypic information, which manifests itself in adaptation, the struggle for survival, and natural selection, travels through the reverse channel. The phenotype (the mature organism) contains more information than the genotype because environmental influences constitute external source information. Since the circulation of information does not take place on just one level, it must undergo a transformation somewhere to allow its "code" to be "translated" into a different one. This occurs during embryogenesis, which is a "translator" from the language of molecules to the language of an organism. This is how microinformation becomes macroinformation.

No genotypic changes take place during the previously described circulation process, which means that evolution is also absent from that process. Evolution develops thanks to spontaneously occurring "lapses" in genotypic transfer. Genes mutate in a directionless, blind, and random way. It is only environmental selection that chooses, that is, fixes in subsequent generations, genes that increase adaptation to the environment—which stand for survival chances. The antientropy activity of the selection process, that is, that which cumulates an increase in order, can be imitated in a digital machine. Given that we do not have such a machine, we must play the "evolution game."

We divide a group of children into groups of equal size. The first group will represent the first generation of organisms. "Evolution" starts the moment each child from the first group receives its "genotype." This is a packet that contains a foil cape and a set of instructions. If we want to be really accurate, we can say that the cape stands for the egg's material layer (plasma membrane) and the set of instructions for the chromosomes in the nucleus. The "organism" learns from the set of instructions "how it is supposed to develop." This consists in a child putting on the cape and having to run along the hallway that has an open window on one of its sides. There is a marksman with a toy gun, filled with peas, outside. A child that has been hit "dies in his struggle for survival" and thus cannot "reproduce." One that completes the run intact places the cape and the set of instructions back in the packet and passes this set of "genotypic instructions" on to the person from the "next generation." The capes have different shades of gray, from very light to almost black, while the hallway walls are painted dark gray. The more the runner's silhouette stands out against the background, the easier it is for the marksman to hit him. Those whose capes are similar in tonality to the hallway walls have the highest chance of survival. The environment acts as a filter here, eliminating those who are the most poorly adapted to it. "Mimicry," whereby an individual's color starts to resemble that of its environment, develops. The original wide range of individual colors narrows down at the same time.

Yet not all individual chances of survival are the result of a given child's "genotype," that is, of the color of his cape. By observing the fate of his predecessors or by simply grasping what is going on, he learns that some forms of behavior (running fast, running while keeping low, etc.) also make it more difficult for the marksman to hit him and thus increase his chances of "survival." In this way, thanks to the environment, an individual obtains nongenotypic information that was not contained in the set of instructions. This is phenotypic information, which is the individual's own personal acquisition. Yet phenotypic information is not hereditary because one passes on to the "next generation" just the "reproductive cell," that is, the packet with the cape and the set of instructions. After a certain number of "crossings" through the environment, only those individuals whose genotype and phenotype (the color of the cape and behavior) offer the biggest chance of survival actually "survive." The initially diverse group becomes more uniform. Only the fastest, the most agile, and those wearing capes in protective coloring survive. Yet each subsequent "generation" only receives genotypic information: it must create phenotypic information by itself.

Let us imagine that, as a result of a factory error, some spotted capes now appear among all the others. The effects of such "noise" stand for a genotypic mutation. Spotted capes can be clearly seen against the background, which is why "mutants" have a very small chance of "survival." They are thus quickly "annihilated" by the marksman with a toy gun—who can be interpreted by us as a predator. Yet if we wallpaper the hallway walls with spotted sheets (environmental change), the situation will change radically: only mutants will be surviving now, and such new "hereditary" information will rapidly eliminate the old information from the whole population.

The act of donning the cape and reading the set of instructions is, as we have already said, equivalent to embryogenesis, in which functions develop alongside the overall form of the organism. All this activity involves the embryogenetic transfer of genotypic information via the first information channel (from biogeocenosis to mature individuals). Learning about the best way of running across the environment stands for acquiring phenotypic information. Each individual who has successfully arrived at the end point is already carrying two types of information: hereditary–genotypic and nonhereditary–phenotypic. The latter disappears from the evolutionary scene for ever with him. Genotypic information, which has passed through the "filter," is transmitted from one pair of hands to another—which constitutes its reverse transfer (via the second channel).

It is thus also in our model that information travels from biogeocenosis to organisms on the "microscopic" level (opening the packet received, familiarizing oneself with the set of instructions, etc.), while traveling from organisms back to biogeocenosis on the macroscopic level (because the packet itself, i.e., the genotype, will not cross the environment, *the whole individual,* who is its "carrier," must do it).

In this game, biogeocenosis refers to the entire hallway together with all the runners (the environment in which a given population exists).

Some biologists, such as Schmalhausen, claim that even though the circulation of information occurs in the way described earlier; the mature organism does not contain more information than the genotype did, which means than an increase in the amount of information caused by the interplay of feedback between the individual and the environment is only apparent in his life and is just a consequence of the regulatory activity on the part of mechanisms produced by the organism thanks to genotypic information. The plasticity of those reactions creates an impression that an increase in the amount of information contained in the organism has taken place.

When it comes to genotypic information, it remains largely unchanged until a mutation occurs. However, the amount of phenotypic information is bigger than the amount of genotypic information; the opposite view contradicts information theory but not biological theories. We have to differentiate between these two issues. If we decide on a frame of reference then the amount of information will be dictated by the trajectory of the phenomenon. We will thus not be able to subtract any of this information from the phenomenon, because we will deem it "superficial." It does not make any difference whether such information emerges as a result of regulatory activity or in any other way—as long as we are investigating its amount in a material object such as an organism, with regard to a given frame of reference.

This is not some kind of academic debate: it is a matter of grave importance for us. The previously cited position suggests that environmental "noise" is only capable of impoverishing phenotypic information (this is what Schmalhausen claims). Yet noise can also be a source of information. Indeed, mutations are an example of such "noise." As we know, the amount of information depends on its degree of probability. The sentence "boron is a chemical element" contains a certain amount of it. Yet if a fly leaves a smudge over the first letter, as a result of which the sentence will read "moron is a chemical element," then on one hand, we will witness a disturbance to the information transfer caused by noise and thus a *decrease* in the amount of information, yet on the other, we will also witness an *increase* in its amount because the second sentence is far less probable than the first one!

What we are presented with here is a simultaneous increase in *selective* information and a decrease in *structural* information. The first type of information refers to a set of possible sentences (of the "x is a chemical element" kind), whereas the second refers to real situations for which sentences only serve as their representations. In this second case, the set of sentences that represent real situations consists of sentences such as "nitrogen is a chemical element," "oxygen is a chemical element," and so on. The number of sentences in this set corresponds to the actual number of chemical elements and thus amounts to approximately one hundred. This is why, if the only thing we know is the set from which our sentence will be selected, the probability of the appearance of a given sentence is 1/100.

The second set contains all the words from a given language that are to replace the x in the sentence "x is an element" ("an umbrella is an element," "a leg is an element," etc.). It therefore contains as many sentences as there are words in that language—which is to say, many thousand. Information is the converse of probability, which means that each one of such sentences is thousands of times more improbable, that is, it contains suitably more information (not thousands of times more because information is a logarithm, but this is of no particular relevance here).

As we can see, one should use the notion of information carefully. In a similar vein, we could also understand a mutation as both a *decrease* in the amount of (structural) information and an *increase* in the amount of (selective) information. The way a mutation is "perceived" is determined by the biogeocenotic environment. Under normal conditions, it will stand for a *decrease* in the amount of structural information pertaining to the actual world; consequently, the organism will become extinct as a result of being less adapted, even though the amount of selective information has increased. Should the conditions change, the same information concerning the mutation will cause an increase in both structural and selective information.

Incidentally, noise can be a source of information only under some very special circumstances: when such information is an element of a set all of whose elements are characterized by high-level organization (complexity). The change of the word "boron" to "moron" as a result of noise involves a shift from one kind of organization to another, whereas the change from the word "boron" to an ink spot involves a complete eradication of any organization whatsoever. A mutation also involves a transformation of one kind of organization into another—unless we are talking about a lethal genetic mutation, which kills an organism in the course of its development.

A sentence can be true or false, while genotypic information is adaptive or nonadaptive. In both cases, we are talking about structural measurement. However, as selective information, a sentence can be only more or less probable, depending on the set from which it has been drawn. Similarly, as selective information, a mutation can be more or less probable (which means that there is more or less of such information).

Phenotypic information is usually structural because it emerges as a result of environmental influences, while an organism responds to those influences through adaptive reactions. We can thus add phenotypic, structural, external-source information to genotypic structural information to obtain a full amount of structural information that a mature individual possesses. Of course, this has nothing to do with the question of heredity: only genotypic information is hereditary.

In practice, determining the balance of information is very difficult for a biologist because it is only on a theoretical level that it is possible to draw a sharp boundary between what is genotypic and what is phenotypic, precisely owing to the presence of regulatory mechanisms. If no external influences were affecting an egg undergoing a division, we could describe its development as "deductive," in the sense that genotypic information undergoes the kinds of transformations whereby no increase in the amount of information occurs. A mathematical system, initially represented by its foundational premises ("an axiomatic kernel") and transformation rules, "develops" in a similar way. The two things together could be called the "genotype of a mathematical system." However, fetal development in this kind of imaginary isolation is not possible because an egg is always subject to some influences—for example, those coming from gravity. We know what a significant effect it has on, say, the development of plants.

Before we start at long last on properly designing an "autognostic" or "cybergnostic" machine, in the final part of my argument, I would like to mention that there exist different types of regulation. There is continuous regulation, which permanently watches the values of controlled parameters, and noncontinuous regulation (breech regulation), which only comes into effect after certain critical values of controlled parameters have been crossed. An organism employs both types of regulation. For example, temperature is mostly regulated in a continuous manner, whereas the sugar level in blood is regulated in a noncontinuous manner. The brain can also be considered to be a regulator that uses both methods. Yet these issues have been discussed so comprehensively by Ashby in his *Design for a Brain* that there is no need to rehearse them here.

Individual development involves a confrontation between two types of information: internal and external. This is how an organism's phenotype develops. However, an organism is supposed to serve itself and evolution; that is, it is to exist and maintain the existence of the species. Information "systems" on the farm are supposed to serve *us*. We should thus replace the law of bioevolution on our farm—a law that says that an individual that is best adapted to the environment survives—with the following new law: "that which *expresses* the environment most adequately survives."

We already know what "expressing the environment" means. It involves collecting structural, and not selective, information. Repetitions are no doubt unnecessary and superfluous, but let me say this once again. A communications engineer examines the probability of the arrival of information in such a way that a hundred-letter-long sentence always contains the same amount of information for him, no matter if it has been taken from a newspaper or from Einstein's theory. This aspect is the most important one in information *transfer*. However, we can also speak about the amount of information in the sense that a sentence describes (represents) a certain situation that is more or less probable. In that case, the information content of the sentence depends not on the probability of the occurrence of letters in a given language or on their overall number but only on the degree of probability a given situation itself manifests.

The relation between the sentence and the real world has no relevance for its transmission via a communications channel, yet it becomes decisive when measuring the amount of information contained, for example, in a scientific law. We shall occupy ourselves with "farming" this second type of information—which we call structural.

"Regular" chemical molecules do not express anything, or they "express only themselves," which is the same thing. We need molecules that would be both themselves and a representation of something outside themselves (a model). This is possible because the designated place of a chromosome not only "expresses" "itself," that is, a DNA particle, but also the fact that an organism that will develop from it will have, say, blue eyes. However, it only "expresses" this fact as an element in the overall organization of the genotype.

But what does it actually mean to say that some hypothetical "organism–theories" "express the environment"? Everything that exists, that is, the world as a whole, is an environment studied by science—yet not everything at the same time. Information gathering consists in identifying systems in the world and then examining their behavior. Certain phenomena, such as stars, plants, and humans, have characteristics that make them "stand out" as systems; others (a cloud, a thunderbolt) are only seemingly granted such autonomy, that is, such relative isolation from the environment. We should reveal now that we are not actually going to start our "information evolution" from scratch; that is, it is not our intention to create something that will first have to reach the level of human knowledge "by itself" and that will be able to develop further only after that point. I do not know whether this would be impossible—most probably not. In any case, such evolution "from scratch" would require an enormous amount of time (perhaps even the same amount that biological evolution needed). But this is not necessary at all. We are going to use our knowledge, also with regard to classification, right from the start (knowledge about what constitutes a system worth studying and what does not). We expect that for some time, we shall be unlikely to make any amazing discoveries, and that they will emerge only after our "farm" has become established. We shall arrive at solution by making a number of approximations. Such a farm can be designed in multiple ways. A gravel riverbed—as a "generator of diversity" and a "selector" that is sensitive to detecting "regularity"—could serve as an early model for it. If the selector consists of a series of barriers with round holes in them, then at the end, we shall have only round stones because no other ones will get through the "filter." We shall have obtained some kind of order out of disorder (from the "noise" of the gravel riverbed), yet the round stones will not represent anything outside themselves. Information, in turn, is a representation. A selector cannot thus act on the basis of "an inherent characteristic" but only on the basis of something outside it. It thus has to be connected, on one side, to the "noise" generator *as a filter* and, on the other side, to a particular section of the external world.

Ashby's idea for constructing an "intelligence amplifier" is based on the notion of the "generator of diversity." Ashby states that any scientific laws, mathematical formulae, and so on, can be generated by a system that acts completely randomly. And thus, for example, the binominal theorem can be "transmitted," in the form of the Morse code, by a butterfly fluttering its wings over a flower, purely by accident. What is more, one does not have to wait long for such unusual events to occur. Since every piece of information—and thus, for example, the binominal theorem—can be transmitted through binary code using a dozen or so symbols, in every cubic centimeter of air, its particles transmit that formula *several hundred thousand times per second* in the course of their chaotic movements. This is absolutely correct: Ashby provides a set of calculations to demonstrate this. Hence we arrive at the conclusion that while I am writing this, the air in my room contains configurations of molecules that express, in binary code, countless numbers of other priceless formulae as well as statements on our topic under discussion that are much more precise and lucid than those I myself have produced. This is still nothing compared with the Earth's atmosphere as a whole! Amazing scientific truths from the year 5000; poems, plays, and songs by Shakespeares who are yet to be born; mysteries from other space systems; and lots of other things come together within it in a split second, and then immediately fall apart.

What does it all mean? Unfortunately, nothing, since those "precious" results of the billionth atomic collisions are mixed up with billions of other, totally meaningless results. Ashby says that new theories count for nothing because they can be produced in great numbers by means of "noise" processes that are as random as atomic collisions in a gas are, while the only things that really matter are elimination and selection. With this, Ashby wants to prove the capabilities of an "intelligence amplifier" as a *selector* of ideas provided by any kind of noise process. Our position on this differs from his. I have referred to Ashby here to demonstrate that it is possible to travel toward a similar goal (although not the same one—an "amplifier" is different from a "farm") via different routes. Ashby thinks that one should start from the greatest diversity and then "filter" it gradually. We, in turn, want to start from significant yet not enormous diversity—a kind of diversity that a self-organizing process (e.g., fertilized egg) represents—and make sure this process "develops" into a scientific theory. Its complexity may then either increase or decrease: this is not the most important thing for us.

In fact, the "generator of diversity" postulated by Ashby already exists.

We can say that mathematics constantly generates countless "empty" structures, while the world, physicists, and other scientists, in ceaselessly combing through that warehouse of diversity (i.e., the different formal systems), from time to time come across something that is useful on a practical level and that "fits" certain material phenomena. Boole's algebra developed before we knew anything about cybernetics; it then turned out that the human brain also uses elements from this algebra—which now provides the foundation for the working of digital machines. Cayley invented matrix calculus several dozen years before Heisenberg noticed that it could be used in quantum mechanics. Hadamard tells a story in which a certain "empty" formal system on which he was working as a mathematician and that he did not think could have anything to do with reality subsequently became useful in empirical research. Mathematicians are thus a generator of diversity, whereas empiricists are the selector postulated by Ashby.

Naturally, mathematics is not a noise generator. It is a generator of order, of various "internal orders." It creates orders that correspond, in a more or less fragmentary manner, to the real world. This fragmentary correspondence facilitates the development of science and technology—and thus of civilization.

It is sometimes said that mathematics is an "excessive" order when compared with reality—which is less ordered than mathematics. But this is not really the case. Notwithstanding its greatness, invariability, necessity, and explicitness, mathematics was shaken for the first time during our century,[6] as a result of the cracks that appeared in its foundations in the 1930s, when Kurt Gödel proved that its fundamental postulate—that of consistency and internal systemic completeness[7]—could not be fulfilled. If a system is consistent, then it is not complete, and if it is complete, then it stops being consistent. Mathematics seems to be as imperfect as any human activity. From my point of view, there is nothing wrong with it—nothing that would downgrade it. But never mind mathematics, since we do not want it anyway. Cannot the mathematization of cognitive processes be avoided? Not the kind that, while doing away with all signs and formalisms, rules over chromosomal and star processes but rather that which, in using the symbolic apparatus and the rules of autonomous transformations, expands through its operations the logical depth that has no equivalence in Nature. Are we doomed to remain within its structure? Let us say first of all, but only as a kind of warm-up, that even though it does not sound very promising, it is extremely easy to start a "mathematical systems farm." It goes without saying that this would be done on the basis of "deductive development" from an "axiomatic kernel," in whose "genotype" are recorded all the rules for any possible transformations. In this way we shall obtain all sorts of "mathematical organisms" one can think of, in the form of complex crystal structures and so on. With this, we would have done the exact opposite to what science used to do. Science used to fill the emptiness of mathematical systems with the material content of phenomena, while we would not be translating such phenomena into mathematics but rather translating mathematics into material phenomena.

We could clearly perform various calculations in this way, or even design various devices, by introducing initial data (e.g., working parameters of a machine that is to be constructed) into the "genotype"—which, in undergoing development, will offer us, in the form of an "organism," an ultimate solution to a problem or a design for a machine. Of course, since we are able to encode given parameter values in the molecular language of the "genotype," we can do the same with a "mathematical organism" and thus translate this crystal or another structure that emerges as a result of "deductive development" back into the language of numbers, design plans, and so on. The solution always "appears by itself" in the course of the already ongoing reactions; we do not have to concern ourselves with the particular stages of that process. The only thing that matters is the final result. It is important for the development process to be controlled by internal feedback so that, once the given parameters have reached certain values, this whole "embryogenesis" can be stopped.

Setting up a "farm that cultivates empirical information" amounts to "inverting" the tree of biological evolution. Evolution started from a uniform system (protocell) and then created millions of branches of phyla, families, and species. A "farm" starts from specific phenomena, represented in their material equivalents, and aims toward "reducing things to a common denominator" so that we obtain a unified theory in the end, encoded via a molecular language into the permanent structure of a pseudo-organism.

Perhaps we should give up on all those metaphors now. We begin by modeling particular phenomena in a given class. We collect initial information ourselves—in a "classic" way. Now we have to transfer it into an information-carrying substrate, which needs to be provided by synthetic chemistry.

Our task consists in representing the trajectory of a system (the course of a phenomenon) by means of a dynamic trajectory of another system. We need to represent processes with some other processes, not with formal signs. A fertilized egg is isomorphic with its "atomic description," sketched on a piece of paper, or with a spatial model made out of spheres that represent atoms. Yet it is not an isodynamic model because a model made out of spheres will clearly not be undergoing any development. The model contains the same information the egg does; yet its information *carrier* is different in both cases. This is why the egg is capable of undergoing development, whereas the paper carrier is not. We need models that are capable of developing further. No doubt, if the signs in the equations listed on a piece of paper wanted to interact with one another, we would not need this whole "information farm." Yet this is, of course, impossible. Setting up a farm, in turn, is extremely difficult and will only be achievable a long time from now, but it is not a complete absurdity, I hope.

The building material for our "information carriers" will be provided, for example, by large molecules of synthetic polymers. They undergo development, grow, and complexify their structure by attaching to themselves particles of "food" that are floating in the medium in which the "carriers" are placed. The carriers are matched so that their development and their subsequent transformations correspond on an isodynamic level to the transformations of a given system (phenomenon) in the external world. Each such molecule is a "genotype," one that develops in accordance with the situation it represents.

First, we introduce a significant number of molecules (several trillion), about which we know that the initial phases of their transformation are progressing in the right direction. "Embryogenesis," which stands for the correspondence between the developmental trajectory of the carrier and the dynamic trajectory of the real phenomenon, begins. The developmental process remains under control, as it is coupled with the situation. The coupling is selective (i.e., the elimination of "those who are not developing correctly" is taking place). All the models together constitute an "information population." This population is gradually moving from one container to another. Each container is a station in which selection takes place. Let us call it a "sieve" for short.

The "sieve" is a device that is connected (e.g., by means of automatic manipulators, perceptrons, etc.) to a real phenomenon. Structural information about the phenomenon's state is encoded in the molecular language by the "sieve," which also creates something like microscopic specks, each one of them providing a "record of the phenomenon's state," that is, an actual section of its dynamic trajectory. Two speck waves then crash together. The specks in the first one, as self-organizing information carriers undergoing development, "predict" the state in which the real phenomenon finds itself by means of the state they themselves have just reached. The second wave contains specks produced in the "sieve," which carry information about the actual state of the phenomenon.

This is a similar reaction to the precipitation of antigens by antibodies in serology. Yet it is the difference between "truth" and "falsity" that provides the rule for precipitation. Here all the specks that have predicted the phenomenon correctly are precipitated because their molecular structure "matches" the gripping molecular structure of the specks sent by the "sieve." The precipitated carriers, as those that have "correctly predicted" the phenomenon's state, move on to the next selection stage, where the process is repeated once again (once again, they crash against the specks that are carrying the news about the subsequent state of the phenomenon; once again, those that correctly "anticipate" this state are precipitated, etc.). Eventually we obtain a certain number of specks that together constitute an isodynamic, specially selected developmental model of the whole phenomenon. Since we are familiar with their original chemical composition, we already know which molecules can be considered dynamic models of the development of the system we have been studying.

This is an introduction to the evolution of information. We obtain a certain number of information "genotypes" that have correctly predicted the development of phenomenon x. At the same time, an analogous speck "farm" is in operation, with the specks modeling phenomena x, y, and z, which belong to the whole class under examination. Let us say we finally obtain carriers for all the seven hundred million elementary phenomena in that class. What we now need is a "theory of that class," which will mean finding its constants or shared parameters within the whole class. We thus have to eliminate all the irrelevant parameters.

We then start farming "the next generation" of carriers, which no longer model the development of the original phenomenon but rather the development of the first generation of carriers. As the phenomenon has an infinite number of detectable variables, the initial selection of significant variables needs to be made. There are very many of such variables, but of course these are not all of them. This initial selection, as already mentioned, takes place in a "classic" manner, that is, it is made by scholars.

The next generation of carriers does not currently model all the developmental variables of the first generation either, yet this time the selection of significant variables happens automatically (via catalytic precipitation). In the course of their development, various individuals from the second-generation carriers leave out different variables manifested by the first-generation carriers. Some of them leave out significant variables, as a result of which their dynamic trajectories diverge from the "true prediction." They are being constantly eliminated by subsequent "sieves." Eventually the second-generation carriers that have "predicted" the whole developmental trajectory of the original ones get selected, despite having left out a certain number of variables. If the structure of the carriers that reach the "finishing line" in the second round is practically identical, this means that we have obtained, or "crystallized," a "theory of the examined class." If there is still (chemical, topological) diversity among the carriers, the selection process has to be repeated to continue with the elimination of insignificant variables.

The "crystallized theories," or, better, "second-generation theoretical organisms" then start "competing" for representation with the analogous specks that form a "theory" of another class of phenomena. We aim to develop "a theory of the class of classes" in this way. This process can be continued as long as we wish, to obtain various degrees of "theoretical generalization." What is unobtainable yet thinkable is some kind of "jewel of cognition," a "theoretical superorganism" located at the very top of this evolutionary pyramid we want to reach: a "theory of everything that exists." It is no doubt impossible to get to; we only mention it to visualize more explicitly the analogy with the "inverted tree of evolution."

The idea we have just discussed, whose explanation sounds rather boring, is also very primitive. We should think about improving it. For example, it would be worth implementing some kind of "actualized Lamarckism" in the "farming" process. As we know, Lamarck's theory about passing on hereditary traits does not in fact correspond to what happens in biology. The method of passing on "acquired traits" could be used in the evolution of information to accelerate the formation of "theoretical generalizations." We have already spoken about "crystallized information," yet it is equally possible that those "theory-carrying" specks will be made of a different substance (polymers). It may also turn out that their similarity to living organisms will be significant in some respects. We should perhaps not start with specks but rather with some relatively large conglomerates, or even "pseudoorganisms," that is, "phenotypes" (which provide an information record of the real phenomenon), and work toward such a "phenotype" developing its own "generalization," its "theoretical outline," or "genotype theory"—once again, contrary to what actually happens in biology.

Never mind such ideas, though, because they cannot be verified anyway. Let us just point out that every "speck theory" is a source of information that has been generalized into a systemic law and that can be coded into a language that will be understood by us. These specks are free from the limitations of formal mathematical systems: they are able to model the behavior of three, five, or six gravitating bodies—which cannot be achieved on a mathematical level, strictly speaking at least.

By initiating the development of the carriers of "the theory of five bodies," we shall obtain data about the location of the real bodies. To do so, we must "circulate them round" in a particular device so that their developmental trajectories become attuned with the trajectories of the system under examination, thanks to feedback loops. This, of course, presumes the existence of self-regulating and self-organizing mechanisms within those carriers. We can thus say that we are like Liao Shi Ming, who offered tuition on how to fight dragons—the only problem being that his student was unable to find a dragon anywhere. Similarly, we do not know how to build an "information carrier" or where to look for the building material. Anyway, I have now briefly explained how we can envisage the remote future of "biotechnology." As you can see, its potential is significant. Having been brave enough to say this, in conclusion, I shall present one more biotechnological possibility.

A separate class would be constituted by "information-carrying sperm," whose task would not be to examine phenomena or devices but rather to produce them. All sorts of desired objects (machines, organisms, etc.) would emerge from such "sperm" or "eggs." Of course, such "working sperm" would need to have at its disposal both encoded information and executory devices (such as biological sperm). A reproductive cell contains information about its final goal (organism) and about the trajectory that will take it to this goal (embryogenesis), but it already has (in the egg) the material needed "to construct a fetus." We can imagine such "working sperm"—which, alongside the information about what kind of object it needs to construct and about how to do it, is equipped with some additional information about how to transform the material from the environment (e.g., on another planet) into the necessary building material. If it also contains an appropriate program, such "sperm," when planted in the sand, will produce everything that can be made from silica. It may just need some additional materials and, of course, an energy source (e.g., an atomic one). However, we should stop on this panbiotechnological note.[8]

LINGUISTIC ENGINEERING

Bodies impact each other on a material, energetic, and informational level. The impact results in a state change. If I throw myself to the ground as a result of someone shouting "Down!," the change in my situation will have been caused by the arrival of information; if I collapse because an encyclopedia falls on me, the change will have been caused by a material action. In the first case, I did not have to fall down; in the second, I did. Material and energy-driven actions are determined, while information-led actions only lead to a change in probability distributions.

This is what the problem looks like roughly. Information-led actions change probability distributions within a range determined by energy and material conditions. If someone shouts "Fly!" at me, I will not do so, even if I want to. Information will have been transmitted but not actualized. It will have changed the state of my brain but not of my body. I will have understood what has been said to me, but I will not be able to actualize it. Language thus has a causal and a comprehending aspect. We shall start from this premise. By language, we shall understand a set of states that has been isolated from the set of "all possible states," that is, a subset of the latter set, in which selection according to a "certain criterion" (criterion x) occurs. For a given language, x is a variable that adopts different values from within a certain range. What kinds of "subsets of states" are we talking about? It will be easier if we use an example. Another such subset, not a linguistic one this time, contains all the possible trajectories of the bodies within the solar system. It is easy to see that even though the number of trajectories is infinite, they are not arbitrary (e.g., square trajectories are impossible). Bodies behave as if certain *restrictions* had been placed on their movement. We say, after Einstein, that the spatial matrix, conditioned by the distribution of bodies, imposes restrictions on their motions. All the possible trajectories of the revolving bodies, and also of those bodies that can be introduced into the system at any time, are not the same thing as the space that has a restricting capability. Similarly, linguistics differentiates between utterances ("trajectories") and language (a kind of "linguistic field"). We can continue with this analogy. Just as the gravitational field restricts bodies in their motions, the "linguistic field" restricts the "trajectories" of utterances. Equally, just as every kinematic trajectory is designated, on one hand, by the field's metric and, on the other, by the limit conditions of the body (its original speed and direction of movement), the shaping of utterances is affected by the conditions of the "linguistic field," such as syntax and semantic rules, and by the "local limit conditions" offered by the diachrony and synchrony of the speaking individual. Just as bodies' trajectories are not a gravitational field, utterances are not a language, even though, if all the masses disappear from the system, restrictions imposed by gravitation will also disappear, and if all Polish speakers die out, the corresponding syntactic and semantic rules, that is, the "field" of our language, will also disappear.

The question then arises as to how "fields"—linguistic and gravitational ones—actually exist. This is quite a difficult question, concerning as it does the "ontological status" of examined phenomena. Bodies' motions and linguistic articulations no doubt exist—but do they exist in exactly the same way that gravitation and language do? In both cases, we shall respond, certain forms of description are used that explain the status quo and allow us to make predictions (only probabilistic ones in the case of language, but this point need not worry us for the moment). Yet we are not obligated to consider such descriptions as ultimate because we do not know whether the pronouncements of Einstein and the linguists about these issues (of gravitation and language) should really count as final ones, permanently valid and unshakeable. These circumstances, however, do not pose any problems for a constructor of interplanetary rockets or talking machines, at least on an ontological level, because in both cases, it is only a technical issue. We shall now present a model distribution of "all the possible languages" on a bipolar scale. We shall call one pole "causal," the other "comprehending." On this scale, natural language finds itself not far from the "comprehending" pole, and physicalist language is somewhere in the middle, whereas the language of heredity is right by the "causal" pole.

The only difference between informational and material "causality" is that effects of purely material causation do not relate to anything; that is, when a certain material phenomenon takes place whereby the role of "information" factors can be considered entirely irrelevant, it is impossible to debate whether this phenomenon is "true" or "false," "adequate" or "inadequate": it simply happens, full stop.

Every linguistic utterance can be considered a form of control program, that is, a transformation matrix. The outcome of all the actualized transformations can be purely informational, or also, simultaneously, material. Control can take place inside the system, when one of its parts (the egg's nucleus) contains a program and its other parts actualize the designated transformations. We can also have intersystemic control—when, for example, two people communicate in speech or writing. Only now and again is it possible to decide whether we are faced with two feedback systems or just one (an important issue in itself but not of utmost significance to us at the moment). A particular utterance, for example, a book, controls the reader's brain processes. Yet, while control programs in the language of heredity are very detailed, utterances that are part of natural language are programs that are full of gaps. The fertilized egg does not adopt any particular strategy with regard to the chromosomes that control its transformations (although it can adopt such a strategy, holistically, toward the environment by opposing the disturbances emanating from it). The receiver can only choose strategies when the oncoming program does not standardize the imposed behavior, for example, when the said program is full of gaps. It then requires filling in, a process that is dictated by the size of the gaps and by the "interpretive potential" of the receiver, which is in turn determined by its systemic structure and preprogramming. Not being controlled in a deterministic way, the reader of novels is in a way forced to make multilevel strategic decisions (regarding what individual sentences, whole transmitted scenes, systems made up of scenes, etc., should refer to). Such a strategy can usually be reduced to information maximization and control optimization (we want to find out as much as possible and in the most coherent and holistic way). The reception of a text *as a program* that requires filling in within the range of acceptable interpretative oscillations is only one element in the hierarchically complex set of proceedings because we do not read to practice a relating or an ordering strategy but rather to learn something. An increase in the amount of information is the correct outcome of a reception that matters to us. As a general rule, transmission activates interpretive decisions and all the other control activities of a syntactic and semantic nature subliminally; that is, "the mental filling in of the fragmentary program" occurs in a way that is inaccessible to introspection. Consciousness only receives the final outcomes of the decision-making processes, already in the form of information—which supposedly transmits the text in an absolutely unmediated manner. It is only when a text is difficult that the hitherto automatic activities are partly "elevated" to the level of consciousness, which joins the process as a superior interpretive authority. It happens in different ways for different individuals because the "difficulty" of a text cannot be measured on a unified scale for all people. In any case, it is impossible to gain full knowledge of the multilevel brain activity through introspection; this inaccessibility is of one of the anathemas of theoretical linguistics. When transmission efficacy turns out to be quite good, that is, the text's constants are being transmitted even though the text as a program of "informational reconstruction" is full of gaps, this can be explained by the fact that the sender's and the receiver's brains are homomorphic systems characterized by a high degree of functional parallelism, especially if they were subject to analogous preprogramming from within the same culture.

The formalization of a linguistic utterance aims at a maximum reduction of interpretive freedom. A formal language does not allow for alternative interpretations—at least, this is how it should be within an ideal limit. In reality, it turns out that this limit does not equal zero, which is why some expressions that are unambiguous for a mathematician turn out to be not so unambiguous for a digital machine. A formal language actualizes—in a noncomprehending (or, at least, "not necessarily comprehending") way—purely informational causation because it is a language without gaps, as all of its elements and all the transformation rules need to be *explicitly* provided right from the start (the fact that there is no room for the receiver's "guesswork" is supposed to prevent embarking on any kind of interpretive strategies). Formal utterances are a step-by-step way of constructing structures that have internal relations but no external ones (i.e., they have no correspondence to the outside world). Nor are they subject to external verification tests: in pure mathematics, truth amounts to the possibility of undertaking a (noncontradictory) construction.

The language of heredity is causal both on an informational and material level. This is a special kind of language because "utterances" generated in it are "tested" after some time with regard to their "biological adequacy," through "natural fitness tests" for living organisms that take place in the organisms' innate ecological environment. The "utterances" of this language should thus meet criteria of "truth" in a pragmatic sense: the efficacy of "causation" is verified or falsified in action, whereby "truth" equals survival, while "falsehood" means "death." To those abstract logical extremes corresponds a really wide continuous spectrum of possible factorizations, since the "genetic sentences" that suffer from "interior contradictions, as they contain lethal genes," cannot even finish the initial embryogenetic phase of their causation, while other sentences are "debunked" only after a certain period of time—for example, the lifetime of one generation or more. Examining that language itself with its individual "sentences," without considering all the "adequacy criteria" entailed in the environment, does not allow us to determine whether and to what extent actualizable causation has been programmed into the egg's nucleus.

The language of causation does not contain any "mental," "emotional," or "volitional" concepts, nor does it contain any general terms.

Despite this, such a language can manifest considerable universalism when we also take into account the fact that, even though the language of chromosomes is entirely nonpersonal and "noncomprehending," since it is not the product of anyone's way of thinking, it does actualize—at the end of the chain of transformations that are controlled by it—the language of comprehending beings. Yet, first of all, the emergence of such a "derivative" comprehending language only takes place at the level of a whole set of human individuals (a single individual will not create a language); and second, the latter does not determine the emergence of a comprehending language but only probabilistically enables such an occurrence.

A pure comprehending language does not exist in reality, yet it could be produced artificially. To do this, it would be necessary to create some isolated systems that would be modifications of Leibniz's "monads," manifesting some determined states undergoing temporal changes. Certain abbreviated names would be assigned to those states. "Communication" involves the transmission of an internal state by one monad to the others. A monad understands another monad because it knows, "from its own inner experience," all the states that it may be informed about by its companions. There is a clear analogy here with a subjective language of introspection, in which one communicates one's emotional states, volitional states ("I want to feel cheerful"), and mental states ("I dream of happiness"). As we already know, "biological adequacy" with regard to the environment is the "X" on the basis of which "utterance" selection takes place in the language of chromosomes. What is this "X" in our monads? Selection takes place on the basis of the adequacy of names with regard to their internal states and nothing else; a pure comprehending language cannot thus serve any purpose in the causal sense we have talked about. No doubt it does not therefore exist in such a "purely spiritual" form. But it does exist in some elementary forms in animals, forms that—owing to the narrowness of their terminology and the lack of syntax—have no right to be called a "language." It is because it is biologically useful when one animal (e.g., a dog) knows about the "internal state" of another animal, while at the same time—since some particular forms of observable behavior correspond to such states—animals are able to communicate such internal states (fear, aggression) to one another in their own "behavioral code" (and do so via sensory channels within a wider range than ours, as a dog can smell fear, aggression, or sexual readiness in another dog).

An expanded, purely comprehending language, such as the language of our "monads," would also be able to produce logic and mathematics because various kinds of operations (addition, subtraction, exclusion, etc.) can be conducted on elementary internal states—provided they are not just experienced at a given moment but can also be recalled later. Please note that such "monads" would not be able to emerge in the natural course of evolution. Yet as soon as they have been built by someone, owing to the lack of any direct contact with the external world, the possibility of creating mathematics and logic would present itself. (We shall assume that these monads do not have senses and that they are only connected among themselves—e.g., by means of wires, along which travel the transmitted and received utterances of the "comprehending language.")

Natural human language is partly comprehending and partly causal. It is possible to use it to say "my leg hurts," so that to understand such a sentence, one must have experienced pain and must have a leg; it is also possible to say "the defeat hurts," because such a language is full of derivatives of internal states that can be projected onto the external world ("the arrival of spring," "gloomy landscape"). It is possible to create mathematics and logic within it and also to actualize various kinds of empirical causation. There is an interesting relation between the causal language of genes and natural language. The language of heredity can be represented, at least partly, if not entirely, in a comprehending human language. Every gene can be marked in a certain way, for example, by being numbered (natural language implies mathematics, together with set theory). Yet natural language cannot be explicitly represented in the language of chromosomes. As we have already said, the language of heredity does not contain any general terms or any names of mental states. If this was just unusual, it would not be worth mentioning. Yet there is something rather instructive about it. A particular chromosomal utterance led to the emergence of Lebesgue, Poincaré, or Abel. We know that mathematical talent is indicated by a chromosomal utterance—even though there are no "mathematical talent genes" that one could just number and isolate. Mathematical talent is preformed by an unknown structural and functional part of the whole genotype. We are unable to foresee to what degree it is to be found in the reproductive cell and to what extent it is "contained" in the social environment.

There is no doubt that the environment turns out to be an "expresser" of talent rather than its creator. Causal language, which does not have any general terms in its vocabulary, is therefore capable of actualizing states in which designates of such terms manifest themselves. A development then takes place from "the particular" to "the general," or from lower-level to higher-level complexity. It is not that the causal language of the genes is a tool that is not general enough, whose examination will not be of much use to the designer, because every "utterance" made in this "language" is "just" a self-actualizing production recipe for a particular copy of a given species—and nothing else. The language of heredity turns out to be surprisingly "excessive" in its universality. This language is a tool for constructing systems that can manage tasks that their very designer (i.e., this language) is not able to cope with because it lacks, for example, an appropriate semantic and syntactic apparatus.

We have thus demonstrated that the effectiveness of causality demonstrated by our language of heredity goes beyond the limits set by our formal and mathematical research. The development of an egg is neither a "tautological" process nor a "deduction" of consequences from the "set of axioms and transformation rules" contained in the cell's nucleus.

While our acts of formalization are always the highest level of enactment, because it is only through such operations that we are able to arrive at the permanent certainty of our findings, evolution's way is an exact opposite of ours. The chromosomal predictive "calculation" cannot afford the "luxury" of enactment because it does not unfold on paper that lasts but rather in real life. This is why all the states in which matter acts causally via information control must be taken into account by it. It is thus possible to say that an organism passes synthetic judgments a priori by means of its reproductive cells because a considerable majority of such judgments turn out to be true (at least on a pragmatic level, as we have already established).

The criteria of such "truth," or at least adequacy, are nevertheless variable, which explains the very possibility of transformism and the evolution of the species. What is most significant for us is that also within the realm of causal language—which is inextricably connected to its material carrier—there are no criteria to assess the "truth," or at least the "competence," of an utterance. Neither a comprehending person nor causal language can develop and act unless it is extralinguistically conditioned and oriented. The criteria of truth, correctness, and effectiveness of languages are to be found outside them—in the material realm of nature. In the absence of such criteria, monsters of nonsense can be produced as much by a comprehending language as by a causal one—the evidence for which is offered by the history of writing as well as the natural history of the species.

We have suggested dividing languages into causal and comprehending ones. In purely causal language, word literally becomes flesh. Such a language does not "explain" anything but only materializes the "content" of its utterances by programming given sequences of actions. It will be interesting to offer a comparison between these two types of languages because causal language starts from a molecular level and then transcends it to reach the macroscopic level of multicellular organisms, while natural language develops on a macroscopic level (of our bodies) and then transcends it "both ways," that is, toward atoms and galaxies. It is possible to detect "connotations" as well as "denotations" of the various "terms" used by them, or even "semantics"—if we are to agree that processes organized by natural language in the "brain environment" correspond to the meanings of that language, while processes that "justify," within the realm of the natural environment, the presence of particular genetically determined traits in an organism correspond to the "meanings" of the language of heredity. In this sense, however, the semantics of causal language is clearly finite because there are no further test "appeals" with regard to the specific function of genetically determined traits outside the ecological environment (a given section of the genotype determines the development of limbs, which "signify" locomotion, another section determines the development of eyes, which "signify" vision, and nothing else), while the "brain environment" of natural language is only an "intermediary test station" because some other such stations exist in the domains of Nature and Culture—which are external to brains.

The problem of general terms, which causal language lacks, is a separate issue. This is why causal language is finite, as demonstrated earlier, while natural language is polyterminous (continuous) thanks to the presence of such terms. To simplify quite dramatically, we could say that the need to produce general terms is generated by the probabilistic nature of the real world, in its most common, stochastic "version." We are referring here to the inseparability of two simultaneous aspects of phenomena: the fact that they can be similar to each other and the fact that they can also differ from each other in some other respects. Every table is in a sense "unique," but it is also an element of the class of "tables." General terms "fix" similarities and minimize differences. Where things and phenomena are unique, general terms would not need to exist—and would not be able to. In reality, there is no order as high as that which is postulated by language: as a description of phenomena, language is usually more ordered than them. Stochastic processes, which are very common in the terrestrial environment, involve the coexistence of random traits ("components of chaos") and ordered traits ("components of order") in the same phenomena (the behavior of humans, animals, society, machines, and complex but nonlinear systems; meteorological and climate changes, etc.). Deterministic frameworks precede nondeterministic ones in the history of science because we are directed, almost automatically and thus immediately, toward the former by language itself. Language ignores random aspects of the stochastics of phenomena, or rather, it squeezes such aspects, in a nonanalytic way, into the corset of general terms. Causal language achieves similar results, in the functional sense, thanks to initiating developmental gradients created by the molecular chains of "conservative" and hence "theological" reactions. This is very important for the designer because it turns out that similar results with regard to invariance can be obtained by means of various techniques. On a molecular level, the "chaotic" overlaps to a shocking degree with the "thermal," that is, with the heat movements of molecules. Conservative reactions, and, thanks to them, developmental gradients, enhance the element of regularity that is present in stochastic processes, which is why the final result is similar—but only on a functional level—to using general terms: in both cases, the order of phenomena is applied, while their randomness is "silenced."

The nature of languages as quantifiable control sequences cannot be grasped in full without taking into account the physical aspects of systems that have produced them. Life is a thermodynamically improbable state in a determined way; that is, it is separated—to a considerable, or, some claim, maximum, degree—from the state of permanent equilibrium. How can a system that is constantly bleeding order as a result of this, and that is destined, from the very beginning, to experience continuous regulation losses, not only gain stability but also reach higher levels of organization, for example, in embryogenesis? It can do this thanks to having circular processes "fitted into it" at every level, starting from the molecular one, processes that—with the rhythm of a bouncing ball being permanently prevented from falling down—must first of all be organized *in time*. There is no mathematical apparatus for a thermodynamic theory of life. When, owing to this lack, Goodwin used the classical apparatus of statistical mathematics in his formal theory of living organisms, the first approximation already showed the leading role that the temporal organization of the cell as a coupled harmonization of molecular vibrations—which biological phenomena are—played in the life process. The cell is a synchronized system of oscillators; this design solution, which emphasizes the periodic regularity of its cycles by constantly regularizing them, is dictated by the input material. Analysis shows us that such a process cannot be stabilized in a nonvibration way: physics, or the properties of the building material, does not allow it. Embryogenesis, metabolism, and morphogenesis are resultant of the collaboration between *temporarily* focused, that is, synchronically attuned, molecular oscillators. This leads to exponential convergence in the stage of fetal development; to pseudo-stability in the mature stage; and then to the loss of balance in the concurrent periodic regularity—which causes old age and death. And thus the phenomenon of oscillation, which is caused within the system by feedback overcorrection, and which is always perceived as negative by cyberneticists, represents, as we can see, the essential dynamic structure of the living process itself. This explains the temporal organization of linguistic utterances as control and regulation programs, that is, as concentrated carriers of order that need to be injected into the system that is constantly losing them. In this light, the "chance" of natural languages is already assumed in the fundamentals of the living processes, just as the elementary reactions of protozoa are the "premise" behind the development of brains. In this sense—provided the evolutionary process takes long enough—they will both be inevitably actualized as rare yet permanent probabilities of a stochastic chain.

Although this may sound like a paradox, we do not actually know that much about the so-called explanatory power of natural language. We shall refrain from a risky debate about "the essence of explanation" and just limit ourselves to the following remarks. Our species departed from the anthropoid stem less than three quarters of a million years ago, under the influence of natural selection—which favored a particular group of systemic parameters, whereby the selection criteria did not cover the ability to construct quantum theories or space rockets. Despite this, the "excess" of the transformation of information in human brains that emerged in the course of this process of selection turned out to be adequate to obtain results that were very remote from Pliopithecus's horizon. At the same time, it would be indeed unusual if it turned out that the set of orders that our mind is able to construct and accept, having as it does a deep sense of "understanding the essence of things," matches precisely the set of all possible orders to be detected in the Universe as a whole. We should admit that this is not impossible, yet it does seem highly improbable. This way of thinking, so modest in its assessment of our abilities, is probably the only way recommended, given our lack of knowledge, because we are not aware of our limitations—which is why it seems more prudent to accept the possibility of their existence rather than be excessively optimistic. Indeed, optimism can blind us, while presupposing some limitations—and, through this, beginning to search for them—may allow us to counter such limitations eventually. This is why we are predicting a still remote state when it will be easier to control phenomena than to understand all the conditions that enable such control—owing to the immense size of the set of variables and parameters that are involved in very ambitious undertakings. The prospect of a definitive bifurcation of the causal and comprehending aspects of natural language—of which human speech functions as a unique amalgamate—seems real to us. If we manage to crossbreed algorithms produced by finite machines with nonalgorithmic streams of information emerging from phenomena supervised by self-organizing gradients, causal language will stop being comprehending, while science will be producing predictions stripped of "explanations" rather than explanations themselves. Comprehending language will continue to function as a language of information campaigns run by gnostic apparatuses, a recipient of the fruits of victory, a transmitter of experiences that cannot be exchanged for anything else and also—which is probably equally important—a generator of axiological standpoints. We should not even think about some kind of sudden revolution, as a result of which such machines will dominate over us on a mental level. The development of causal generators will accelerate the changes we are already experiencing today because the symbiosis, or synergy, of scientists and information machines is taking place more and more frequently. We have to admit that the extent to which control will remain in human hands partly depends on a point of view. The fact that man is able to swim by himself does not mean that he is capable of crossing an ocean without a ship—not to mention jets and space rockets in this context! A similar evolution is starting, in a kind of parallel way, in the information universe. Man is capable of directing a gnostic machine toward a problem that he could perhaps solve by himself (either he himself or his great grandchildren), but in the process, the machine may open his eyes to a problem whose existence he did not even suspect. Who actually has the lead in the latter example? It is difficult to imagine, even vaguely, both the degree of functional uniformity that the "cognitive man–machine tandem" can represent and all those degrees of freedom that the human brain working in such conditions is enriched by. We should emphasize that this only refers to the early stages of the linguistic "dichotomy." It is difficult to say anything specific today about its future stages.

But what will a philosopher say about this approach? The designer will display in vain the fruits of his labor to the philosopher, pointing out the extent to which answers to classical philosophical questions depend on the technologically constructed limit conditions (whether *nihil est in intellectu quod non prius fuerit in sensu*[9] also depends on the specific characteristics of chromosomal preprogramming in the brain so that a certain excess of preprogramming can make "a priori synthetic knowledge" available to it). For the philosopher, the success of the designer of gnostic machines will represent the failure of thinking as such, including practical thinking, since in this way it will have ousted itself even from the creation of instrumental truths. The designer, whose great grandfather despaired over sailboats yet built steamers at the same time, shares the philosopher's concern—but not his reservations.

The philosopher will remain deaf to the designer's arguments because, feeling contempt toward spontaneous, nonhuman, externalized thinking, he wants to think through everything that exists himself—by creating the right kind of system or a meaningful structure. Yet what is the relation between systems that differ from one another to a greater or lesser degree? Each one can arbitrarily decide that it occupies a singularly valid, unique "meta"-position with regard to the others. We thus find ourselves amid a series of circular processes. Even though such logic is fascinating, it serves us no good if every position turns out to be possible to defend, as long as there are no internal contradictions within it. A thought that wants to reach the paradise of certainty locates such a paradise in various places, while the map of such localizations, that is, the history of philosophy, is a language-based search for what is located outside it, if anywhere at all. There is also a position according to which metaphysical systems are hyponoic creations, produced in the subconscious and then dressed by consciousness in language. Here metaphysics turns out to be the most logical and the most persistent of our dreams. The philosopher will debunk this point of view—embarrassing as it is from a psychoanalytic standpoint, which turns the philosopher's labor into a dream—by strongly debasing psychoanalysis. This is great fun, somewhat futile yet beneficial in terms of developing some new cultural frameworks;

it consists of shifts within the system, in which an arbitrary change of position involves the transformation of both evaluative and cognitive perspectives. The mode of thinking that does not penetrate us to become a causal (or evaluative) action turns out to be rather powerless. That which is eternal in classical metaphysical systems draws its nourishment from the roots located in the semantic looseness of language—which, in its diachronic universality that binds together various cultures, dispenses the blurriness that facilitates adaptability and stretchiness, while giving the impression of being a solid foundation. It is because not only can one draw established meanings from it but one can also insert some new ones into it, by means of acts that are supposed to be discoveries yet that are really just arbitrary—with the kind of arbitrariness that is more menacing the more unintentional it is. This kind of philosophizing is thus a way of reinforcing bottomlessness because the "bottoms" that are being strengthened in the linguistic material are not needed: each one can be penetrated through to "move on"; each one can be questioned. Where does the fierceness of this way of proceeding come from? It is not necessary to evoke any new beings to explain it: just like with every act that neither initiates nor supports life, this is pure decadence. Yet it is a more noble kind of decadence than some others, one in which language plays the role of a partner who is ready for anything, logic plays the role of Kama Sutra, and the absolute plays the role of sexual pleasure, ever more sophisticated because it is always infertile. No doubt we cannot live without philosophy, and it is not true that we should *primum edere, deinde philosophari,*[10] because the apparently simple activity of eating already implies a whole bunch of directions, from empiricism through to pragmatism. There is a difference between the philosophical minimum, which every system makes internally coherent and then directs outward, and the distillation and maceration of language, so that it evaporates until we are left just with the certainty that can satisfy the passion for learning. It is impossible to barricade oneself in language consistently and absolutely. There are two ways out of it: one into the world of real actions, the other into the world of beings, which are supposedly not produced by language but only detected by it. Phenomenologists have warned us that giving up on the sovereignty of the world of logical and mathematical truths will push man into animalistic randomness, yet we do not have to choose between parts of the supposedly indestructible alternative offered by them—"man as a random being" versus "man as fundamentally a being of reason"—because both occur at the same time. As we already know, it is impossible to construct without language, even if one is a nonhuman constructor. Therefore, if rules, which are a molecular transposition of syntax and logic, have even amino acids and nucleotides subjugated to them in their embryogenetic discourses, does it not seem highly probable that, as an adaptive terrestrial phenomenon and thus a random one, language is at the same time a universal phenomenon, even on a cosmic scale? The reason for saying this is that the similarity of planetary environments leads to the antientropy systems that emerge in them having to produce approximate representations of those environments so that language, logic, and mathematics turn out to be far-flung derivatives of Nature itself: otherwise, it would not be possible to counter Nature's fluctuations, which destroy any kind of order. The causal language of bioevolution creates, as its first derivative, the language of the nervous system, the neural codes of singular control (in the "organism–organism" and "organism–environment" relations), and, as its second derivative, natural language, thanks to the symbolized exteriorization of the neural codes, which uses any number of specially adapted sensory canals (vocal, visual, or tactical speech). Logic thus turns out to be relative not to the *Homo sapiens* species but rather to the material Universe, where a whole set of functionally, if not structurally, similar logics—which function well because they have been selected in the process of evolution—can exist. Such comparative linguistics, when extended across the Universe, leads to the conclusion that all languages—chromosomal, neural, and natural—are mediated, that is, "ontologically nonautonomous," because they are systems for constructing structures by means of the selection and organization of elements, structures that can only be acknowledged as real or debunked as false by the real world. It is only the *degree* of mediation, or falsifying generality, that is different. As a result, the range of utterances is the widest in natural language because the criteria of correct functioning are much stronger in the language of embryogenesis or neuroregulation than in natural speech. In other words, all languages are "empirically subversive," but, alongside the criteria of empirical and logical "survival," natural language also has cultural criteria—which is why biological monsters are not capable of survival, unlike the monsters of linguistic nonsense or cultural delusion. Disability or death are a punishment for poor design in the language of biology. Metaphysicians do not have to pay such a high price for an analogous sin because the environment of human minds is incomparably more liberal toward meanings that are vegetating (or serving penance) in it than the natural environment is toward living organisms.

We can only address such an inevitably general program to future generations. The scale presented at the beginning of these deliberations needs to be enclosed in a circular shape. The process of language formation initiates the emergence of hereditary information. The language of its causality—on its first, apsychic level—is a consequence of the accumulation of knowledge gained in the process of "penetrating," through trial and error, the area between physics (including quantum physics) and the chemistry of polymers and colloids in a given set of elements, within a narrow range of temperatures and energies. After several billion years, it leads to the emergence, on the level of social sets, of natural language— which is partly comprehending and partly causal. To overcome the formal limitations to which it is subject in aiming to achieve the precision that every designer needs, this language should in turn develop, with the help of tools that have become autonomized on an informational level through being located in extracerebral material systems, a causal language "of the next kind"—which, almost by accident, will cross the threshold of "understanding" or "comprehensibility." This will be the price that may be able to allow it to reach the level of creative universality. That level will be placed higher than the first, chromosomal level—which started all the information changes. This new language will be semantically and syntactically richer than both its predecessors, just like natural language is richer than the language of heredity. This whole evolution is an informational aspect of a process in which complex systems emerge from simple ones. We do not know anything about the systemic laws of such processes because, when faced with phenomena characterized by an antientropic growth gradient, physics and thermodynamics have so far maintained a kind of "reluctant neutrality." And because it would be imprudent to say anything else about an area so obscure, we shall have to fall silent now.

THE ENGINEERING OF TRANSCENDENCE

We mentioned earlier that alongside "information farming," there is also another possibility of taming the information deluge. We shall reveal it now. We are going to do this by using a unique, even ontological, example. In this way, we shall present to the reader the very core of future possibilities. This does not mean that we actually consider the plan we are going to reveal worth actualizing. Yet it is worth presenting, even if just to show the wide range of possible pantocreatic activity.

It is often said that the separation of the present reality from transcendence, so commonplace today, is pernicious in that it undermines the universe of fixed values. Because life on Earth is the only thing that exists, because it is only in this life that we can seek fulfillment, the only kind of happiness that can be offered to us is purely carnal. Heavens have not revealed anything to us; there are no signs that would indicate the need to devote ourselves to some higher, nonmaterial goals. We furnish our lives ever more comfortably; we build ever more beautiful buildings; we invent ever more ephemeral trends, dances, one-season stars; we enjoy ourselves. Entertainment derived from a nineteenth-century funfair is today becoming an industry underpinned by an ever more perfect technology. We are celebrating a cult of machines—which are replacing us at work, in the kitchen, in the field—as if we were pursuing the idealized ambience of the royal court (with its bustling yet idle courtiers) and wished to extend it across the whole world. In fifty years, or at most a hundred, four to five billion people will become such courtiers. At the same time, a feeling of emptiness, superficiality, and sham sets in, one that is particularly dominant in civilizations that have left the majority of primitive troubles, such as hunger and poverty, behind them. Surrounded by underwater-lit swimming pools and chrome and plastic surfaces, we are suddenly struck by the thought that the last remaining beggar, having accepted his fate willingly, thus turning it into an ascetic act, was incomparably richer than man is today, with his mind fed TV nonsense and his stomach feasting on delicatessen from exotic lands. The beggar believed in eternal happiness, the arrival of which he awaited during his short-term dwelling in this vale of tears, looking as he did into the vast transcendence ahead of him. Free time is now becoming a space that needs to be filled in, but it is actually a vacuum, because dreams can be divided into those that can be realized immediately—which is when they stop being dreams—and those that cannot be realized by any means. Our own body, with its youth, is the last remaining god on the ever-emptying altars; no one else needs to be obeyed and served. Unless something changes, our numerous Western intellectuals say, man is going to drown in the hedonism of consumption. If only it was accompanied by some deep pleasure! Yet there is none: submerged into this slavish comfort, man is more and more bored and empty. Through inertia, the obsession with the accumulation of money and shiny objects is still with us, yet even those wonders of civilization turn out to be of no use. Nothing shows him what to do, what to aim for, what to dream about, what hope to have. What is man left with then? The fear of old age and illness and the pills that restore mental balance—which he is losing, in being irrevocably separated from transcendence.

Irrevocably? Yet transcendence could certainly be created. No, not in a metaphorical sense, to practice some beliefs the way one exercises in the morning for health reasons. Belief must be true: let us thus create some unshakeable foundations for it. Let us construct immortality and eternal justice, which will start dispensing punishments and awards. Where are we supposed to construct all this? In the next world, of course...

I am not joking. It is possible to construct "the next world." How? With the help of cybernetics...

Imagine a highly complex system that is larger than a planet. We only program it in a sketchy manner. Thanks to ongoing evolution, it should be able to develop landscapes and seas more beautiful than ours, as well as intelligent organisms. They should have an environment at their disposal—within that system, of course. We spoke earlier about the origins of this process: machinic processes were then divided into two parts—one constituted by organisms, the other by their environment.

The new machine is immense. It also possesses an additional third part: the Next World. When an individual, an intelligent being, dies, when his existence on Earth comes to an end, when his body decays, his personality gets inside that third part via a special passage. Justice reigns there; there is award and punishment; there is Paradise, and, somewhere, there is also a mysterious unfathomable Creator of Everything. It can also go another way. This third part need not have exact equivalents in any earthly religion. In any case, the possibilities are limitless. We want to connect with "our dear departed ones" Out There? No problem. Enlighten the spirit in the domain of eternal existence and expand individual capacity for cognition and experience? Nothing could be simpler than that: a personality on its way to the "Next World" will develop relevant "emotional and intellectual subsystems." Perhaps we prefer Nirvana? An after-death union of all personalities into one mindful Being? This can also be achieved. We can construct many worlds like this. It is possible to construct various types of them and then examine in which one the "sum of happiness" will be the greatest. The values of the "felicitological index" will serve as our design guide. It is possible to create for any created beings any cybernetic paradises, purgatories, and hells that will await them. Appropriate "selectors," performing to some extent St. Peter's role, will suitably direct the condemned and the blessed at the entrance to that "next world." It is also possible to construct the Last Judgment. In one word, Everything.

Fine, we shall say, let us accept that this crazy experiment can indeed be conducted, but what good is it to us? And why should we do it in the first place?

But this is only a preliminary phase...

Imagine that a generation of intelligent beings who are similar to us will be able to construct such a machine in a thousand or a hundred thousand years. In any case, I only keep saying "machine" because we do not have any other word for it. What would a skyscraper be for a caveman? An underground cave? A mountain? Imagine an artificial park. All the trees are real, but they have been transplanted from afar. Or think of an artificial sea or a satellite. They are made of metal. Yet if something is built from the same material as the Moon, and if it is the same size as the latter, then how are we going to recognize its "artificiality"? When we say "artificial," we all too frequently mean "imperfect." But this is how things are just now. Instead of saying "machine," perhaps it is better to call it "the created." This will be a total world, with its own laws, and it will be indistinguishable from the "real" one owing to the skill of its Designers. Anyway, when it comes to the technical aspect of Creation, I would like to refer the reader to the next section ("Cosmogonic Engineering").

The creators of that world will thus say to themselves, The creatures that live there do not know anything about us, about our fragile corporeality—which comes to an end so rapidly and irreversibly—and they are so much happier than us! They believe in transcendence, and this belief is fully justified. They believe in afterlife—and rightly so! They also believe in the Next World, in Reward and Punishment, in Infinite Mercy and Loving Omnipotence. While, after their death, both they and the nonbelievers find out that this was all indeed real... unfortunately, our children will not have a chance to live in such a world. Although... wait a second! Could we not just simply transfer them there? Because who are children really? They are creatures similar to us—in their looks, minds, and emotions. How do they come into being? We "program" them in a way given to us by Evolution, by means of sexual intercourse. This is probabilistic programming, which is in line with the rules of Mendelian inheritance and population genetics. We already know our human genotype really well. Instead of conceiving children the way we have done so far, we shall transfer the very same traits that are in us as potential mothers and fathers, already recorded in the cells of our ovaries and testes, to the very heart of the "the created," which we shall specially design for this purpose. It will be a Promised Land, while our act will serve as a great Exodus into it. In this way, humanity in its future generations will conquer the Next World, Transcendence, and everything it has dreamed of for centuries...and all this will be true rather than imagined; it will be a reality awaiting us after death rather than a myth which is to compensate to some extent for our biological frailty!

Is this really impossible? I think that it *is* possible, at least in principle. What we are calling "the created," that world with its eternal level of transcendence, from now on will become a dwelling place for a happy humankind...

Yet this is deception, we say. How can you make someone happy by means of deception? The Designers are amused by this accusation. Why is this "deception"? Is it because that world has different laws than ours—that is, compared with ours, it is enriched by this whole actualized transcendence?

No, we reply, it is deception because this world is not true. You created it. Yes, we did. Yet who created your "real" world? If it did indeed have a creator, does this not mean it is a similar kind of "deception"? No? Where is the difference then? We both created a civilization; does this mean it is also a deception? Finally, as biological organisms, we are a product of the natural process, which shaped us through millions of games of chance. Why is it a bad thing that we want to take this process into our own hands?

But this is not the point, we say. Those creatures are going to be enclosed, trapped in that world of yours, in that crystal palace of absolute fulfillment—which cannot be found outside it.

Yet this is a contradiction, they reply. Indeed, we did build "total fulfillment" into that world, which is why it is richer—rather than poorer—than the "natural" one. It does not pretend; it does not imitate anything else: it is itself. Life and death are the same there as they are in our world, except that they are not the end..."Enclosed...?" What do you know about its size? Perhaps it is as big as the Metagalaxy? Do you consider yourselves to be trapped in it, prisoners of the stars that surround you?

But this world is not true! we cry out loud.

What is truth? they reply. That which can be verified. And it is possible to verify more things there than here because here everything ends at the limits of empiricism and falls apart together with it, while there even belief comes true!

Fine, we reply, we have one last question. In its mortality, that world is equivalent to ours, is it not? Yes, it is. So, at the end of the day, there is no difference between them! In your world, it is equally possible to doubt, and to develop a conviction about the pointlessness of all Creation, as it is in this ordinary one. The fact that doubt disperses after death cannot in any way affect mortality itself. And thus a similar hedonistic, consumption-driven and disoriented civilization can emerge in your new wonderful world that developed in the old one ... so why are you constructing this world? Only to create the possibility of some "pleasant after-death disappointment ... ?" Because surely you must understand that any eternal mysteries will be occurring in the third, "transcendent" part of your world—which is not in the least going to affect its everyday course. For things to be different, that world of yours, with its mortality, would have to bear clear signs and indications of the existence of its metaphysical extension. And thus, in its mortality, it cannot be identical with ours.

This is true, the Designers reply.

Yet our world can also have a "metaphysical extension"; it is just that our present civilization does not believe in its existence! we scream. Do you realize what you have done? You have replicated, atom by atom, what already exists! So now, in trying to avoid such pointless plagiarism, not only must you add "the next world" to your design, but first of all, you must change its material basis—its mortality! You thus have to introduce miracles into it, that is, to change the laws of nature, which means physics, which means everything!

Yes, indeed, the Designers reply. Belief without after-death fulfillment means incomparably more to mortals than fulfillment or transcendence not preceded by belief ... this is a highly interesting problem. It is real, that is, soluble, only for an observer who is positioned outside that world—or rather outside both worlds, the real and the transcendental. Only such an outside observer will be able to know whether belief is justified. When it comes to your suggestion that we introduce miracles into this "new world," we have to reject it. Does this surprise you? Miracles are not a confirmation of belief. They are a transformation of the latter into knowledge because knowledge is based on observable facts—which the "miracles" would then become. Scientists would make them part of physics, chemistry, or cosmogony. Even if we were to introduce prophets who are able to move mountains, this would not change anything. There is a difference between receiving messages about such acts and events in holy scripts, in the glory of legends, and experiencing them directly. It is possible to create a world that either has the knowledge about transcendence outside of it or has a capacity for *believing* in transcendence—whereby this transcendence either exists or not, yet we cannot find out and prove it one way or another. It is because to prove belief is to destroy it as it only exists in its full absurdity and groundlessness, in its rebellion against empiricism, in its pious hope shaken by spells of doubt, in anxious expectation and unfulfilled certainty, guaranteed by "visual aids" such as miracles. Put briefly, a world in possession of the current knowledge about transcendence and what it looks like is a world without belief.

This is where the dialogue comes to an end. The conclusion is that the source of Great Anxiety, and of the equally dangerous mindlessness, lies not in the "amputation" of transcendence by materialism in man but rather in the present social dynamics. It is not a renaissance of transcendence that is needed but rather a renaissance of society.

COSMOGONIC ENGINEERING

We have demonstrated earlier the fruitlessness of a pantocreatic undertaking, whose aim would be to fulfill the dream of eternal existence in the Next World. However, we have to remember that this fruitlessness does not refer to the technical aspect of the plan but rather to the fact that the presence of "transcendence"—which cannot be empirically verified—has exactly the same effect on the future of the inhabitants of this world as its posthumous absence. It therefore does not really matter whether "the other side" exists if we cannot find out about it right here. And if we can, then transcendence stops being what it is and turns into a continuation of existence that eradicates any belief.

Pantocreatics focused on creating "here-and-now" worlds makes more sense to me and is definitely worth pursuing. People who devote themselves to such tasks are called Cosmogonic Engineers. The word *cosmogonic* derives from the term *cosmogony*. There is a similarity here with *electronics*—since both stand for designer activity.[11] The cosmogony expert examines the creation of worlds; the Cosmogonic Technologist creates worlds. But we are talking about creation proper, not just replicating Nature in one way or another.

When starting to construct a world, the Cosmogonist needs to decide first of all what this world is to be like: strictly deterministic or indeterministic, finite or infinite, surrounded by certain prohibitions—that is (since this is one and the same thing), manifesting regular behaviors that can be called its laws or allowing those laws to undergo some changes. Unfettered variability would (unfortunately) mean chaos, as we said earlier; it would mean an absence of causal links, an absence of connections, and thus freedom from any kind of regulation. Incidentally, chaos is one of the things, or rather states, which is most difficult to create because the building material (which we of course take from Nature) is characterized by order. The remnants of this order are likely to penetrate into the foundations of what is being constructed. Anyone can prove this, even by means of a very simple experiment such as programming a digital machine so that it provides us with a long sequence of numbers that is entirely random, that is, chaotic. It will be more random than a sequence that a person could draw "from his head" because the regularity of his mental processes does not allow for any kind of "empty," that is, completely random, action. Yet even the most perfect machine, if ordered to behave randomly by us, would not do so infallibly. Otherwise, authors of random number tables would not have all those problems they have been experiencing.[12]

Our Designer starts work by placing some constraints on diversity. His creation needs to have spatial and temporal dimensions. He could in fact give up on time, yet this would limit him unnecessarily: where there is no time, nothing happens. (Strictly speaking, we should put it the other way around: where nothing happens, there is no passage of time.) It is because time is not a quantity that is introduced into the system (the world) from outside, but rather its immanent characteristic—which results from the nature of the ongoing changes. It is possible to create several times, running in different directions—some reversible, some not. Of course, from the point of view of the observer—who is external to this world—there is only one time within the latter. The reason for this is that the observer is measuring that time with his own clock, and also because he has placed all those various other times in the only time he has—the one given to him by Nature. Our Cosmogonist is unable to go beyond Nature; he is making his construction within it and from the materials provided by it. However, because Nature has a hierarchical design, he can locate his activity on some of its levels. His systems can be open or closed. If they are open, that is, if it is possible to observe

Nature from them, their subordinate status with regard to the Big Thing within which the construction project is located becomes apparent. This is why he is most likely to devote himself to constructing closed systems.

Before we say a few words about the aims of such a construction project, we should ask about its stability. Yet the notion of stability is relative. Nature's atoms are relatively stable—relatively, because a great majority of elements tend to decay after a longer or shorter period of time. There are no transuranium elements on Earth anymore (although it is possible to synthesize them) because our planetary system has existed long enough for those unstable transuraniums to have decayed. Stars are also unstable: none of them can exist longer than a dozen or so billion years. The Cosmogonist Engineer has at his disposal cosmogonic knowledge that is incomparably more extensive than ours. He therefore knows—either exactly or approximately—how things were, how they are, and how they are going to be, that is, whether the Universe is an entity that pulsates as a finite yet limitless body, and that changes every twenty billion years or so from "blue" contractions (when the light of the centripetal converging galaxies turns blue) to "red" ones (when the light, with its waves "stretched" as a result of the Doppler effect, moves in the opposite direction in spectrograms), or whether it behaves differently altogether. In any case, I think the duration of one phase, that is, those twenty billion years, is, practically speaking, the temporal limit of his design calculations, because even if during that time the "blue compression"—which would annihilate both life and everything it has created with its enormous temperature—was not to take place, the atoms that would be his building material (the way bricks are for us) would not actually cope with such "duration." Pantocreatics does not thus create eternity because this is impossible. Fortunately, it is also unnecessary. We have nothing in common with a rather peculiar being who would want to exist for billions of years on an individual level and who would realize what such an existence means. (No human being is ever going to be able to imagine that.)

We spoke about stability, starting from atoms. Then we moved, quickly and prematurely, to the Universe as a whole. Atoms are stable. Stars and planets are less stable than atoms. The duration of geological periods is even shorter. The life of mountains is really limited because it is only calculated in tens of millions of years. Mountains fall apart within that time span, washed away by rain and stream water, and end up as a relatively smooth layer covering the continents and ocean bottoms.

Oceans and continents themselves constantly change shape—and they do it relatively quickly (according to our timescale), that is, only over the period of several million years. In that case, if the Cosmogonist designs his project to last approximately the same amount of time that evolution took, that is, some three or four billion years, we must conclude that even if it is not a very modest undertaking, neither is it particularly bold. What would be truly bold would be his attempt not to build from Nature and within its limits but rather to *direct* Nature, that is, to take evolution into his own hands. Yet it would not just be a biological or homeostatic evolution but rather the evolution of the whole Universe. Such a plan to become a steersman of the Great Cosmogony, and not just a creator of the lesser one we are debating here, would indeed be an astounding act of boldness. We are not going to discuss it any further. Is it because it is completely and absolutely impossible, now or in the future?

Maybe it is, but it is certainly also interesting. Despite everything, questions arise with regard to where the energy that will be capable of pointing the ongoing changes in the desired direction may come from; what kind of feedback should be embedded into the process; how can Nature be made to control Nature so that it shapes and leads itself—through regulatory and not energetic intervention—where the real, or rather ultimate, engineers of the Universe will want it to go? Yet we shall not be discussing any of this. We shall return to our lesser worlds, constructed from natural ingredients—not against but rather within Nature.

Our Cosmogonist (who no doubt feels closer to us, now that we have understood, thanks to the preceding digression, that he is not completely independent and does not thus have any imaginary control over the course of Everything) can actualize worlds from different philosophies. We have already discussed what would happen if he were to create a "binary" world that would include transcendence. Yet he can also construct the world of Leibniz's philosophy, with its preestablished harmony. Please note that a person constructing such a world can introduce the infinite speed of signal distribution into it because in that Universe, everything that happens has been preprogrammed in advance. We could explain the mechanism of this phenomenon in more detail, but it is probably not worth it.

Imagine that our Designer now wants to turn his world into a habitat for intelligent beings. What would present the greatest difficulty here? Preventing them from dying out right away? No, this condition is taken for granted. His main difficulty lies in ensuring that the creatures for whom the Universe will serve as a habitat do not find out about its "artificiality." One is right to be concerned that the very suspicion that there may be something else beyond "everything" would immediately encourage them to seek exit from this "everything." Considering themselves prisoners of the latter, they would storm their surroundings, looking for a way out—out of pure curiosity, if nothing else. *Just* preventing them from finding the exit would amount to offering them knowledge about their imprisonment, while simultaneously taking away the keys. We must not therefore cover up or barricade the exit. We must make its existence impossible to guess. Otherwise, the inhabitants will start feeling like prisoners, even if that "prison" was actually to be the size of the whole Galaxy.

Infinity is the only solution. It would be best if some universal force could enclose their world so that it looked like a sphere—which would enable them to travel its length and breadth without ever reaching "the end." Some other technical applications of infinity are also possible: if we make sure that the force is not universal but rather peripheral, as a result of which approaching "the end of the world" will result in the shrinking of all the material objects without exception, it will be impossible to reach that end, just as it is impossible to reach absolute zero in mathematics. Each new step will require even more energy, while also itself being smaller than the previous one. In our world, this happens in various "places," for example, when a body accelerates to the speed of light. The amount of energy grows to infinity, while the material object to which this energy has been applied is not going to reach light velocity anyway. This kind of applied infinity is an actualization of a decreasing sequence whose limit is zero. But perhaps we have had enough of such cosmotechnological deliberations. Do we really believe they can be actualized? Even if no one is going to embark on this task, it will be out of choice rather than impotence.

If this is the case, then we should demonstrate, by means of an example, how many possible things that could be constructed with the right amount of will and support are actually not being constructed—or, more broadly, achieved.

Let us imagine (but only for illustrative purposes; otherwise, we are not going to be able to visualize anything) that we have a large, complex system the size of ten Moons that is a homeostatic pyramid consisting of closed, descending feedback systems. It resembles a digital machine that is self-repairing, autonomous, and self-organizing. Among the hundred trillion of its elements are "planets," suns around which those planets revolve, and so on. Whole swarms, whole countless floods of impulses, are constantly traveling inside this enormous body (which is likely to be connected to a star cluster as its energy source), in the form of starlight rays, motions of atmospheric surfaces of planets, organisms of native animals, ocean waves, waterfalls, forest leaves, colors and shapes, smells and flavors. All the inhabitants of the "machine" who are its parts are experiencing all this. These parts are not mechanical but rather constitute the machine's processes. The processes manifest a special kind of coherence and gravity and special kinds of relations: they are capable of producing an intelligent personality and sentient senses. They therefore experience their world in the same way we experience ours, because ultimately what we perceive as smells, fragrances, or shapes is nothing else than a hustle and bustle of bioelectric impulses in brain folds, in a location supervised by the global receiver, that is, consciousness.

The Cosmogonist's undertaking is considerably different from the phantomological phenomena described earlier because phantomatics is a way of deceiving the natural brain by introducing impulses into it that are identical with the impulses that would be entering it if the person were really located in Nature's material surroundings. The Cosmogonist's world, in turn, is a domain to which *Homo naturalis,* a corporeal being such us ourselves, has no access—just as a sun ray has no access inside the electric processes thanks to which a digital machine examines optical phenomena. We do in fact know a similar kind of "local noninvasion-ism" from our own world because it is impossible to enter someone else's dream or reality, that is, his consciousness, to participate directly in its experiences.

Unlike in a phantomatic situation, in cosmogonics, both the world and its inhabitants are "artificial" (if this is the word we want to use for what is created). None of the inhabitants know about it—or *can* know about it—though. They experience exactly what a person in real life or in the phantomat does (because we already know that those two experiences cannot be told apart by the person undergoing them). Just as we cannot get out of our skin or see someone else's consciousness, the inhabitants of that cosmic creation cannot find out about its hierarchical nature in any way, that is, about the fact that it is a world contained in another world (namely, ours).

Equally, they are unable to establish whether someone created them and the cosmic habitat they freely roam, and, if so, who it was. Yet even though we have not been created by anyone (or at least by any personified being), there are plenty of philosophies that claim the contrary, declaring that our world is not all there is. But people who have pronounced such things have the same senses and brains that we do—sometimes rather sharp brains! We can therefore conclude that also in that other world, there will emerge various philosophers who will declare similar theses—except that they will be right. However, because there will be no way of confirming that they are right, the empiricists of that world will shout them down for being metaphysicians and spiritualists. It is also possible that a physicist from that world who will be working on matter will announce to his fellow men, "Listen! I've discovered that we are all built from electric impulses, which are in constant motion!" And he will be right, because this is precisely how those creatures and their world had been constructed by the Cosmogonist Engineer. Yet this discovery is not going to change the general conviction that existence is material and real. Again, rightly so—because they are constructed from matter and energy, just like we are constructed from vacuum and electrons, but this does not make us doubt our materiality.

But there is a difference on the design level, namely, both that other world and its inhabitants are material processes (such as, e.g., processes in a digital machine, which are being utilized by it to model the development of a star). Yet in a digital machine, bunches of impulses that form a model of a star are at the same time electric charges traveling in transistors' crystals, in the vacuum of cathode lamps, and so on. The physicists from that world are going to figure out that the electrical impulses from which they themselves and their world are constructed consist of some subordinate elements: they will thus arrive at the existence of electrons, atoms, and so on. But this is not going to mean much as far as their ontology is concerned because when we ourselves discovered that atoms consisted of mesons, baryons, leptons, and so on, this did not lead us to any ontological conclusions about our supposed "artificial" origin.

The physicists from that world would only be able to discover the fact of creation (or rather of "having been created") by *comparing* our real world with their own. It is only then that they would be able to see that our world is one level of Reality short when compared with theirs (since they are made of electrical impulses and it is only those impulses that are made of the same material that our world is). Figuratively speaking, a created world is perhaps like a very stable, very long, and internally coherent dream that no one is dreaming but that rather "is dreaming itself"—inside a "digital machine."

Let's now go back to the question about the causes that can set intelligent beings on the path of creative activity in the Universe. There can probably be a number of them, of various kinds. I would not like to speculate over the kinds of causes that are going to point a particular cosmic civilization in that direction. It is enough if we discuss the scope of its technological activity; the motivation is then going to emerge in the course of that civilization's development. Perhaps it will be a defense against the avalanche of information. In any case, a descendant civilization (i.e., one that has been preprogrammed and enclosed in the preceding way) is going to become "encysted" with regard to the rest of the Universe and will become inaccessible for any external activity (e.g., signals). It is rather amusing to think that this civilization will itself be able to construct within its world (provided it is vast and diverse enough) some further, hierarchically subordinate worlds—which will be embedded into one another the way nested dolls are.

So that all of this does not look like some kind of insane daydreaming, we should point out that the complexity of any system must decrease slowly yet gradually if it is not being supplied to the system from outside. (In other words, systemic entropy increases.) The bigger the system, the more possible states of equilibrium it has, and the longer it can seem to be able to defy the law of increasing entropy. A local decrease in entropy can occur, for example, in the course of organic evolution, whose thermodynamic equilibrium remains negative on a global scale because an increase in the amount of information has taken place over several billion years. Naturally, the ratio of the system as a whole must be positive (an increase in solar entropy is incomparably bigger than its decrease on Earth). We have mentioned before the idea of "connecting" a cosmogonic creation to a star as a source of desired order. Equally, the whole surface of the "external sphere" of such a world could be turned into an "absorbent" for the energy coming from the natural Universe. This would be the only chance for the creatures living there: either they decide that entropy of a really large system—in this case, their own—does not have to increase, or they figure out that their "everythingness" is having energy supplied to it from outside.

Let's look again at the hierarchy of the nested world, which is a result of a decision made by a cosmic civilization for which our world was not perfect enough. That civilization is then going to create "encysted world no. 2." Yet its inhabitants, dissatisfied as they will be with their living conditions after several million years, and desiring a better future for their offspring, will create world no. 3 for them—inside their own world and using its materials. All those subsequent worlds are what we can call "cosmomeliorators," "evil rectifiers," "ontological repairers," or whatever other name we want to give them. It is possible that in one of them, there will eventually emerge an adequate perfection of existence to warrant the termination of any further creative activity in the Universe. This activity will have to come to an end anyway because members of civilization no. 100,000 cannot plant their sons and daughters on the surface of an atom.

Someone could ask whether I believe it is at least a little bit likely that humans will one day make such—or similar—plans.

Asked directly like this, I should give a direct answer. I do not think so. Yet, if we take into account all those countless intelligent worlds revolving inside those enormous galaxies, while the number of those galaxies is incomparably higher than the number of dandelion seeds floating in the air above the wide meadows, and also than the number of the grains of sand in a desert, this very number makes any improbability possible—as long as it can be actualized (even if in one out of every million galaxies). Yet for no one in this whole abyss of stardust to arrive at an idea about such an undertaking, to try to bite more than it seems possible to chew—this indeed seems rather improbable to me. Before someone categorically disagrees, please take a moment: July nights, during which the sky is so starry and beautiful, really encourage such contemplation.

8

A LAMPOON OF EVOLUTION

Several million years ago, the cooling of the climate began. It was a harbinger of the approaching ice age. Mountains grew, continents rose, jungles gave way to grass planes owing to the rising drought. The formation of steppes resulted in the shrinking of the living environment for the four-pawed wood animals. The latter's in-the-air existence among the tree branches made their hand movements much more precise and positioned their thumbs opposite their other fingers, while turning their eyes into their main orientation sense. This particular environment required them to adopt a vertical posture, perhaps more frequently than any other. Various tribes descended from the trees—which were ever thinner and thus offered less shelter—to try their luck on the wide steppe planes. The baboon then developed as a result of giving up on the vertical posture and face and through the secondary development of a doglike muzzle. Apart from the baboon, there is just one more experimenter from among those who left their tree residences that has remained among the living.

It would be futile to try to identify man's linear genealogy because attempts to descend to earth and walk on two feet were made over and over again a countless number of times. Anthropoids, with their uncertain gait but already adjusted on a neural level to adopting a posture that had been formed in the thick of the jungle, eventually entered the steppes—this pre–ice age ecological niche where four-legged herbivores were grazing. They already had human hands and eyes but lacked a human brain. Competition privileged their growth. Living in groups, these animals competed against one another. Owing to some unique shifts in their inner secretion, their childhood, that is, the period of collecting experiences under the group's watch, extended significantly.

Their facial expressions and the sounds they made were used as a mode of communication—which was subsequently to develop into speech. It was probably then that primitive man acquired considerable longevity when compared with the anthropoids. Indeed, individuals from groups that had the most experience, that is, the oldest and the longest living, won the struggle for survival. It was the first time during evolution that a species that was capable of longevity had been selected because, for the first time, that particular trait turned out to be biologically valuable as a treasury of information.

This "human prologue" involved a shift from an accidental, "monkeylike" use of tools to their production. The latter developed from continuing with "monkey" technology, that is, throwing stones or a sharp pole—which represented the beginning of action at a distance. A shift to the Paleolithic era involved the development of simple machines and the exploitation of processes from the surrounding world: fire as a tool of homeostasis, which freed man from his climate, and water as a means of transportation. The mode of living changed from hunting and gathering to nomadism and then sedentarization, whereby man moved from eating plants to cultivating them. It was already a million years after the beginning, in the Neolithic period, that the latter events took place.

It seems we did not actually descend from the Neanderthal man; rather, we destroyed that particular living form, which was our close kin. This is not to say we did it as murderers or voracious eaters, as the struggle for survival can take many different forms. The Neanderthal man was so close to primitive man, *Homo primigenius,* that those tribes might have crossbred—which is probably what happened. Yet, even though the Neanderthal man, so mysterious to us owing to the considerable size of his skull—which was bigger than ours—created his own culture, he died out together with it. Primitive man created a new culture. A short period of time (in geological terms) then passed before the first proper stage of technological development began: several thousand years of various civilizations, mainly based in the subtropics. Yet it was merely an instant compared with the million years that the formation of man and the social group took.

During the first stage, "natural" sources of both nonhuman energy (draught animal) and human energy (slave) start to be used. The discovery of the wheel and rotary movement, which even highly developed civilizations (e.g., those of Central America) skipped, forms the basis of the construction of machines with a narrow scope of activity, which are incapable of self-adaptation. Soon after, energy from man's environment—wind, water, coal—is exploited, and after that, electricity. As well as putting machines in motion, the latter allows for information to be sent over long distances. This facilitates a dynamic coordination of activities and a faster progress in transforming a natural environment into an artificial one.

The shift to the second stage begins with some significant techno-logical changes. Obtaining engine power that matches the power of Nature allows man to overcome gravitation. Alongside atomic energy, cybernetic construction is enabled, which involves replacing the me-chanical construction of machines with programming their development and activity. This is an obvious consequence of trying to imitate life's phenomena—which are treated, although not always consciously, more as an example, or a directive for action, rather than just as an object of desperate admiration evoked by their undisputed superiority.

Constructing systems of ever greater complexity gradually fills a large gap in theoretical knowledge that separates the already quite extensive knowledge about simple machines, such as the steam or electric engine, from knowledge about complex systems, such as evolution or the brain. In its full swing, such development aims toward "general imitology" because man is learning to create everything that exists—from atoms (antimatter, synthetically produced in laboratories) through to equivalents of his own nervous system.

The rapid increase in information that follows indicates to man that manipulating information is a separate branch of technology. Significant help is offered by studying methods used in this respect by bioevolution. The prospect of overcoming the information crisis arises, thanks to the automatization of cognitive processes (e.g., via "information farming"). This may result in the perfection of action, based on the principle of constructing reliable systems of any complexity from unreliable elements. Again, this will happen after knowledge about the analogous technology of biological phenomena has been acquired. A complete separation of the production of goods from its human supervision begins to look possible; "hedonistic technologies" (such as phantomatics) emerge. The limit of this sequence of developments lies with cosmogonic engineering, that is, the creation of artificial worlds which are nevertheless so sovereign and independent from Nature that they replace its worlds in every respect. The difference between "the artificial" and "the natural" thus begins to blur because "the artificial" is capable of exceeding "the natural" within any range of parameters that are of importance to the Designer.

This is what the first stage of man's technical evolution looks like. It does not represent the end of development. The history of civilization, with its anthropoid prologue and the possible extensions we have outlined here, is a process of expanding the range of homeostasis, that is, of man changing his environment, over the period of between a thousand and three thousand years. This ability, which penetrates the micro and macro universe with its technical tools, all the way to its furthest visible "pantocreatic" limit, does not touch the human organism itself. Man remains the last relic of Nature, the last "authentic product of Nature" inside the world he himself is creating. This state of events cannot last for an indefinite period of time. The invasion of technology created by man into his body is inevitable.

RECONSTRUCTING THE SPECIES

This phenomenon, which is central to the second stage of a civilization's development, can be analyzed and interpreted in various ways. It can take many different directions and forms—within certain limits. Since we need some kind of schema to continue with our discussion, we shall resort to the simplest one, but we have to remember that it is just a schema—and thus a simplification.

First, it is possible to perceive the human organism as given and thus fixed in its overall design. In that case, biotechnology's tasks will consist in eliminating diseases or preventing them and also in replacing waning functions or defective organs either with their biological substitutes (such as transplants, tissue grafts) or with technical ones (prostheses). This is the most traditional and shortsighted approach.

Second, while doing everything as described earlier, it is possible to combine those actions with a superior one, which will involve replacing Nature's evolutionary gradients with man's purposeful regulatory activity. Such regulation can in turn have various goals. It may focus on eliminating all those harmful consequences caused by the absence of natural selection, which destroys the inadequately adapted, from the artificial environment of that civilization. Alternatively, it may replace a modest program with a comprehensive one: that of biological autoevolution. The aim of the latter is to form an ever greater number of perfect human types (through significantly changing hereditary parameters, e.g., mutability, susceptibility to cancers, body shape, inner- and cross-tissue correlations, and, last but not least, parameters that regulate life span or even the size and complexity of the brain). In other words, this would be a plan for creating "the next model" of *Homo sapiens,* extended in time over hundreds or perhaps even thousands of years. It would take place through slow and gradual changes rather than a sudden leap—which would smooth out intergenerational differences.

Third, perhaps the whole problem should be treated in a far more radical way. We can consider as inadequate both Nature's design solution to the problem of "What is an Intelligent Being to be like?" and the solution that could be reached by autoevolutionary means learned from Nature. Instead of improving or "patching up" the model that exists within a certain set of parameters, it is possible to set some new arbitrary values for them. Instead of a relatively modest biological life span, we could demand quasi-immortality. Instead of strengthening the design provided by Nature within the constraints set by its building material, we could demand the highest strength that the existing technology can offer. In other words, we could replace reconstruction with a total rejection of the existing solution and then design a completely new one.

The latter solution to the dilemma seems so absurd and unacceptable to us today that it is worth listening to the argument that its supporter might present.

First of all, he will say, the path of "preventative and prosthetic" solutions is necessary and inevitable. The fact that we are already on this path serves as the best evidence for it. There already exist prostheses that temporarily change our heart, lungs, and larynx; we have synthetic blood vessels, a synthetic mesentery, synthetic bones, synthetic lining of the pleural cavity, and artificial joint surfaces made of Teflon. We already have hand prostheses that are directly activated by means of the bio-electricity of the muscle stumps in the pectoral girdle. There are designs for a device that could record nerve impulses moving someone's limbs when he walks. Someone who is paralyzed as a result of the damage to his spinal cord will be able to walk thanks to a machine that will send appropriate impulses "recorded" by a healthy person to his legs and that he will be able to regulate in any way he wants. At the same time, transplant opportunities are increasing: after the cornea, bone parts, and red marrow, the time has come for transplanting life-sustaining organs. Experts claim that a lung transplant is likely to take place in the not-too-distant future.[1] Overcoming the organism's biochemical defense against foreign proteins will allow us to transplant hearts, stomachs, and so on. Whether we use transplants or substitute organs made from abiological substances will always be decided by the state of the knowledge and technology at a given time. It will probably be easier to replace certain organs with mechanical ones, while others will need to wait until transplant technology has developed far enough. Importantly, any further development of biological and abiological prosthetics will be dictated not only by the needs of the human system but also by the needs of new technologies.

Thanks to the research conducted by American scientists, we already know today that the strength of muscle contractions can be significantly increased by inserting an electronic amplifier of impulses between a nerve and a muscle. A model for such a device is able to collect nerve stimulations sent to muscles from the skin, amplify them, and direct them to appropriate effectors. Russian scientists working in the area of bionics—a science that studies effectors and receptors in living organisms—have independently constructed a device that significantly reduces reaction time in humans. This time is currently too long in space rocket rudders or even in supersonic airplanes. Nerve impulses travel at the speed of merely a hundred meters per second, yet they have to get from a sensory organ (e.g., the eye) to the brain and, from there, via the nerves, to the muscles (effectors)—which takes several tenths of a second. The scientists collect the impulses coming from the brain and traveling along the stems and direct them straight to the mechanical effector. It is therefore enough for the pilot to *want* to move the rudder for it to shift. The situation that will ensue once such technologies have developed far enough will lead to some paradoxes. Equipped with a prosthesis, an individual who suffers from an injury as a result of an ill-fated accident or illness will exceed a normal person's capabilities. It is because it would be hard not to provide him with the best prosthesis available—and the available ones will be faster, more efficient, and more reliable than the natural organs!

When it comes to the "autoevolution" discussed here, it is to be limited to systemic transformations that still remain within the limits of biological plasticity. This reservation is not necessary though. An organism is unable to produce through programming the genotypic information of diamonds or steel because it would need a high temperature and a high pressure for that—and these cannot be achieved in embryogenesis. Yet it is already possible to create, for example, prostheses, permanently set in the jaw's bones, whose teeth components are made from the hardest materials possible—materials that an organism itself is unable to produce and that are practically indestructible. What is thus clearly important in an organism is its perfect design and functioning, not its genesis. When using penicillin, we are not interested in whether it was obtained in a laboratory, in test tubes, or as actual fungus in a growth medium. When, in planning human reconstruction, we limit ourselves to the means whose development will facilitate the information transfer of the human genotype, we simultaneously give up, completely unnecessarily, on supplying the body with some enhanced systems and some new functions that would be very useful and valuable for it.

In reply, we say that the supporter of the revolution in human redesign does not probably realize what the consequences of his postulates may be. We are not just talking about some narrowly conceived attachment to man's present body. The whole of culture and art, including some of its most abstract theories, is saturated with corporeality the way it was formed and shaped by Nature. Corporeality has informed the canons of every historical aesthetic, of every existing language, and, through that, of the totality of human thought. Our spirit is corporeal too: it is not accidental that this word derives from respiration. Contrary to what it may seem, there are no values that could have emerged without the presence of the corporeal factor. Love itself is entirely corporeal—in its least physiological sense. It would be an act of extreme madness if man really was to undergo a transformation owing to the technologies that he himself has created and if he was to consider a robot with a perfect crystalline brain his successor. It would actually amount to a collective suicide of the human race, even though such a suicide would be covered up by the apparent continuation of humanity in thinking machines— which are part of the technology created by man. In this way, man would ultimately allow the technology he himself has brought about to push him out of his place of existence, of his ecological niche. Having removed a less adapted species from the stage of history, technology would thus become a new synthetic species.

Such arguments do not convince our opponent. He says, I am perfectly familiar with the corporeality of human culture, yet I do not consider everything that exists in it worthy of eternal preservation. You must surely know the terrible influence that random occurrences—such as the location of reproductive organs—have had on the development of certain concepts or on the emergence of social and religious conventions. The economy of action and the indifference toward what we consider to be aesthetic reasons resulted in the proximity and partial merging of the ducts that excrete the products of metabolism and the sex ducts. This proximity, justified on a biological level—which is in fact an inevitable consequence of a design solution implemented during the amphibian and reptilian stage, that is, hundreds of millions years ago—led people to the conclusion that the sex act was shameful and sinful, once they had examined and investigated their own organic functions. The impurity of this act seemed almost natural to them because it was being carried out with organs that were very closely linked to excretory functions. The organism should avoid the final products of excretion: this is important from a biological point of view. Yet at the same time, it should aim at sexual conjunction, which is needed for evolutionary purposes. The concentration of two such diametrically opposite yet important imperatives must have contributed to the wide emergence of myths about the original sin and about the natural impurity of sex life and its manifestations. Torn between genetically programmed repulsion and attraction, the mind would produce either civilizations based on the idea of sin and guilt or civilizations based around shame and ritualized debauchery. This is my first point.

Second, I am not advocating any kind of "roboticization" of man. When I discussed various electronic and other prostheses, it was only with a view to drawing on the currently available specific examples. By a robot, we understand a mechanical blockhead, a roughly humanoid machine equipped with human intelligence. It is thus a primitive caricature of man, not his successor. Systemic reconstruction does not have to involve giving up on any valuable features—only an elimination of those features that are considered imperfect and primitive in man. Evolution acted with great speed when forming our species. Its unique tendency to preserve the design solutions of the original species for as long as possible endowed our organisms with a whole series of faults that are not seen in our four-legged ancestors. Their bone pelvis does not support the weight of their entrails. Yet since in man the pelvis is supposed to carry and support him, muscle membranes developed as a result—but they significantly impede labor. Vertical posture was also harmful for hemodynamics. Animals do not have varicose veins, but varicose veins are one of the plagues of the human body. The sudden increase in the size of the skull resulted in the tilting of the pharyngeal cavity at the right angle (where it turns into the esophagus), introducing air turbulence into that space and leading to the sedimentation of the considerable amount of aerosols and germs on the pharyngeal wall. As a result, the throat became an exit point for a great number of infectious diseases. Evolution tried to prevent this by surrounding that area with a ring of lymphatic tissue, yet this improvisation was not successful: it actually became a source of some new ailments, since those tissue clusters developed into centers of focal infections.[2] I am not saying that man's animal ancestors represented ideal design solutions. From an evolutionary point of view, any species is "ideal" as long as it is capable of continued existence. I am only saying that even our incredibly limited and incomplete knowledge allows us to imagine solutions—which cannot be actualized as yet—which would free man from considerable suffering. All kinds of prostheses seem inferior to us to natural limbs and organs, because at present, such prostheses do indeed come second in terms of their execution. I do understand that, where technology creates no obstacles, conventional aesthetic can be satisfied. The outer surface of the body does not seem beautiful to us when it is covered with shaggy fur; the same would apply if it was to be covered with steel sheeting. Yet it may, of course, not differ from the skin at all as far as the eye or the other senses are concerned. It is a different matter with sweat glands. We know that civilized people are really concerned about reducing the effects of their glandular activity—an activity that causes great problems to some in maintaining their personal hygiene. Never mind such details though. We are not talking about what may happen in twenty or a hundred years but about what is still thinkable. I do not believe in any final solutions. It is very probable that "superman" will be considered an imperfect creation after some time, because new technologies will enable him to achieve what seems to us a fantasy that cannot ever be actualized (e.g., "switching from one personality to another"). Today we believe that it is possible to create a symphony, sculpture, or painting via a conscious mental effort. At the same time, the thought of "composing" a successor for ourselves, with any kind of orchestration of his spiritual and physical traits we want, seems like a terrible heresy. Yet the desire to fly or to study the human body, machine building, or examining the origins of life on Earth also used to be seen as heresies in the past. Barely several hundred years separate us from those times. If we are to behave like intellectual cowards, we can, of course, remain silent on the topic of any probable future developments. But in that case, we should at least make it clear that we are behaving like cowards. Man cannot change the world without changing himself. We can take the first steps on a given path while pretending we do not know where it leads. Yet this is not the best strategy.

This pronouncement by the supporter of species reconstruction deserves to be at least considered, even if not accepted. Any principled opposition may arise from two different standpoints. The first one is more emotional than rational, at least in the sense that it withholds its consent to revolutionize the human organism without accepting any "biotechnological" arguments. It considers the human form the way it is today untouchable, even if it admits that this form suffers from various weaknesses. It is because even those weaknesses, both physical and spiritual, have turned into values in the course of historical development. No matter what form it takes, the outcome of the autoevolutionary activity would dictate that man is to disappear from the surface of the Earth. In the eyes of his "successor," he would become a dead zoological term, just as the Australopithecus or the Neanderthal are for us today. For an almost immortal creature, which would be in command of both its body and its environment, the majority of eternal human problems would not exist. A biotechnological revolution does not just therefore mean annihilating the *Homo sapiens* but also its spiritual legacy. Unless we treat it as a figment of our imagination, this position seems rather scornful: instead of solving his problems and finding answers to the questions that have preoccupied him for centuries, man is to hide away from them in some kind of materialist perfection. What a shameful escape and an abandonment of responsibility—when, by means of technology, the *Homo* is becoming transformed into the deus ex machina! The second standpoint does not exclude the first one: it probably shares its argument and emotional content, yet it does so in silence. When it does announce itself, it poses questions. What kinds of specific improvements and reconstructions does the "autoevolutionist" propose? Is he refusing to provide any detailed explanation because it seems premature? But how does he know if the perfection of biological solutions, unmatched thus far, is going to be exceeded one day? What facts does he base his belief on? Is it not rather that evolution has already reached the limit of its material capabilities? That the complexity represented by the human organism is a critical value? Of course, it is already known today that within the limits of individual parameters, such as the speed of information transfer, the reliability of *local* action, the stability of a function as a result of copying its actuators and controllers, machinic systems can beat man—yet there is a difference between separately multiplying power, efficiency, speed, and strength and integrating all those optimal solutions within *one* system.

The autoevolutionist is prepared to pick up the gauntlet—and to offer counterarguments to the arguments presented. Yet before he moves on to discuss the position adopted by his rationalist opponent, he reveals that the first standpoint is actually not that alien to him. It is because, deep down, he feels the same strong objection to any plans for species reconstruction that the person who has condemned it in absolute terms does. But the autoevolutionist sees such future transformation as inevitable, which is why he is looking for all kinds of reasons that would support it, so that the necessary action overlaps with the outcome of the decision made. He is not an a priori opportunist: he does not think that what is necessary *must* at the same time be good. At the same time, he hopes that it at least *may* turn out to be good.

CONSTRUCTING LIFE

To design a dynamo machine, one does not need to know the history of its invention process. A young engineer can do very well without it. The historical circumstances that shaped the first generators are, or at least can be, completely irrelevant to him. In any case, a dynamo machine as a device for transforming kinetic or chemical energy into electrical energy is rather outdated. When, in the not-too-distant future, electricity will be produced directly, for example, in a microfusion cell, thus overcoming the troublesome circularity of the present transformations (of coal's chemical energy into thermal energy, then of thermal into kinetic energy, and only then of kinetic energy into electrical energy), only a historian of technology will remember the design principle of the earlier generators. This kind of separation from developmental history is unknown in biology. We say this because we are about to embark on a critique of evolutionary solutions.

Yet this can only be a critique of design solutions and not of any of the prior stages. People are actually more inclined to notice the perfect side of biological solutions because their own capabilities lag far behind those biology offers. For a child, every act performed by an adult is magnificent. One has to grow up to notice a weakness in what used to be seen as perfection. But this is not all. Loyalty toward the designer—who gave us death together with life, and who bestowed on us more suffering than delight—nevertheless prompts us to evaluate biological achievements in a way that is not limited to a lampoon of the designer's activity. It is because the evaluation proper should see him the way he really was. And he certainly was not almighty. At the moment of its inception, evolution was a Robinson placed on an empty planet—lacking as it did not just tools and any kind of help, not just knowledge and the ability to make predictions, but also itself, that is, a mind capable of making future plans, because apart from the hot ocean, the thundery lightning, and the oxygen-free atmosphere with the burning sun, there was no one around. By saying that evolution started in such and such way, that it did this and that, we are thus personifying the early unfolding of the process of self-organization—which did not just lack agency but also a goal.

This process was a prelude to a great opus, without itself knowing either the opus as such or even its first opening notes. The only thing molecular chaos had at its disposal, apart from its individual material potential, was a large degree of freedom: time.

Barely a hundred years ago, the Earth's age was estimated at forty million years. We now know the Earth has been in existence for at least four billion years. I remember learning that life on Earth had existed for several hundred million years. The remnants of organic substances known to us today that belonged to those early living creatures are two billion seven hundred million years old. If we count backward from today, then 90 percent of evolution's time passed before the first vertebrates developed. The latter took place over 350 million years ago. After another 150 million years, the successors of those osseous skeleton fish came out onto the dry land and conquered the air. Then, after the emergence of the mammals—which are fifty million years old—man arrived one million years ago.

It is easy to juggle with billions. Yet it is difficult to grasp the design significance of such numbers and such spans of time. As we can see, the acceleration of subsequent solutions is not just a characteristic of technical evolution. Progress is accelerated by the accumulation of not only theoretical, socially assembled knowledge but also genetic knowledge recorded in the genetic code.

During two and a half billion years, life developed solely in the oceans. The air and land were dead in those times. We know around five hundred fossil species from the Cambrian (over half a billion years ago), yet it was only possible to discover a handful from the pre-Cambrian period. This surprising gap has not been explained until this day. It seems that the number of living forms increased dramatically within a short period of time (i.e., over millions of years). Pre-Cambrian forms almost exclusively include plants (algae): there are hardly any animals.

The latter can be counted on the fingers of one hand. In the Cambria, though, they appeared in great numbers. Some scientists are inclined to accept a hypothesis about some kind of radical global change to the Earth's conditions at that time. This could have been an increase in the intensity of cosmic radiation, as is claimed by Shklovsky's hypothesis we discussed earlier. Whatever happened, this unknown factor must have acted on the scale of the whole planet, because the pre-Cambrian gap applies to all the paleontological data. Conversely, it is not like, for some unknown reason, the ocean waters from the early Cambrian days contained a relatively small amount of living organisms on the whole, and the emergence of numerous new species in the Cambria had been preceded by a sudden increase in the number of earlier forms. There had already existed many living forms in the Archeozoic: geological data reveal that long before the Cambria, the oxygen/nitrogen ratio in the atmosphere was similar to what it is today. And because oxygen in the air is a product of the activity of living organisms, their overall mass cannot have been significantly smaller than it is today. The lack of fossil forms has been at least partly caused by their fragility: pre-Cambrian organisms did not have a skeleton or any mineral shells. How and why such "reconstructions" occurred in the Cambria remains unknown. Perhaps this problem will never be solved. Yet it is also possible that a more thorough examination of biochemical kinetics will help us solve this puzzle, were we to establish, on the basis of the current structure of protein homeostasis, what kinds of more primitive forms were most likely to have preceded it—of course, as long as solving such a puzzle is limited to inner systemic factors rather than to some kind of one-off consequence of cosmic, geological, or climatic changes at the dawn of the Cambria.

We are discussing the preceding because "the Cambrian revolution" may have been caused by some kind of "biological invention" on the part of evolution. Even if that was the case, it did not change the originally accepted fundamental principle of design, based on cellular building blocks.

The evolution of life was no doubt preceded by the evolution of chemical reactions. The first cells did not therefore have to feed on dead matter as their source of order. In any case, they would not have been able to solve right away the difficult task of being able to synthesize organic bodies from simple formulae, such as carbon dioxide, using the energy of solar photons. This accomplishment was only achieved by plants, which developed the ability to produce chlorophyll and the whole enzyme apparatus that catches radial quanta. Fortunately, in the early days, the first organisms must have had at their disposal some organic substances, which they were able to assimilate easily and which were a remnant of the excess that had produced them. The latter emerged through processes such as electric explosions in the atmosphere of ammonia, nitrogen, and hydrogen.

But let us return to the fundamental dynamic problem of the basic cell. It must control the significant parameters of its transformations so that they do not escape from the range of fluctuations beyond the limit of reversibility—into entropy, and hence death. In a fluid colloid medium, such control must take place with limited speed. Fluctuations caused by the statistical nature of molecular motions cannot occur faster than the circulation of information across all the cells. Otherwise, the central regulator, the nucleus, would lose control over the local processes: the information about the need for an intervention would then normally arrive too late. This would be the start of irreversible changes. The size of the cell is thus dictated, in the last instance, by the parameters that regulate the speed of information transfer from any location in the cell to the regulators as well as the speed of the local chemical processes. In its early stages, evolution tended to produce cells that would greatly differ in size. Yet, owing to the preceding limitations, it is impossible to obtain a cell the size of a pumpkin or, indeed, an elephant.

We should point out that for a human technologist, a cell is a more than unusual device, one that can be admired rather than understood. A "simple" organism, such as *E. coli*, undergoes division every twenty minutes. During that time, the bacterium produces proteins at the rate of one thousand molecules per second. Since a single protein particle consists of around one thousand amino acids, which all have to be properly "distributed" in space and "adjusted" to the emergent molecular configuration, this is not a mean task. The most cautious estimate indicates that the bacterium produces *at least* one thousand bits of information per second. This number will seem really striking if we compare it with the number of bits of information a human mind can cope with: about twenty-five bits per second. A printed page with *little* information excess contains around ten thousand bits. As we can see from this, the cell's information potential is the largest within its inner processes, that is, those that serve to maintain its dynamic existence. The cell is a "factory," in which the "raw material" is everywhere—alongside, above, and below the "production machines." Cell organelles, ribosomes, mitochondria, and similar microstructures that, in terms of their size, are situated in the middle of the scale between the cell and the chemical particle are such "machines." They consist of ordered and complex chemical structures, with "machining tools" such as enzymes "attached" to them. It looks like the "raw material" is passed on to the "machines" and their "tools" not by some special directional forces that would attract the necessary material but repel the unwanted or inappropriate one but simply by regular thermal motions of molecules. Those "machines" are thus kind of bombarded by streams of particles cavorting about in molecular suspension. It is up to their specificity and selectivity to pick the "correct" elements from all this apparent chaos. As all those processes without exception are of statistical nature, general thermodynamic considerations lead to the conclusion that in the course of such transformations, mistakes, that is, errors, should occur (e.g., placing "false" amino acids where protein's molecular spiral is emerging). Such errors must nevertheless be rare, at least normally, because it is impossible to detect proteins that have been "falsely" synthesized by the cell. In recent years, much research into the kinetics of life's chemical reactions has been conducted, whereby the latter have been seen not as rigidly repeated cyclical processes but as a malleable whole—which, while being supported in its ongoing development, can be directed quickly and easily to achieve the currently desired goal. After working out the "output parameters" of the model cell, a large computing machine took thirty hours to calculate the most optimal harmonization of reaction speeds and their individual links in the cell. This is where the formalization of the problem, so essential in science today, leads us: a bacteria cell solves the same problems in fractions of a second—naturally, without an electronic or neural brain.

Cellular uniformity is both real and apparent. It is real in the sense that the protoplasm is a colloid solution of large-sized molecules made up of proteids, proteins, and lipids; that is, it is a "chaos" of molecules suspended in a liquid medium. It is apparent, because the cell's transparency defies any attempts to observe its dynamic microstructures, while cutting and preserving them with colorants causes changes that destroy their original organization. The cell, as it was discovered thanks to laborious and difficult research, is not even such a metaphorical "factory," as the preceding descriptions might suggest. The processes of osmosis and diffusion between the nucleus and the protoplasm do not just occur as a result of some physical mechanisms, according to the gradient of differences in osmotic pressure, but those gradients themselves are principally controlled by the nucleus. In the cell, we can distinguish microcurrents, that is, particle microstreams that are miniaturized equivalents of the circulatory system. Organelles are the nodal points of those currents. They function both as "universal machines" equipped with sets of appropriately distributed enzymes and as accumulators of the energy thrown in the right direction at the right moment.

It is relatively easy to imagine a factory that would contain machines and raw materials floating around one another, yet it is difficult to envisage one that would constantly keep changing its shape, the mutual coupling of its aggregates, or their production capacity. The cell is a system of watery colloids, with multiple streams of enforced circulation and with a structure that is not only mobile on a functional level but whose transformations are also chaotic (in the sense that it is even possible to shift the location of the protoplasm: as long as this does not destroy some basic structures, it will still work and hence exist). It is constantly shaken by Brownian motions and suffers constant deviations from stability. It is possible to control cellular processes in a given way only in a statistical sense, on the basis of the probabilistic tactics of instant interventionist and regulatory decisions. Oxygenation processes occur in the cell by means of electron transfer via a "liquid pseudocrystalline semiconductor." They display specific rhythms, which are caused by this ongoing regulatory intervention. The same applies to some other processes, for example, energy cycles involving the accumulation of energy in adenosine triphosphate.

All higher organisms are actually just combinations of this elementary building material; they are a way of "drawing conclusions and ramifications" from the results of data encoded in every cell, all the way up from the bacterial one. Also, no tissue organism shares the cell's universality, although the latter is to some extent replaced by the plasticity of the central nervous system. Yet at the same time, such universality is manifested by every amoeba. It is no doubt really convenient to have a leg that, if need be, can become a tentacle and that can be replaced if lost. I am referring here to pseudopodia in amoeboids. Equally useful is the ability "to open one's mouth in any place." An amoeba can do this too, as it sprays food remnants and engulfs them with protoplasm. It is here that the system of original premises manifests itself for the first time. Cells, becoming conjoined into tissue, may create a macroscopic organism that will have a skeleton, muscles, vessels, and nerves. But even the most perfect regeneration is already not as extensive as the universality of function, which has been lost together with protozoon characteristics. The building material places a restriction on creating "reversible organs." The protoplasm is to some extent capable of contracting, conducting impulses, and digesting the ingested food, yet it does not contract as efficiently as a specialized muscle cell does; it does not conduct impulses as well as nerve fibers do; and neither can it bite into its food or chase it successfully, especially if the latter is full of energy and in the process of fleeing. While specialization is a unidirectional magnification of particular traits of cellular universality, it also involves giving up on this universality—the outcome of which, a rather important one, is an individual's death.

A critique of the "cellular premise" can be approached in two ways. First, it can be approached from a genetic standpoint: we take as a given the liquid (water) environment of bodies such as amino acids and other organic compounds, which are products of the chemical activity of the ocean and the atmosphere. It is only there that those bodies came together; it is only there that they were able to react with one another in a way that brought them to self-organization under the Earth's conditions "merely" a billion and a half years ago. Taking those initial conditions into account, we could ask about the possibility of actualizing a "prototype" that would be different from the one offered by the evolutionary solutions. Second, we could wonder—notwithstanding the necessity of the preceding situation—what would be the most optimum solution that would remain independent from the preceding limitations. In other words, would the developmental prospects of organization not look better if a Designer were to start it in a solid or gaseous environment?

There is absolutely no way for us to be able to compete, even on a hypothetical level, with the colloid version of the homeostatic solution produced by Evolution. This is not to say that this solution cannot be surpassed. Who can tell if the absence of certain atoms, certain elements in the raw material that formed the first cells and that evolution had at its disposal, did not close, right from the start, the road to some other, perhaps more energy efficient and more dynamically stable states and types of homeostasis? Evolution made do with what it had; it made probably the best possible use of its materials. Yet, because we accept the omnipresence of self-organizing processes in the Universe, which is to say, we do not think they occur only in some very special circumstances, as a result of some kind of unique and special coincidence, we also accept the possibility of the emergence, within liquid phases, of some other types of self-organization than the protein-based one, and maybe even the colloid-based one—whereby those other variants may be both "worse" and "better" than the "terrestrial" option.

But what does this "worse" or "better" actually mean? Are we not trying to sneak in some kind of Platonism with those terms, some kind of entirely arbitrary valorization? Progress, or rather the possibility of progress, is our criterion. By this we understand introducing such material homeostatic solutions that can not only continue despite some inner and outer disturbances but that can also develop, that is, increase the range of homeostasis. These are systems that are perfect not only on the level of their adaptation to the current state of the environment but also in their capacity for change, so that any such changes both need to correspond to demands from the environment and enable some further transformations in the future. This is to avoid barricading off the road of subsequent existential solutions or getting stuck in a developmental blind alley.

Seen in this way, human evolution deserves both a positive and a negative evaluation. Negative, since—as we shall see later—through its original selection (of its building material) and the subsequent methods of its design activity, it deprived its final and highest products, that is, us, of the opportunity to continue in a steady manner the work of progress on the biological level. Biotechnological as well as moral aspects stop us from simply continuing with the evolutionary methods: biotechnological because we are too *determined* as a particular design solution by Nature's causal forces, and moral because we reject the method of blind trials and that of blind selection. At the same time, it is possible to evaluate the evolutionary solution positively because, despite our biological limitations, we have the freedom of action, at least with regard to the future, thanks to the social evolution of science.

It seems rather likely that "the terrestrial option" is neither the worst nor the best of all possible ones—judging by the previously established criteria. Deliberations of statistical nature are actually not allowed within the solar system because it consists of just a few planets. Yet, were we to use such a limited reference point, we would arrive at the conclusion that protein homeostasis in cells is still better on average, because the other planets of the same age within the system have not produced any intelligent forms. Yet, as I say, this is quite a risky argument, because time scales and speeds of change may differ on different planets: methane–

ammonia planets may have belonged to a different evolutionary sequence, one in which our centuries would count for millions of years. This is why we shall prevent ourselves from any further speculation on this topic.

Let us now move from "liquid" homeostats to solid and gaseous ones. We want to ask about the developmental prospects of organization—should some kind of Designer initiate such a process in gaseous or solid concentrations of matter.

This issue has real rather than just academic significance, because a reply to the question raised in the previous paragraph can refer both to potential engineering activity and to the probability of the emergence of some other, noncolloid but "solid" or "gaseous" "evolutionary processes" on some other bodies in the Universe that do not resemble the Earth. As we know, the speed of reactions is paramount here. Of course, it is not the only factor, because the course of each reaction must be kept within certain limits; it must be strictly controlled and repeatable. The creation of circular processes signifies the emergence of the earliest initial forms of automatism on a molecular level. They are based on feedback, which partially frees the central regulator from the need to have to watch constantly over everything that is going on in its realm. And thus on to gases. Reactions can take place faster in them than they do in a water environment, but temperature and pressure are very important factors here. Evolution used "cold" technology on Earth, that is, technology based on the catalysis of reactions with a view to starting and accelerating them, and not on high temperatures. This round-about method was the only one possible. This is because even though the complexity of a system that produces high pressure and high temperatures can be lower than the complexity of a catalytic system, evolution could not, of course, have created such a system from nothing. It thus became "Robinson the chemist." Here it is not the "absolute" information balance that is decisive, that is, the fact that less information is needed to build the right sorts of pumps or to couple certain reactions (e.g., those involving the focusing of solar rays) to create conditions that will enable reactions between bodies; rather, the best kind of information is that which can be used and activated at present. The Earth did not offer similar opportunities within the realm of solid bodies or in the atmosphere. Yet could a propitious state have emerged under some other circumstances? We are unable to answer this question: we can only hypothesize here. Without a doubt, we are already capable ourselves of building homeostats from solid bodies, even though they are still primitive (e.g., electronic machines). Yet such solutions, burdened as they are with a number of basic inadequacies, can only be considered a prologue to design proper.

First, the models constructed by us are "macro-homeostats," that is, systems in which no direct relationship exists between their molecular structure and their functions. This relationship is not just about its usefulness for performing those functions, something that is no doubt also required in an electronic machine, in which the wires must have particular conductivity, and transistors or neuromimes—particular characteristics. It is also about the fact that a complex system, subject as it is to a great number of elements whose behavior it cannot always control, needs to be built on the principle of "obtaining certain effects by means of using uncertain parts." Those parts must therefore be equipped with self-repairing autonomy that compensates for any internal or external damages. The machines constructed thus far do not have this characteristic (although some new ones that are in the pipeline will have it, at least partially).

Second, this state of events has its consequences. A digital machine may need to have some of its parts (e.g., lamps) cooled and will thus require a pump to be used to maintain the circulation of the liquid coolant. Yet this pump itself is not a homeostat. It is in fact much simpler on the level of its homeostatic design, yet, should it get damaged, the machine as a whole will most probably come to a halt in no time. But a pump in an organic homeostat, for example, in the heart—although designed for purely mechanical action (blood transfusion)—is a multilevel homeostatic system. It is, first of all, part of a superior homeostat (the heart plus the vessels plus neural regulation). Second, it is a system with local autonomy (the autonomy of the regulation of heart contractions, which is built into its neural nodes). Third, the heart itself consists of many million microhomeostats such as muscle cells. The solution is very complex, yet it also demonstrates multifaceted protection against disturbance.[3] Evolution, as we said earlier, has solved this task by using the "cool" technology of molecular catalysis in a liquid medium. We can imagine working toward a similar solution while using some solid building material—for example, to construct some crystalline homeostats. Molecular engineering and solid body physics are going in this direction.

It is too early to think about constructing a "universal homeostat" such as the cell. We are taking the opposite trajectory to the one evolution took, because paradoxically, it is easier for us to construct narrowly specialized homeostats. The equivalents of the neuron can be found, for example, in neuristors, neuromimes, and artrons—from which certain systems, such as MIND (Magnetic Intergrator Neuron Duplicator), are constructed. The latter performs the logical function of recognizing various complex images that consist of sequences of information signals. Cryotron-type systems can almost compete on size with the neural cell (still a decade ago, such elements, i.e., cathode lamps, were a *million* times bigger than the neuron!), while they are ahead of it when it comes to speed. For the time being, we are unable to re-create self-repairing tendencies. Incidentally, the tissue of the central nervous system does not undergo regeneration either. Yet we do know of crystal systems that emerge as a result of the trace contamination of their atomic network by atoms of certain elements in a way that, depending on its original design, makes the whole unit behave as if it were a cascade amplifier, a heterodyne.receiver, a transmitter, a rectifier, and so on. It is possible to construct, for example, a radio set from such crystals. A further step will involve not just putting together any functioning entity from crystalline blocks but rather building a radio (or an electronic brain) from one block of crystal.

Why are we interested in the preceding solution? The uniqueness of such a system lies in the fact that a crystal radio, cut in two, constitutes *two independent* and still working—even if at half speed—radio sets. We can cut those parts in two again and again, and every time we shall obtain a "radio set," as long as the last remaining part still contains its necessary functioning elements, that is, atoms. We are approaching here the limits of using the parameters of the building material that had been reached by evolution on a different material front, so to speak: in colloids. Evolution also uses "molecular engineering," from which it started its whole design project. Molecules had always been its building blocks; it was capable of selecting them according to their dynamic usability and information capacity (enzymes are a source of universal solutions: they can serve any functions that involve synthesis and decomposition, together with functions connected with the transfer of inner-cell and hereditary information, as elements of chromosomal genes). The systems created by evolution are capable of functioning within a narrow range of temperatures (between 40 and 50 degrees Celsius) and not below the freezing point of water (in which all life's reactions take place). Low temperatures, even those around absolute zero, are more beneficial for molecular microminiaturization because, thanks to the phenomenon of superconductivity, the system gains an advantage over biological systems under those conditions (even though it is actually far from maintaining such an advantage on the level of all the parameters considered by life).

Thanks to the systemic equilibrium created by low temperature—which is bigger than the one determined by a drop of the protoplasm—the need for self-repairing interventions diminishes. Instead of solving this problem, we thus evade it in a way. We do actually know that crystals manifest "self-repairing tendencies," because a damaged crystal, when plunged in a solution, autonomously repairs its atomic network. This opens up certain prospects, although we are still unable to actualize them. A far more difficult problem is caused by "gaseous homeostasis." As far as I know, this issue has not yet been discussed in scholarly literature. The science fiction novel *Black Cloud* (1957) cannot be considered to belong to the latter, even though it was authored by a renowned astrophysicist, Fred Hoyle. Yet the "organism" he presented in it—a large nebula, an ensemble of cosmic dust and gases that has a dynamic structure stabilized with electromagnetic fields—can be constructed, I think. Another matter, of course, is whether such "organisms," made from electricity and gases, can emerge in the course of interplanetary "natural evolution." This seems impossible for many reasons.

It may look like we are discussing entirely imaginary issues, having crossed the boundary of what is permissible a long time ago. Yet this is not quite the case. We can formulate, as a general law, the following statement: only those homeostats are actualized by natural forces whose final states are achieved in the course of their gradual development, in accordance with the direction of the general thermodynamic probability of phenomena. Many unwise things have already been said about the Queen of the Universe, Entropy, about "the rebellion of living matter against the second law of thermodynamics"—which is why we have to emphasize clearly how incautious such semimetaphorical theses are and how little they have to do with reality. As long as it is a cold atomic cloud, the original nebula has less order than the Galaxy—which is carefully arranged like a disc with segregated stellar material. Its apparent early "disorder" nevertheless contained a source of high order in the form of nuclear structures. When the nebula falls apart into protostellar whirls, when gravitational forces squeeze those gas spheres hard enough, then the door of atomic energy suddenly "breaks open." The expelled radiation, in its struggle against gravitation, begins to form stars and their systems. To generalize, even though large material systems always aim at states with maximum probability, that is, with the highest entropy, they do so via so many intermediary states and so many varied routes that eventually, over a period of time that can be as long as tens of billions of years, not just one and not even ten types of self-organizing evolution may emerge "on the way" without violating the second law of thermodynamics but rather their countless myriad. There thus exists a large but still seemingly empty (since we do not know its elements) class of homeostatic systems *that can be built*—be it from solid, liquid, or gaseous building material. This class contains a unique subclass: a set of homeostats that can only emerge without the external interference of a personified Designer and only thanks to the causal forces of Nature.

This clearly shows that man can beat Nature: the latter is only capable of constructing some of the possible homeostats; we, in turn, on gaining the necessary knowledge, will be able to build all of them.

Such optimism when it comes to space design has to be coupled with some reservations, full of ifs and buts. We do not know whether humanity is going to gain all the information that is necessary for carrying out the previously described "construction tasks." There may exist "an information acquisition threshold," just as there exists a limit to the speed of light, yet we do not know anything about it. Besides, we should remind ourselves of the actual proportions involved in this "man against Nature" task. In the face of such a problem, we are like ants who are promising themselves to carry the Himalayas on their backs to another location. Perhaps I am exaggerating in favor of the ants. Perhaps their task would actually be easier. This is when we compare all of contemporary technologies to the tools at ants' disposal, that is, their own jaws and backs. There is only one difference—the fact that ants can only develop their tools in the course of biological evolution, while we can start an information evolution, as already discussed. This difference may decide in favor of man's victory one day.

CONSTRUCTING DEATH

Living organisms have a limited life span as well as experiencing processes of aging and death. Yet these processes are not inseparable. Protozoa are final as individuals, but they do not die; rather, they divide into their offspring. Some metazoa, such as hydras—which reproduce by budding—can live under laboratory conditions for a very long time without any signs of aging. It is therefore not true that all protoplasm in a metazoon has to age, which is why the aging of colloids (their thickening, sol-gel transformation from a liquid to a jelly state) cannot be seen as equivalent to the old age in our lives. Colloids in the plasma do age in a similar way to abiological colloids, but what may look like a cause of this is actually its consequence. The aging of cellular colloids is caused by their loss of control over their life processes, not the other way round.

Renowned biologist J. B. S. Haldane has outlined a hypothesis in which an individual's death is seen as being caused by hereditary factors, that is, lethal genes, which manifest themselves in the life of an organism so late that they are not subject to elimination through natural selection anymore. It is difficult to accept this hypothesis. Not just immortality but even Methuselah-like longevity does not pay off in evolution. An organism, even if it does not age on an individual level (i.e., "does not break"), ages within the evolving population, just as an excellently pre-served model of a Ford from 1900 is completely outdated today, as a design solution that is incapable of competing with modern cars.

Yet at the same time, protozoa cannot *not* go on dividing for an unlimited period of time. It is actually possible to "force them" to live to an old age—several dozen times longer than the life of an average individual—by keeping them on a diet so "meager" that it will just barely sustain their systems' life functions but that will not supply any substances that would increase them, with a view to producing two descendant or-ganisms. The old clones (populations) of the simplest organisms do age in a sense: their individuals begin to die out and are only brought back to life by the process of conjugation, in which an exchange of hereditary information takes place. This issue is rather difficult to comprehend. The problem of death can be analyzed in different ways. Was it "embedded" into organisms by evolution? Or is it rather an accidental phenomenon, a secondary consequence of decisions made by the designer with regard to issues other than individual existence? Is it therefore an equivalent of an act of destruction, with which the designer cancels the prior solution and starts working on a new one, or is it rather an unintended consequence of some kind of "fatigue" of material?

It is not easy to find a clear answer to this question. We should distinguish between two things here because longevity is a different problem than mortality, which requires a different solution. As we have already said, longevity becomes biologically significant when the offspring requires a longer period of care before it becomes independent. These are, however, exceptions. Normally, once natural selection has been made and offspring has been produced, the fate of parent organisms becomes the latters' "individual" problem—which, in Nature, means no one's problem. Whatever degenerative processes accompany old age, they do not affect the further evolution of the species. The tusks of old mammoths crisscrossed, making mammoths die from hunger gradually, yet selection was unable to eliminate this phenomenon as it occurred after their sex life had stopped. The old age of animals or plants, shifted beyond the selection threshold, is thus outside the latter's jurisdiction. This applies not only to degenerative changes but also to longevity. Unless it is biologically useful (as it was in man's early days) for a future generation, prolonged existence, if it emerged randomly as a result of a particular mutation, will also be sentenced to random disappearance because there is no selection factor that would fix it genetically. We can actually see it in the distribution of longevity in plant and animal kingdoms. If the selective genes that are randomly significant become coupled with those that warrant longevity, this will be the only chance for longevity to emerge. This perhaps explains why tortoises and parrots live for a long time. There is no explicit correlation between an animal type and longevity: other birds have quite short life spans. Sometimes it is the environment that is beneficial for longevity, which is why sequoias are the longest living organisms (five to six thousand years).

Procreation is without a doubt a necessary factor in evolution; limiting individual existence in time is just its consequence. An organism must arrive at procreation completely fit; we can say that its continued existence results from "inertia," that is, the "dynamic push" initiated by embryogenesis. Evolution is like a shooter who is trying to reach a particular target, for example, a flying bird; yet what is going to happen to the bullet when it has reached that target, where it is going to go, whether it will keep flying for eternity or immediately fall to the ground does not have any significance either for the bullet or for the shooter. Of course, we should not simplify the issue too much. It is difficult to compare organisms as diverse as sequoias and hydras with vertebrates. We know that complexities differ and that dynamic laws of such organisms have their own hierarchy. That a hydra is almost immortal actually means very little for man as the "interested party." The constant maintenance of the inner-systemic correlation of processes must be harder the higher the mutual dependence of the building parts, that is, the stricter the organization of the whole system. Every cell commits "molecular errors" in the course of its existence, the sum of which it is no longer able to compensate for after some time. Or rather, it cannot do it while it exists in its current form. Division is a form of regeneration, after which the processes start again as if from scratch. We do not know why this is the case. We do not even know whether it has to be this way. Neither do we know whether these are inevitable phenomena because evolution never showed any "ambition" to solve the problem of homeostatic regulation *over any period of time*. Its efforts were entirely focused on another issue, which it approached up front: species longevity, the supraindividual immortality of life as a sum of homeostatic transformations on the scale of the planet. It did manage to solve this one.

CONSTRUCTING CONSCIOUSNESS

Anyone who has spent some time observing the behavior of an amoeba, going on a hunt in a droplet of water, must have been surprised by its similarity to intelligent, not to say human, action, demonstrated by this little drop of protoplasm. We can read accounts of such hunts in an old but excellent and worthy book by H. S. Jennings, *Behavior of the Lower Organisms* (1906). Crawling at the bottom of its droplet, an amoeba encounters a smaller one, which it starts to engulf by protruding its pseudopodia. The other one is trying to break free, but its attacker is holding on tight to a part of its body. The victim's body keeps stretching—until it breaks. The remaining part of the saved victim departs with some reasonable speed, while the attacker sprays what it has devoured with its plasma and then sets on its way. In the meantime, the part of the victim that has been "devoured" is moving vigorously. Swimming inside the protoplasm of the "predator," it suddenly reaches its epineuston, breaks through it, and crawls outside. "Surprised," the attacker at first allows its prey to escape but then immediately begins to chase it. A series of rather grotesque situations ensues. The attacker catches up with its victim several times, but the latter always manages to escape. After many futile attempts, the attacker "gives up," abandons the chase, and slowly departs in search of some better luck on his hunt.

What is most astonishing about this example is the degree to which we are capable of anthropomorphizing it. The motivation behind the droplet of protoplasm's actions is entirely comprehensible to us: a chase, the devouring of a victim, initial stubbornness in running after it, and eventually, giving up on the chase after "realizing" that it is not worth it.

It is not an accident that we are discussing this in a section devoted to the "building material of consciousness." We ascribe consciousness and intelligence to other humans because we ourselves possess both.

To some extent, we also ascribe both to animals that are close to us such as dogs or apes. However, the less an organism resembles us in its design and behavior, the more difficult it is for us to accept that it may also experience emotions, anxieties, and pleasures. Hence the quotation marks that I have used in the story of the amoeba's hunt. The building material from which its organism is "made" may be incredibly similar to that of our bodies, yet what do we know, what can we guess, about the experiences and sufferings of a dying beetle or snail? A situation in which an "organism" is a system that consists of some kinds of cryotrons and wires maintained at the temperature of liquid helium, or is a crystalline block, or is even a gas cloud held together by electromagnetic fields, evokes more resistance and more stipulations.

We have already discussed this issue when speaking about "consciousness in an electronic machine." Now we only need to generalize what was said earlier. If the fact that x is conscious is only determined by the behavior of this x, then the material from which it is made does not matter at all. It is therefore not just a humanoid robot or an electronic brain but also a hypothetical gaseous–magnetic system, with which we can have a chat, that belongs to a class of systems endowed with consciousness.

The general problem can be formulated as follows: is it really possible that consciousness is a state of a system that can be reached via different design methods and by using different materials? So far we have claimed that not everything that is alive is conscious—yet everything that is conscious must be alive. But what about consciousness manifested by systems that are clearly lifeless? We have already encountered this snag and have managed to deal with it. It is not so much of a problem when the human brain, which is to be reproduced in some other material, is used as a model. Yet the brain surely is not the only possible solution to the problem of "how to construct an intelligent and sentient system." When it comes to intelligence, our resistance will not be too strong because we have already constructed prototypes of intelligent machines. It is a little trickier with "sentience." A dog reacts to a hot object it has touched; does this mean that a system with a feedback loop that gives out a cry as soon as a burning match approaches its receptor is also sentient? Not at all, this is just mechanical imitation, we hear. We have heard this plenty of times. Such stipulations assume that alongside intelligent actions and reactions to stimuli, there also exist some "absolute beings," such as Intelligence and Sentience, united in the Duality of Consciousness. But this is not the case.

Physicist and science fiction writer Anatoly Dneprov has described an experiment in his novella, whose aim was to debunk a thesis about "infusing with spirituality" a language-to-language translation machine by replacing the machine's elements such as transistors and other switches with people who have been spatially distributed in a particular way. Performing the simple functions of signal transfer, this "machine" made of people translated a sentence from Portuguese into Russian, while its designer asked all the people who constituted the "elements" of that machine what this sentence meant. No one knew it, of course, because the language-to-language translation was carried out by the system as a dynamic whole. The designer (in the novella) concluded that "the machine was not intelligent." But a Russian cyberneticist issued a reply in a journal that had published the story, pointing out that if the whole humanity was distributed so that every person corresponded on a functional level to one neuron in the brain of the designer from that novella, then the system as such would be able to think, but only as a whole, and none of the persons participating in that "human brain game" would understand what that "brain" was thinking. However, this does not of course mean that the designer himself lacks consciousness. A machine can even be constructed from pieces of string or slightly rotten apples, from gas atoms or pendulums, from small flames, electric impulses, radiating quarks, and anything else we want, as long as it is a dynamic equivalent of the brain on a functional level, and then it will behave in an "intelligent" way—if by "intelligent" we mean being able to act in a universal manner while aiming at the goals chosen in a comprehensive selection process and not preprogrammed in advance (such as instincts in insects). Only a technical difficulty can disable any of those actualizations (there will not be enough people on Earth to enable "repeating" with them—as neurons—the human brain; it would also be difficult to avoid connecting them by some kinds of telephones, etc.). Yet these problems do not really apply to any of the reservations raised against "machinic consciousness."

I once said (in my *Dialogues*) that consciousness is a system's characteristic one learns about by being this system oneself. But we are not talking about just any system. It is not even that this must be a system located outside our body. In each of its eight trillion cells, there are at least several hundred enzymes that are sensitive to the concentration of a particular chemical product. The enzyme's active group is a kind of "input." The enzymes thus "sense" the lack or excess of the product—

which starts their own proper reactions—yet what do we as owners of all those cells and enzymes know about this process? As long as only birds or insects could fly, "flying" was equivalent to "living." Yet we know too well that devices that are completely "dead" are also able to fly today. It is the same thing with the problem of intelligence and sentience. The conclusion that an electronic machine will eventually be capable of intelligent thinking but not of feeling, of experiencing emotions, is a product of the same misunderstanding. It is not like some neural cells in the brain have properties of logical switches, while others just occupy themselves with "experiencing emotions." Both types are very similar and only differ when it comes to the *place* they occupy in the neural net. Similarly, the cells of the visual and auditory areas are actually identical. It is therefore entirely possible that crossing the neural pathways so that the auditory nerve connects with the occipital lobe, and the optical nerve with the auditory cortex, as long as such an intervention is performed early enough (e.g., in an infant), would lead to successful seeing and hearing, even if one was to "see" with the auditory cortex and "look" with the visual cortex. Even quite simple electronic systems already have "reward" and "punishment" types of connections, which are functional equivalents of "pleasant" and "unpleasant" experiences. This binary-value mechanism is very useful because it accelerates the process of learning—which is, of course, why evolution developed it. Generally speaking, we can thus say that a class of "intelligent homeostats" contains living brains as one of its *sub*-classes, the other ones being made up of homeostats that are completely "dead" in a biological sense. However, this "deadness" only stands for its nonprotein character as well as for the lack of many individual parameters possessed by us, living cells and organisms. It would be quite difficult to classify a system which—even though it was built, for example, from electromagnetic fields and gas—would be capable not only of conducting thought operations and reacting to stimuli but also of reproducing, drawing "nourishment" from its surroundings (e.g., from an electric socket), moving in any direction, growing, and subjugating various functions to its existence as its overarching principle.

In other words, when it comes to consciousness in homeostats, we do not so much need "in-depth" answers as solutions. Does this mean we have returned to our starting point to engage in a tautology? Not at all. We have to decide empirically which parameters of the system must remain constant so that consciousness can manifest itself in them. Since the boundaries between "transparent" and "opaque," or "clear" and "fuzzy," consciousness are fluid, the boundaries of such states will have to be drawn arbitrarily, just as it is only in an arbitrary manner that we can decide whether our friend Mr. Smith is already bald or not yet. In this way, we shall obtain a set of parameters that is needed to form consciousness. If they are all displayed by a random system (e.g., one built from old iron stoves), we shall ascribe consciousness to it. Yet what if they are different parameters, or parameters with slightly different values than the preselected ones? Then, according to our definition, we shall say that the system does not have a human type of consciousness—which will be absolutely true. But what if, even if it does not display any of those parameters, the system behaves like a genius who can understand all the people at the same time? This does not change anything, since, clever as it is, it lacks human consciousness: no human is equally smart. Is this not sophistry? someone may ask. It is, of course, possible for a system to display a consciousness that is "different" from human consciousness, like that "genius" mentioned earlier does, or one that claims to draw the greatest pleasure from bathing in solar radiation.

Here we transcend the limits of language. We do not know anything about the capabilities of this "other consciousness." Naturally, if it was to turn out that "human-type" consciousness was characterized by parameters A, B, C, and D, to which would correspond, respectively, the values 3, 4, 7, and 2, and if some kind of system displayed the values of those parameters as 6, 8, 14, and 4, if it manifested what we would perceive as a rather unusual intelligence, we would have to wonder whether the risk of extrapolation (that would consider it as possessing some kind of "double consciousness") would be permissible. What I have just said sounds both naïve and crude. What is at stake here is that neither those parameters nor their values would probably be isolated but would rather function as nodes in the "general theory of consciousness," or rather, in the "general theory of intelligent homeostats whose degree of complexity is not lower than the complexity of the human brain." It will be possible to perform some extrapolation within that theory, which will naturally carry a certain risk. How should extrapolation hypotheses be verified? By constructing "electronic attachments" for the human brain? Yet we have already said a lot here, perhaps even too much; the most sensible thing will thus be to fall silent. We should only add—which probably goes without saying—that we do not believe in the possibility of constructing intelligent agents from pieces of string, rotten apples, or iron stoves—just as it is hard to construct palaces from birds' feathers or soap foam. *Not every* kind of material is equally useful as an ingredient for a design in which consciousness is to be "started." But this should be self-evident enough, which is why we are not going to say any more about it.

ERROR-BASED CONSTRUCTS

The thermodynamic paradox concerning the bunch of monkeys blindly hitting typewriters' keys until they accidentally produce *Encyclopaedia Britannica* was actualized by Evolution. A countless number of external factors can increase the mortality of a population. Selection for high fertility is an answer to this problem. It is a directional consequence of a directionless action. Through the superimposition of two systems of transformation, each one of which is random with regard to the other one, an ever more highly organized order is produced.

Sexes exist because they are beneficial from an evolutionary point of view. The sex act enables the "confrontation" between two types of hereditary information. Heterozygosity is an additional mechanism that both popularizes "construction innovations," that is, mutations, within a population and protects organisms against the harmful consequences of the manifestation of those "innovations" in the development of an individual. A zygote is a cell that develops from the fusion of two sex cells, a male and a female, while the genes responsible for particular traits, that is, alleles, can be dominant or recessive. Dominant ones manifest themselves in the course of development, recessive ones only when they encounter their recessive partners. This is because mutations are usually harmful, and thus an individual that has been designed according to the new mutated genotype plan typically has less of a chance of survival than a normal individual. Yet at the same time, mutations are inevitable as attempts to find the way out of a critical situation. Flying insects sometimes produce wingless offspring—which usually perish. When the land sinks or the sea rises, what used to be a peninsula can become an island. Wings carry flying insects to waters in which they die. It is then that wingless mutants become a prospect for the continuation of the species. Mutations are therefore both harmful and useful. Evolution combined the two sides of this phenomenon. A mutating gene is usually recessive. On encountering a normal one—which is dominant—it does not manifest itself in the design of a mature organism. Yet individuals born in this way carry a latent mutated trait, which they pass on to their offspring. Early recessive mutations must have occurred with the same regularity as dominant ones did, yet the latter were eliminated by natural selection because all the traits—including the very mechanism of heredity, with its tendency toward mutation ("mutability")—are subjected to it. Recessive mutations have become more numerous, creating within a population its emergency unit or evolutionary reserve.

This mechanism, mainly based on errors of information transfer—which mutations are—is not a solution to which a human designer would resort. Under some circumstances, this mechanism allows for new design features to manifest themselves while selection is not at work. This takes place in small and isolated populations, in which, owing to the frequent crisscrossing of individuals descendant from the same parents, and owing to the ensuing uniformity of the genome, mutated recessive traits can encounter one another so often that suddenly we end up having a significant number of phenotypic mutants. This phenomenon is called "genetic drift." Certain forms of organisms that cannot be explained otherwise may have emerged in this way (e.g., the gigantism of deer's antlers). But we do not actually know whether it was this particular factor that formed the great bone "sails" on the backs of Mesozoic lizards. We cannot solve this problem because sex selection may also have been the crucial factor, but we do not know anything about the tastes of lizards from millions of years ago.

The fact that the frequency of mutations is itself a hereditary trait, that certain genes increase or decrease it, casts some rather peculiar light on this issue. Mutations are thought to be accidents that change the text of the genetic code and thus lose control over its transfer. Even if they were indeed random in the past, selection evidently could not eliminate them. From a design viewpoint, it is very important to know whether it could not because it "did not want to," that is, because a nonmutating species loses its evolutionary plasticity and, with the changes taking place in the environment, becomes extinct, or because the benefits coincide with an objective necessity, that is, with the fact that mutations are inevitable as a result of the statistical motions of molecules, which cannot be controlled.

From an evolutionary point of view, such a distinction does not matter, yet it can be significant for us because, if the reliability of information-carrying molecular systems such as genes is inevitable, how is it possible to design unreliable systems whose whole level of complexity is equal to an organic one? Let us imagine we are trying to create "cybernetic sperm," which, by digging into the crust of a foreign planet, are to create the machine we need from that planet's building material. A "mutation" can lead to this machine becoming useless. Evolution copes with this problem since, as a statistical designer, it never places its bets on singular solutions: its stake is always a population. This is a solution an engineer cannot accept: is he supposed to grow "a forest of developing machines" on the planet and only then choose the most efficient one from them? But what if the task involved designing systems with a higher degree of complexity than the genotypic one—for example, systems that are to program "hereditary knowledge" in advance, as previously discussed? If, beyond a certain threshold, an increase in complexity automatically increases mutability, instead of an infant who is an expert at quantum mechanics, we may obtain someone who is mentally disabled. We are currently unable to solve this problem: some further cytological and genetic research is needed.

The problem of tumors is linked to the question of the control of information transfer and of intercellular correlation. Cancer most probably results from a chain of successive somatic mutations. The literature on the subject is so extensive that we are unable to go deep into it. Let us just say that there are no data that would discredit this thesis. Cells divide in tissues throughout the course of their whole lives. As every division may be accompanied by a mutation "slip," the possibility of tumors emerging is proportional to the number of divisions and thus to the length of an individual's life. In reality, susceptibility to tumors increases exponentially as an organism ages. This most probably results from the fact that certain somatic mutations are a form of preparation for precancerous mutations that follow—which, after a series of further divisions, end up producing tumor cells. An organism can defend itself to some extent against the invasion of cancerous growth, yet its defense mechanisms weaken with age—which is another factor that affects the emergence of tumors. There are a number of carcinogenic factors, including some chemical compounds and ionizing radiation: what they have in common is their harmful effect on chromosomal information. Carcinogenic factors thus act in a nonspecific way, at least partly: they constitute "noise"—which increases the probability of further errors occurring during the process of cell division. Not every somatic mutation leads to cancer. There are also some nonmalignant tumors that develop from certain mutations; a cell needs to be damaged in a way that does not lead to its death but only to its nucleus, as a regulator, escaping from under the control of the organism as a whole.

Does this ultimately mean that mutations are an inevitable phenomenon? This is a disputable point because it is equally possible that we are dealing with a remote consequence of an initial design premise adopted by Evolution. A somatic cell does not contain more genotypic information than does the sex cell from which the whole organism developed. As the latter allowed for mutability, the somatic cell—which is its derivative—will also inherit this trait. Cells in the central nervous system are not subject to cancer, and neither do they divide—while a transformation can only occur during the divisions that follow. In a way, cancer would thus be a consequence of a "mutation decision" made by Evolution in its earliest stages.

The virus hypothesis can be reconciled with the mutation one owing to the significant biochemical kinship between viruses and genes. A "cancer gene" can in a sense be a "cancer virus." A virus is a name we give to a system that is foreign to an organism and that enters it from outside. This is actually the only difference.

Matters are complicated by the great diversity of tumors and their various types, such as sarcomas, which mainly occur in young individuals. Besides, cancer is not some kind of fatalistic necessity because people who live until a very advanced age do not have to succumb to it. An explanation that just refers to complete randomness is not satisfactory because it is possible to identify (e.g., in mice) pure lines that differ significantly with regard to their susceptibility to tumors, which means that cancer is a hereditary trait. Such hereditary tendencies have not really been identified in humans. However, it is very difficult to differentiate between the decreased frequency of cancer-inducing mutations and the possible high systemic resistance because we know that an organism can destroy cancerous cells, provided they are not numerous.

No matter how these as yet unresolved problems are going to be explained, we believe that while cancer therapy—despite its relatively modest achievements so far (especially in preventative care)—can count on some significant advances in the domain of pharmacology (thanks to highly selective cytostatic drugs), the complete elimination of cancer susceptibility does not seem possible. It is because cancer is a consequence of one of the principles of the cell's functioning that lie at the very foundations of life.

BIONICS AND BIOCYBERNETICS

We have discussed both the dynamics of information transfer and the technology of its hereditary inscription (the latter one in the introduction to "The Creation of Worlds"). Together they form a method by means of which evolution combines the maximum stabilization of genotypes with their essential plasticity. Embryogenesis consists not so much in initiating selected programs of mechanical growth as in setting in motion regulators endowed with significant autonomy and only equipped with some "general instructions." Sexual development is not therefore just a "race" between biochemical reactions involved in procreation but rather a constant collaboration and coemergence between them as a whole.

In a mature organism, an ongoing interplay between the hierarchies of regulators from which this organism is built also takes place. When providing reaction variants that have not been preprogrammed too rigidly, a logical extension of the principle "let them get by in any way they can" equips an organism with an autonomy of the highest order, thanks to constructing the "regulator of the second kind"—the nervous system.

An organism is therefore a "multistat": a system with so many possible states of equilibrium that in the course of an individual life, no doubt only some of them can be actualized. This principle applies to both physiological and pathological states. The latter count as specific kinds of states of equilibrium, despite the unusual values that certain parameters adopt in them. An organism also "gets by in any way it can" when harmful reactions start to reoccur in it. This tendency to enter into a regulatory circle that is full of errors is one of the consequences that emerge from the activity of the highly complex homeostatic pyramid with multilevel stability—which every metazoon is.

The latter cannot be thrown off this state by the antagonism of moderately efficient superior regulation, which is normally based on a one-dimensional scale of oscillation between two values (hindering excitation; increasing or lowering blood pressure; increasing or decreasing its acidity; accelerating or slowing down one's pulse, bowel peristalsis, breathing, or glandular secretion, etc.). There exists fully local regulation—at the limits of the brain's supervisory range (wound healing), which diminishes in old age ("the anarchy of the organism's peripheries": local degenerative changes that can be easily observed, e.g., on the skin of individuals of advanced age), as well as organ regulation, systemic regulation, and finally, holistic regulation. Two methods of transferring control and reflexive information crisscross here: discreet and analogue. The first one is normally used by the nervous system, the second by the system of endocrine glands. Yet even this is not a diametrical split because signals can be addressed through a cable connection (like in a telephone), or they can travel along all the information channels while only the addressee proper responds to them (like in the case of sending radio signals, which—even though they can be received by anyone—are directed only at one select ship in the sea). When "the matter is of vital importance," the organism begins a double information transfer: a threat immediately leads to enhancing the readiness of tissues and organs via neural pathways, while the hormone of "analogue activity," adrenalin, gets pushed into the bloodstream. This multiplicity of information channels guarantees its proper functioning even when some signals fail to arrive.

We have already discussed bionics, a science that is involved in actualizing on a technical level solutions observed in the kingdom of living organisms. The study of sensory organs—which are usually far inferior to technological sensors—has been particularly productive. Bionics is an activity undertaken by a biotechnologist who is interested in short-term results. The modeling of living organisms (especially of the nervous system, its parts and sensory organs)—which is close to bionics but which does not have as its goal such technical immediacy but rather focuses on studying the function and structure of organisms—belongs to biocybernetics. In any case, the boundaries between these new subdisciplines are fluid. Biocybernetics has already made some significant inroads into medicine. It includes the prosthetics of organs and functions (artificial heart, heart–lung machine, artificial kidney, under-skin impulse generators for the heart, electronic limb prostheses, reading and orientation devices for the blind—work is even being conducted on introducing impulses into blind people's nondamaged optic nerve, outside the eyeball, which ties in with the phantomatics we have proposed) and diagnostics as a way of introducing the "doctor's electronic helpers" both in the form of diagnostic machines—two versions of which already exist (a general diagnostic device and a specialist one)—and machines that directly extract the necessary information from the patient's organism (a device that autonomously performs an ECG or an EEG test and that undertakes an automatic preselection, thus eliminating irrelevant information and providing final results that have a diagnostic value). We also have "electronic control attachments," among which we can list an autonomous narcotizer—which both examines a number of parameters in an organism (e.g., bioelectricity in the brain, blood pressure, the degree of its oxygenation, etc.) and, when necessary, increases the supply of the narcotic or sobering agent, or one that increases blood pressure as soon as it has dropped; as well as equipment whose task is to keep permanent watch over certain parameters of the patient's organism, for example, a device that he always carries with him and whose task, if he suffers from hypertension, is to offer a continuous supply of the necessary agent that will keep his blood pressure at a normal level. This overview is both brief and incomplete. We should point out that medicine's traditional means, medication, belongs to a group of "analogue informers," since it is typically administered "in a general way," by being introduced into the body's orifices, internally, or into the blood channel, where this medication is supposed to find its correct systemic or organic "addressee" "by itself." Acupuncture, in turn, can be considered to be a method of introducing "discrete" information by irritating neural stems. While pharmacopoeia is an activity aimed at changing the homeostat's inner state directly, acupuncture is a way of acting on the homeostat's "inputs."

Like every designer, evolution cannot expect to arrive at just any outcome. For example, the mechanism of "reversible death" that various spora, algae, or even small metazoa experience is amazing. Body temperature maintenance in mammals is also very useful. Combining these traits would be a perfect overall solution, but it is not possible. Even though hibernation in certain animals is close to it, it is not an actual case of "reversible death." Life functions, blood circulation, breathing, and metabolism do slow down, but they do not stop. Besides, such a state exceeds the regulatory capacity of the phenotype's physiological mechanisms. The possibility of its occurrence needs to be programmed in advance. Yet this state is particularly valuable, even more so in the era of cosmonautics—and especially in the form it takes in bats.

Prior to their formation, all our ecological niches were seemingly full. Insect-eating birds thus filled both daytime and nighttime (the owl); there did not seem to be any room for any new species—either on Earth or in the trees. Evolution thus led bats into the twilight "niche," in which day birds were already beginning to fall asleep, while night birds had not yet gone hunting. Under those circumstances, the eye was rendered helpless by the changing light conditions—which were poor. Evolution therefore developed a supersonic "radar" in bats. Finally, bats found a hiding place in caves' ceilings—which, prior to that, also used to be an empty ecological niche. But the hibernation mechanism of those flying mammals is the most amazing thing. Their body temperature can drop to zero. Metaplasia practically comes to a halt then. The animal does not look like it is asleep but rather like it is dead. Awakening starts from an increase in muscular change. After several minutes, blood circulation and breathing are already working, and thus the bat is ready to fly.

It is possible to induce a human being into a very similar state of deep hibernation—by means of an appropriate pharmacological technique and a cooling procedure. This is extremely interesting. We know of some cases in which congenital diseases caused by a mutation, which consist in an organism being unable to produce some vital bodies, can be compensated for by introducing such bodies into tissue or blood. Yet in this way, we are only restoring, temporarily, the physiological norm. Hibernation treatment, in turn, exceeds this norm, and thus also the possible systemic reactions that have been genotypically preprogrammed. It turns out that while regulatory potential is limited by heredity, it can be expanded by certain procedures. We return here to the question of the "genetic littering" of humanity—which is caused, indirectly, by natural selection coming to a halt in a civilization and, directly, by the products of civilization that increase mutability (ionizing radiation, chemical factors, etc.). Medical prevention of genetic diseases and ailments turns out to be possible, without changing the deficient genotypes, since it is not the genotype itself but rather a maturing or an adult organism that is being influenced medically. Yet such treatment has its limits. Deficiencies caused by an early manifestation of the damage to the genotype—for example, thalidomide-caused disability—would of course be incurable. In any case, medical and pharmacological treatment seems to us today to be most natural because it remains within medicine's remit. However, the removal of the "lapses" of genetic code may turn out to be a simpler procedure (although by no means an innocent one), and also a more radical one in its consequences than belated therapy applied to deficient systems.

It would be difficult to overestimate the prospects of this "antimutational and normalizing" autoevolution: changes to genetic code would first lead to a reduction in, and then elimination of, congenital somatic and psychological defects, thanks to which masses of unfortunate disabled people—of whom there are millions today and whose number is still on the increase—would disappear. And thus the results of gene therapy, or rather their bioengineering, would prove to be salutary. Yet every time it turns out that it is not enough to remove the mutated gene, and that this gene has to be replaced by another one, the problem of "trait design" presents itself before us, with all the dangers it entails. One of the Nobel laureates, who received the prize precisely for his studies on heredity, and thus may be said to be directly interested in similar achievements, declared that he would not want to live to see them actualized owing to the terrifying responsibility man would then have to take on.

Although creators of science deserve the greatest respect, it seems to me that the preceding point of view is not worthy of a scientist. One cannot simultaneously make discoveries and avoid taking responsibility for their consequences. We know the outcomes of such behavior in some other, nonbiological disciplines. They are pitiful. The scientist tries in vain to narrow down his research so that it takes the form of information gathering, which is protected with thick walls against problems to be covered by its application. Evolution, as we have already demonstrated it, explicitly and implicitly, acts ruthlessly. In gradually getting to know its engineering activities, man cannot pretend that he is gathering solely theoretical knowledge. Someone who gets to know the consequences of his decisions, who acquires the power to make them, will carry the burden of responsibility—something Evolution as an impersonal designer coped with easily because it did not carry this burden.

IN THE EYES OF THE DESIGNER

As a creator, Evolution is an unbeatable juggler, who is performing his acrobatics in an extremely difficult situation, owing to the narrow technological scope of permissible maneuvers. No doubt it deserves more than admiration—it deserves to serve as a lesson to others. But if we turn our eyes away from the particular difficulties of its engineering activity and focus solely on its results, we are prompted to compose a lampoon of evolution. Here are our accusations, from less general to more general ones.

1. There exists nonuniform excess of information transfer and organ construction. According to the principle detected by Dancoff, Evolution maintains the excess of genotypically sent information at the lowest possible level that can still be reconciled with the continuation of the species. It is thus like a designer who is not so bothered about all his cars reaching the finishing line: he will be entirely happy if just most of them do. This principle of "statistical design," in which success is determined

by the majority and not the totality of results, is alien to our mentality.[4] This is especially the case when the price to be paid for the low level of information excess is not defects in machines but rather in organisms, including human ones: each year, 250,000 children are born with serious congenital defects. The minimum level of excess also applies to the design of individuals. As a result of the nonuniform wearing out of functions and organs, an organism ages unevenly. Deviations from the norm occur in various directions; they usually take the form of a "systemic weakness," for example, in the circulatory, digestive, or joint system. Ultimately, despite the whole hierarchy of regulators, the obstruction of one little blood vessel in the brain or the failure of one pump (the heart) causes death. Mechanisms that are supposed to prevent similar disasters—such as anastomosis[5]—tend to fail in the great majority of cases. Their presence rather resembles the "formal execution of the regulations," like in a factory in which fire equipment does find itself in the right place but is so inadequate or is attached in such a makeshift manner that, in case of need, it will actually prove useless.

2. The principle of the nonelimination of unnecessary elements from individual development contradicts the first principle—that of information economy, or even miserliness. The relics of the long extinct forms that preceded the present species are carried through, almost mechanically, as a result of inertia. And thus, for example, during embryogenesis, the fetus repeats the developmental stages of the long-gone embryonic developments, subsequently developing, the way a human embryo does, a gill, a tail, and so on. Those organs are used for some other purposes (bronchial arches develop into a jaw, a throat, etc.)—so on the surface, it looks like it does not matter. Yet an organism is a system so complex that any unnecessary excess of complexity increases the possibility of discoordination, of the emergence of pathological forms that lead to tumors, and so on.

3. The existence of biochemical individualism is a consequence of the second principle—that of "unnecessary complication." Cross-species transferability of genetic information is understandable, since pan-hybridization, that is, the possibility of crisscrossing bats with foxes, or squirrels with mice, would collapse the ecological pyramid of living nature. Yet this mutual foreignness of genotypes in different species also continues within a single species, in the form of the individual dissimilarity of systemic proteins. Even a child's biochemical individualism is different from such individualism in his mother. This has some serious

consequences. Biochemical individualism manifests itself in a fierce defense on the part of an organism against every protein that is different from its own, thus making any life-saving transplantations (of skin, bone, organs, etc.) impossible. To save the lives of people whose bone marrow was incapable of producing blood, the whole defense mechanism of their bodies had to be paralyzed—only then was it possible to undertake the transplantation of the tissue coming from human donors.

In the course of evolution, the principle of biochemical individualism was not disturbed, that is, it was not subject to selection for the uniformity of systemic proteins in a species, since an organism is designed to rely purely on itself. Evolution did not take into account the possibility of any support intervention from outside. We can therefore understand what caused the present situation, but this does not change the fact that medicine, in bringing aid to an organism, must at the same time fight against the "unreasonable" tendency of that organism to defend itself against salutary measures.

4. Evolution is unable to arrive at solutions through gradual changes, if every single one of such changes is not useful *immediately,* in the current generation. Similarly, it cannot solve tasks that require a radical reconstruction rather than just small changes. In this sense, it is "opportunistic" and "shortsighted." Very many systems are characterized by a complex design that would otherwise be avoidable. We are talking about something else than the "unnecessary complication" discussed in point 2. There we criticized evolution's excess as a result of *the road it had chosen to travel to reach its final stage* (egg cell–fetus–mature organism). In point 3, we demonstrated the harmfulness of unnecessary biological complexity. Now that we are becoming even more iconoclastic, we are criticizing the core design of certain individual solutions in the system as a whole. Evolution was unable, for example, to produce mechanical devices such as the wheel because the wheel must be itself from its very beginning, that is, it must possess an axis of rotation, a hub, a disk, and so on. It would thus have needed to have developed in a staggered fashion because even the smallest wheel is already a finished wheel—and not some kind of "transition" stage. Though a great demand for such a mechanical device never really existed in organisms, this example clearly indicates the kinds of tasks that Evolution is unable to solve. Many mechanical elements in an organism could be replaced by nonmechanical ones. Blood circulation, for example, could be based on the principle of an electromagnetic pump, if the heart was

an electric organ producing suitably changing fields, whereas blood cells would be dipoles or would possess a larger ferromagnetic insert. Such a pump would maintain a more steady blood flow, using less power, independently from the elasticity of the organ's walls—which have to compensate for the fluctuations of pressure with every push of the blood into the aorta. Since the organ transmitting the blood would base its activity on the direct transformation of biochemical into hemodynamic energy, one of the more difficult and still unsolved problems—that of properly nourishing the heart when it needs it the most, that is, during the contraction—would disappear altogether. In the situation executed by Evolution, the contracting muscle crushes to some extent the light of the vessels that nourish it, as a result of which the flow of blood, and thus of oxygen, to muscle fibers temporarily decreases. The heart is of course still able to cope with this state of things, yet such a solution is all the worse because it is entirely unnecessary. The small excess reserve of blood supply results in a situation in which coronary disease is one of the main causes of death worldwide today. The "electromagnetic pump" solution has never been actualized, even though Evolution is capable of creating both dipole molecules and electric organs. However, such a project would require an entirely improbable yet simultaneous transformation in two systems that are almost entirely isolated from each other: blood-producing organs would need to start producing the postulated "dipoles" or "magnetic erythrocytes," while the heart would have to stop being a muscle and become an electric organ. As we know, such a coincidence of blind mutations is a phenomenon one might await in vain for as long as a billion years—which is actually what has happened. In any case, Evolution did not even perform a much more modest task, such as sealing up the opening in the interventricular septum in reptiles. It does not mind a decreased hemodynamic productivity because it leaves its creations with the most primitive biochemical and organic setup, as long as they are able, with its help, to keep working toward the preservation of the species.

We should note that at this point in our critique, we are not postulating any solutions that would be impossible in evolutionary, that is, biological, terms, such as certain material transformations (say, replacing bone teeth with steel ones or cartilage joint surfaces with synthetic Teflon ones). It is impossible to imagine any reconstructions of the genotype that would enable an organism to produce Teflon (fluorocarbon). However, it would be possible, at least in principle, to program in advance within the human genotype organs such as the previously mentioned "hemoelectric pump."

Evolution's opportunism and shortsightedness, or rather blindness, in practice means using solutions that first emerge randomly and replacing them only when an accident opens up a different possibility. Yet when the once obtained solution *blocks* the way for any others, no matter how perfect and how much more efficient they might be, the development of a given system comes to a halt. And thus, for example, the jaw of predator reptiles remained a very primitive system mechanically for tens of millions of years; this solution "was dragged" through almost all of the reptile branches (provided we do come from the same ancestors); a change for the better was "successfully carried out" only in mammals (predators such as the wolf)—and thus extremely late. As biologists have correctly observed a number of times, Evolution is a diligent constructor only in working out absolutely vital solutions, as long as they serve the organism during the phase of its full vitality (for sexual reproduction purposes). But everything else that does not have such critical significance is more or less abandoned, left to the devices of accidental metamorphoses and completely blind strokes of luck.

Evolution is, of course, unable to predict any consequences of its present activity—even if it was to lead the whole species down a developmental blind alley, while a relatively small change would be able to prevent it. It actualizes what is possible and convenient right away, without caring about the rest. Larger organisms have larger brains, in which the number of neurons exceeds the increase in mass. This explains its apparent fondness for "orthoevolution," a slow yet continuous enlargement of the body size—which nevertheless very often turns out to be a real trap and a tool of future extinction: not a single one from the old giant branches (Jurassic reptiles) has survived up to this day. And thus, with all its miserliness, which it manifests in undertaking only the most inevitable "alterations," Evolution is the most extravagant designer of all.

5. Evolution is also a chaotic and illogical designer. You can see this, for example, in the way in which it distributes regenerative potential among the species. An organism is not based on the principles of human technological design—*macroscopic* spare parts. An engineer designs things in a way that makes whole blocs in devices replaceable. Evolution designs a "microscopic principle of spare parts," which manifests itself all the time, since organ cells (of skin, hair, muscles, blood, etc.—apart from a few, such as those of neurons) are being constantly exchanged—

through the division of other cells. The descendant cells are these very "spare parts." This would be a perfect principle, better than the engineering one, if it was not so often defied in practice.

The human organism is built from billions of cells. Every one of them contains not only genotypic information that is necessary for the functions it performs but also complete information—the very same that an egg cell possesses. Theoretically, it would thus be possible to develop, say, cells of the tongue's mucous membrane into a mature human organism. In practice, this is not possible because such information cannot be activated. Somatic cells lack embryogenetic potential. We do not really know why this is the case. Some inhibitors (growth restraints) may be at work here, in response to the principle of intertissue collaboration. Incidentally, recent studies have demonstrated that cancer growth supposedly consists in the atrophy of those inhibitors (histones) in cells that have undergone somatic mutation.

In any case, it would seem that all organisms, at least at the same stage of development, should undergo regeneration roughly in the same way, since they all possess similar excesses of cellular information. Yet this is not the case. There is in fact no close relation between the place occupied by a species in the evolutionary hierarchy and its regenerative potential. The frog is a very poor regenerator—almost as bad as man. Not only is this disadvantageous from an individual point of view, it is also illogical from a design perspective. No doubt this state of events was caused by something that happened in the course of evolution. Yet we are not occupying ourselves at the moment with finding excuses for evolution's weakness as a creator of organic systems. The final stage of each evolutionary branch, that is, the currently living "model" that has been put into "mass production," reflects, on one hand, the current conditions that it is expected to cope with, and on the other, the billion-year-long road of blind trials and searches that all its ancestors had traveled. Present-day solutions, which are inevitably compromised, are also tarnished with the inertia of all the previous designs—which also involved compromises.

6. Evolution does not accumulate its own experiences. It is a designer that forgets about its past achievements. Every time it has to seek them anew. Reptiles undertook an invasion of the air twice—once as naked-skinned lizards; the second time, they had to develop organic, executive, and neural adaptation to flight conditions from scratch. Vertebrates left the ocean for the land and then returned to water—at which time

"aquatic solutions" had to be embarked on anew. The curse of every perfect specialization lies in the fact that it represents only an adaptation to the current conditions; the better the specialization, the easier it is for a change to those conditions to lead to extinction. Yet sometimes the best design solutions are crammed into various side lines—lines that are characterized by extreme specialization. The spectacled cobra's sensory apparatus that reacts to infrared radiation can detect differences in temperature to the level of 0.001 degree. The electric sense of some fish reacts to differences in voltage of 0.01 microvolt per millimeter. The auditory organ of moths (which are eaten by bats) reacts to the vibrations of supersonic echolocation of those "flying mice." The sensitivity of the tactile sense in some insects is already at the limit of being able to receive molecular vibrations. We know how developed the olfactory sense is in macrosomatic creatures, which include certain insects. Dolphins have a system of hydrolocation. The concave front section of their skull, cushioned in fat, is a reception screen for the bunch of transmitted vibrations. It acts as a converging reflector. The human eye reacts to individual quanta of light. When a species that has developed such organs becomes extinct, "evolution's inventions," similar to the one just described, die out together with it. We do not know how many of them have died out over the past millennia. If they do continue to exist, there is still no way of popularizing those "inventions" beyond the species, family, or variety in which they developed. Consequently, an old person is a toothless creature, even though this problem has already been solved, dozens of times—each time somewhat differently (in fish, sharks, rodents, etc.).

7. We know very little about the way in which Evolution makes its "great discoveries," its revolutions. They do happen: they consist in creating new phyla. It goes without saying that also here evolution proceeds gradually—there is no other way. This is why we can accuse it of complete randomness. Phyla do not develop as a result of adaptations or carefully arranged changes but are a consequence of lots drawn in the evolutionary lottery—except very often, there is no top prize.

We have already said much about the evolution of genotypes, which is why what I am going to present now, following Simpson,[6] will probably be understandable without any further explanation. In large populations, under low selective pressure, a reservoir of latent genetic variability develops (in recessively mutated genotypes). In small populations, in turn, the emergence of new genetic phyla can accidentally come to a halt. Simpson calls it a "quantum evolution" (this leap is nevertheless less revolutionary than the one previously postulated by Goldschmidt[7]—who called the results of hypothetical genotypic macroreconstructions "promising monsters"). The way it happens is that an abrupt transition from heterozygosity to homozygosity takes place in mutants. Traits that have so far remained latent suddenly manifest themselves within a wide range of genes at the same time. (A phenomenon of this kind must be exceptionally rare—occurring, say, once or twice per every quarter of a billion years.)

The isolation and shrinking of a population takes place most frequently at times of sudden increase in mortality such as during disasters and catastrophes. It is then that evolutionary radiations sporadically emerge from among the millions of organisms that are dying out. They are the new "test models" that come into being abruptly, in the way described earlier, and that had not been preselected in any way. Only the further course of evolution puts them to some "practical tests." Since Evolution's method is always random, the circumstances that are propitious for "great inventions" do not have to lead to the latter in a necessary way but only just in a probable one. It is true that an increase in mortality and isolation facilitates the "surfacing" of a greater number of phenotypic mutants from the "emergency" reserve that has so far remained hidden in the gametes, yet the reserve itself may turn out to be not so much a lifesaving invention, a new systemic form, as a patchwork of pointless and harmful traits. It is because, direction-wise, selection pressure does not have to coincide with mutation pressure. Land can turn into an island, while wingless insects will become winged purely by accident—which is going to make their situation even worse. Both are equally possible; it is only when vectors of mutation pressure and selection pressure point in the same direction that some considerable progress will be possible. Yet this phenomenon as we understand now is extremely rare. In the eyes of the designer, this situation equals one involving stocking supplies on rescue boats on a ship in such a way that, after a disaster has struck, the castaways are met by some surprises—they wonder what is inside the container with "iron ratio" written on it on their boat: is it freshwater or maybe hydrochloric acid, cans full of food or full of stones? And even though it sounds grotesque, the preceding picture precisely matches Evolution's method and the circumstances in which it arrives at its greatest achievements.

The monophyletic character of the emergence of reptiles, amphibians, and mammals proves that we are right: they all developed only once;

each class appeared only a single time across all the geological epochs. It would be very interesting to know the answer to the following question: what would have happened if, 360 million years ago, early vertebrates had not come into being? Would it have been necessary to have waited another "hundred million years"? Or would the reoccurrence of this mutation-based creation have been even less probable? And did this invention not eliminate another construct that was potentially possible?

These questions cannot be answered, since what happened, happened. However, as we have already said, mutation almost always involves a change of one type of organization into another—although frequently a "pointless" one in adaptation terms. The high level of the genotype's organization thus creates conditions in which a sequence of random draws makes constructing a more progressive variation or branch into a phenomenon whose probability is always close to 1—provided this sequence is very long. (By a "progressive" form, I understand a form which, following Huxley, not only dominates with its organization the present ones but also functions as a potential transition point on the way toward some further developmental stages.) Using the example of "great evolutionary breakthroughs," we have once again clashed, rather dramatically, with the evidence-based statistical character of the natural designer. An organism reveals how a certain system can be built from uncertain parts. Evolution, in turn, demonstrates how, with gambling—where there are just two stakes, life and death—one can practice engineering.

8. Given that we are moving toward a more fundamental critique of Evolution, we should censure, in passing, its control method. The feedback that controls genotypes is full of errors, which explains the "genetic littering" of the population. We shall now focus on one of Evolution's original and most fundamental premises: its choice of building material. Tiny gluey protein droplets are Evolution's test tubes and laboratories. From them it produces skeletons, blood, glands, muscles, furs, carapaces, brains, nectars, and venoms. The narrowness of this "production bottleneck" is surprising when compared with the universality of its final products. However, if we ignore the restrictions imposed by the cold technology, and if it is not the perfect skills of molecular and chemical acrobatics but rather the general principles that lie behind the rational design of optimal solutions that are of interest to us, then we are ready to make a number of accusations.

How can one imagine an organism that would be more perfect than a biological one? As a determined system, which is similar in this respect to natural systems, it will be able to maintain its ultrastability thanks to the supply of the most efficient, that is, nuclear, energy. If we give up on oxygenation, then the cardiovascular and lymphatic systems, the lungs, all the central breath regulators, the whole chemical apparatus of tissue enzymes, muscle change, and the relatively low and significantly limited muscle strength will become redundant. Nuclear energy will facilitate universal transformations. A liquid medium is not the best carrier—yet even such a homeostat could be built, if someone really wanted it. This opens up possibilities of acting remotely in various ways—either in a wired and thus discreet way (by using "cables" such as nerves) or in an analogue way (in which case, e.g., radiation becomes an equivalent of information-carrying analogue hormonal compounds). Radiations and fields of forces can also act on the homeostat's environment—in which case the primitive mechanics of limbs, with their plain bearings, will become redundant. No doubt a "nuclear energy-powered" organism looks both ridiculous and nonsensical in our eyes, yet it is worth imagining the situation of a person in a spaceship just taking off to evaluate properly the whole fragility and narrowness of the evolutionary solution. With increased gravitation, the body—which consists mainly of liquids—experiences a sudden hydrodynamic overload: the heart fails, blood either drains away the tissues or tears apart the vessels, effusions and edemas appear, the brain stops working a short while after oxygen supply has stopped, and even the bone skeleton turns out to be too weak a construct at that point to cope with the forces acting on it. Today man is the most unreliable element in the machinery he has created—he is also the weakest link, mechanically, in the processes at work.

Yet even giving up on nuclear energy, fields of forces, and so on, does not necessarily take us back to biological solutions. A system that is more perfect than a biological one is a system that has one more degree of freedom—on the level of materials. It is a system whose shape and function are not predetermined, and that can produce, if need be, a receiving organ or an effector, a new sense or a new limb or a new way of moving. In other words, it is a system that is able to accomplish directly, thanks to the control it has over its "soma," what we ourselves do in a roundabout way—by means of technologies, through second-degree regulators, that is, brains.

This roundabout character of our activity could nevertheless be eliminated. If you have as much time as three billion years, you can explore matter's mysteries so that it becomes redundant.

The problem of the building material can be approached in two ways: from the point of view of the organism's temporary adaptation in Nature, in which case the solution adopted by Nature has many positives, or from the point of view of its future prospects, in which case all its limitations come to the fore. The temporal limitation is the most significant one for us. If you have billions of years at your disposal, you can construct quasi-immortality—provided this is what you want. Evolution could not have cared less about it.

Why are we discussing the problem of aging and death in the section devoted to the weaknesses of the building material? Is this not rather a problem concerning the organization of the material? We have said ourselves that the protoplasm is, at least potentially, immortal. It is a permanently self-renewing order—which means that there is no need for processes to stop as a result of them falling apart at the very core of its design. This is a difficult issue. Even if we do have an idea about what happens in an organism within seconds or hours, we know almost nothing about the laws to which it is subject over the years. Our ignorance is successfully obscured by terms such as *growth, maturation,* and *aging*—yet they are just metaphorical and fuzzy names given to certain states, not precise descriptors.

Evolution is a designer–statistician; we already know this. But it is not only its species-building activity that has an averaging statistical character: the design of an individual organism is based on the same principles. Embryogenesis is a centrally controlled chemical explosion with a teleological precision, which is also informed by statistics—since genes do not determine the number or location of particular cells in the "final product." Taken in isolation, none of the tissues in a metazoon have to die; such tissues can be grown for years on artificial surfaces, outside the organism. The latter is therefore mortal as a whole, while its component parts are not. How are we to understand this? The organism undergoes various disturbances and injuries in the course of its lifetime. Some of them come from the environment; others are caused by the organism itself. The latter point is perhaps the most significant one. We have already discussed certain ways of derailing life processes—which, in a complex system, primarily stands for the loss of correlative balance. There are several types of such processes: the stabilization of pathological equilibrium, as in the case of a stomach ulcer; a vicious circle, as in the case of high blood pressure; and last but not least, snowballing reactions (epilepsy). Cum grano salis, cancer growth can be counted among such reactions. All those disturbances accelerate aging, but aging also takes place in individuals that never fall ill. Old age seems to be a consequence of the statistical nature of life processes, a very primitive representation of which is offered by a shot fired with pellets. No matter how precise the barrel is, the pellets disperse more widely the farther they have traveled. Aging is a similar distribution of processes that involves their gradual escape from central control. Death arrives when the distribution of those processes has achieved a critical value and when the reserves of all the compensatory apparatuses have become exhausted.

Such statistics—which are reliable as the originally accepted premise of the emergence of flow equilibrium (Bertalanffy's *Fliessgleichgewicht*, 1952) as long as the organisms being constructed from the approved elements are simple—are likely to fail once we have crossed a certain threshold of complexity. From this perspective, a cell is a more perfect creation than a metazoon, no matter how paradoxical this sounds. We have to understand that in saying this, we are using an entirely different language, or that we are dealing with entirely different issues than those with which Evolution was concerned. Death is its multiple outcome that results from constant change. It is a consequence of growing specialization and also of Evolution having used this and not that material at the beginning—the only material that it was possible to create at the time.

In reality we are thus not lampooning this nonhuman creator for real. We are rather interested in something completely different. We just want to become more perfect designers than evolution ever was—and we must be careful not to repeat its "mistakes."

RECONSTRUCTING MAN

What is at stake for us is trying to improve on man. Various approaches are possible in this respect. We can practice "maintenance engineering," that is, medicine. In that case, the norm, that is, an average state of health, becomes a model for us, so that certain actions need to be performed to enable every person to achieve this state.

The range of those actions is slowly increasing. It may even include embedding into the organism parameters that had not been anticipated on a genetic level (such as the possibility of hibernation, as mentioned earlier). We can gradually move to ever more universal prostheses and to overcoming the system's defenses to facilitate successful organ transplantation. All of this is being actualized now. First kidney and lung transplants have been carried out; transplants in animals are conducted on a much wider scale (the "spare" heart). In the United States, there even exists a "spare parts" association, which coordinates and supports research in this area. It is therefore possible to retune the organism gradually by changing its particular functions and parameters. This is probably going to be a dual-track process, which will occur under the pressure of objective necessities and technological possibilities. It will take the form of *biological* transformations (the elimination of defects, disabilities, etc., via transplants) and *prosthetic* transformations (when a "dead" mechanical prosthesis will be a better solution for the user than a natural transplant). It goes without saying that within this range, prosthetics must not lead to man's "roboticization." This whole phase, which will surely extend not just to the end of this century but also to the beginning of the next one, is premised on the acceptance of the basic "design plan" provided by Nature. The directives for designing the body, its organs and functions, together with the original premise to use protein as building material—with all its inevitable consequences such as old age and death—remain unchanged.

Expanding the human life span beyond the period of one hundred years *statistically* (i.e., so it becomes an average life span for an individual existence), without interfering with man's genetic information, seems unrealistic to me. Many learned gentlemen have already revealed to us that "as a matter of fact" and "in principle," man could live to be 140 to 160, since some individuals can live that long. This is an argument worthy of the line of thought that says that "as a matter of fact," every one of us could be a Beethoven or a Newton because those characters were also human. Of course they were human, just as long-lived highlanders from the Caucasus are human, but this is really of no consequence to the average population. Longevity is a product of particular genes; whoever distributes them across the population will make it statistically long-living. Any program for some more radical changes is impossible to actualize and will probably still be impossible in the next hundred years. We can only keep imagining a revolutionary program that would institute a reengineering of the organism—in a primitive and naïve way, for sure, but we can certainly do it.

First of all, we need to think what it is that we actually want.

Just as there is a scale of spatial quantities—from metagalactic clouds through to galaxies, local stellar systems, viruses, molecules, atoms, and quanta—there is a scale of temporal quantities, that is, of various lengths of time. The latter one roughly matches the former. Individual galaxies are the longest-lived (a dozen or so billion years), then we have stars (around 10 billion years), biological evolution as a whole (3–6 billion years), geological epochs (50–150 million years), sequoias (around 6,000 years), man (around 70 years), a one-day fly, bacteria (around fifteen minutes), viruses, cis-benzene, and meson (millionths of a second).

Designing an intelligent being whose individual life span will match that of geological epochs seems entirely implausible. Such a person would have to be the size of a planetoid—or give up on his permanent memory of past events. Of course, a field opens up here for some grotesque ideas straight from sci-fi: we can imagine some long-living creatures whose memory is located, for example, in the city's gigantic underground "mnemotrons," connected by means of UV waves with reservoirs of their youth memories from one hundred thousand years ago. The biological threshold seems to be the limit point of a realistic increase in life span (if it is a sequoia, then it is six thousand years). What should be the most important characteristic of such a long-living creature? Yet longevity cannot be a goal in itself. It has to serve some other purpose. Without a doubt no one can predict the future in a surefire way, either today or in a hundred thousand years. Autoevolutionary potential should thus be the main feature of the "improved model," so that it can develop in a way and direction that will match the requirements of the civilization created by it.

What is therefore possible? Almost everything, with just one exception. Having conspired in advance, people could decide one day, many thousands of years from now, "Enough! Let things be the way they are now; let them remain like this forever. Let us not change, seek, or discover anything new, since things cannot be better than they are now, and even if they could, we do not want it."

Even though I have outlined many unlikely things in this book, this one seems to me to be the most unlikely of them all.

CYBORGIZATION

Separate consideration needs to be given to the only project of human reconstruction proposed by scientists with which we are familiar today— a project that is still purely hypothetical. Yet this is not a project for universal reconstruction. It is supposed to serve some particular goals, that is, an adaptation to the Universe as an "ecological niche." It goes under the name of "cyborg" (which is an abbreviation of the term *cybernetic organization*). "Cyborgization" consists in removing the digestive system (except for the liver, and perhaps also sections of the pancreas), as a result of which the jaws and their muscles, as well as the teeth, will also become redundant. If the problem of speech is to be solved in a "cosmic" way—through the regular use of radio communication—the lips will disappear too. The cyborg has a number of biological elements, such as the skeleton, the muscles, the skin, and the brain, but this brain consciously regulates what used to be involuntary body functions. In the key points of its organism, we can find osmotic pumps, which, in case of need, inject nutritious or activation agents—medication, hormones, impulse-inducing substances—or agents that lower basic metabolism, or even agents that induce hibernation. Such readiness for self-hibernation can significantly increase survival chances in case of a breakdown.

Even though the cyborg is capable of running in anaerobic conditions (naturally, with an oxygen supply in its suit), its circulatory system is designed in quite a "traditional" way. The cyborg is not a partly prosthesized human; it is a human that has been reconstructed in part and equipped with an artificial system of nutritional regulation—which facilitates adaptation to various cosmic environments. But such reconstruction has not taken place on a microscopic level; that is, living cells are still the main building material of the cyborg's body. Besides, changes to its organization cannot be carried through to its offspring (they are not hereditary). Presumably "cyborgization" could be accompanied by the reconstructions of biochemistry. Thus it would be very desirable to make an organism independent from having to rely on the constant supply of oxygen. Yet this is a path to the "biochemical revolution" we have discussed. In any case, we know that there is no need to go as far as looking for bodies that store oxygen more efficiently than hemoglobin does to manage without the air supply for a relatively long period of time. Whales can dive for over an hour—something that is enabled by more than just the enlargement of their lungs. They have specially developed groups of organs for this purpose. We could thus also borrow some elements for our reorganization "from the whale."

We have not yet expressed our opinion on whether cyborgization is desirable. We are discussing this issue just to show that problems of this kind have actually been addressed by scientists. We should notice, however, that such a project would probably be impossible to actualize today (not only because of medical ethics but also owing to the low chances of survival after such a cumulated surgical intervention and as a result of replacing vital organs with various "osmotic pumps"), even though, as a matter of fact, this project is rather "conservative."

The main source of criticism arises not so much from the set of proposed operations as from their final result. Contrary to what it may seem, the cyborg is by no means a more universal human being than the "current model." It is rather man's "cosmic variant"—aimed not so much at all the celestial bodies as at those that resemble the Moon or Mars. The rather cruel procedures we have discussed actually yield a rather poor result with regard to the cyborg's adaptive universality. The strongest objection is raised by the very idea of "degeneralizing man," that is, producing various human types, more or less in the likeness and image of various kinds of ants. Perhaps those analogies did not occur to the original designers, but they do impose themselves even on the nonprejudiced. It is possible to perform hibernation without osmotic pumps, just as it would be possible to equip an astronaut with a number of microattachments (automatic ones or ones controlled by him) to introduce certain substances into his system. The cyborgian lack of lips therefore seems to be an effect aimed at the wide public rather than at biology experts. I admit that when it comes to similar reconstructions, it is easier to talk in general terms about a future necessity than to offer some improvements that seem convincing from a design point of view—even if they were to be still unachievable today on a technical level. Industrial chemistry is still hopelessly behind systems biochemistry, while molecular engineering, together with its information-carrying applications, is still in its infancy when compared with the molecular technology of organisms. Yet the measures that Evolution used—"out of despair," so to speak, rather than out of choice—when it was limited by the objective conditions of the "cold technology" and the very narrow selection of elements (practically just carbon, hydrogen, oxygen, sulfur, nitrogen, phosphorus, or, in trace amounts, iron, cobalt, and other metals), are not really an achievement in the field of constructing homeostats that would satisfy the Universe's high standards. When synthetic chemistry, information theory, and general systems theory develop much further, the human body will turn out to be the least perfect element in such a world. Human knowledge will exceed the biological knowledge accumulated in living systems. At that point, plans that, for the time being, are seen as aspersions cast against the perfection of evolutionary solutions will actually become realized.

THE AUTOEVOLUTIONARY MACHINE

Given that the possibility of man's reconstruction seems unlikely to us, we are inclined to think that techniques used for this purpose must also be unrealistic. Brain surgery, "bottled fetuses developing under the control of genetic engineering"—these are the kinds of images offered by the science fiction literature on this topic. Yet the procedures that will be used may actually be invisible. For several years now in the United States, digital machines programmed to arrange marriages have been in operation—although they are not yet commonly used. A "machinic matchmaker" selects couples that are best matched physically and intellectually. According to the (so far limited) results, the stability of machinically arranged relationships is twice as high as that of regular marriages. In recent years, the average age when people marry has become lower in the United States, while 50 percent of the marriages split up within the first five years—as a result of which there are many twenty-year-old divorcés and many children lacking proper parental care. A replacement for the family model of child-rearing has not been invented yet because it is not only a question of finding the means to support appropriate institutions (orphanages); parental feelings also have no equivalent, and their continued lack early on leads not just to a negative childhood experience but also to the development of often irreversible defects in the domain of emotions we describe as "complex." This is the situation at present. Humans form couples in a random, or we could say "Brownian," way. They get together after a certain number of transitory relationships, once they have encountered the "right" partner—the state of events which seems to be confirmed by mutual attraction. Yet this form of assessment is really quite random (since, in 50 percent of cases, it turns out to be erroneous). "Machinic matchmakers" change this. Research results provide the machine with the knowledge about the candidates' psychosomatic characteristics, which allows it to search for perfectly matched couples. The machine does not eliminate freedom of choice because it does not lead to just one candidate. Acting in a probabilistic manner, it offers choices from among a preselected group, which belongs to the confidence interval. The machine can select such groups from *millions* of humans, whereby an individual, acting traditionally, using "chance," can only come across several hundred humans in his or her lifetime. In this way, the machine actually fulfils the age-old myth about men and women being destined for each other yet looking for each other in vain. The important thing is for social awareness to accept this fact permanently. However, these are purely rational arguments. The machine expands the range of options, yet it does so by averaging things, above the head of an individual—thus depriving us of the right to errors and sufferings and to all the other afflictions of life. Yet someone may desire precisely such afflictions, or at least want to maintain the right to risk. Although there is a general conviction that people get married to stay married, someone may prefer to experience the twists and turns with a frivolously chosen partner in a marriage that ends terribly than to "live happily ever after" as part of a harmonious married couple. However, on average, the advantages of arranging marriages from the position of "better knowledge" at the machine's disposal are so much greater than the disadvantages that there is a significant chance for technology of this kind to be widely adopted. When it becomes a cultural norm, the marriage discouraged by the "machinic matchmaker" will perhaps be a forbidden and thus tempting fruit, while society will surround it with an aura that used to accompany, for example, mésalliances in the old days. It may also happen that a similar "desperate step" will be considered in some circles to be a "sign of exceptional courage," a way of "living dangerously."

"Machinic matchmakers" may turn out to have some very serious consequences for our species. When individual genotypes have been deciphered and then entered, next to "psychosomatic personality profiles," into the machine's memory, the matchmaker's task will lie in making a choice that not only matches humans with other humans but also genotypes with other genotypes. It will thus be a two-level selection process. First, the machine will isolate a class of partners who are matched on a psychosomatic level, after which it will submit this class to the second-level elimination process to reject candidates who are highly likely to produce offspring that would be considered undesirable for various reasons—for example, disabled (which will be a correct decision), or endowed with a low level of intelligence, or with unstable personalities (which raises some doubts, at least for the time being). Such behavior seems desirable as a way of stabilizing and protecting the genetic material of the species, especially at a time when the concentration of mutagen bodies in the environment of a given civilization is on the increase. It is a short way from stabilizing a population's genotypes to controlling their future development. We are entering here the domain of planned control, which offers a fluid transition to controlling the evolution of the species as such—since matching people's genotypes amounts to controlling evolution. This technology seems the least drastic one as it remains invisible—yet this also brings up a difficult moral dilemma. According to the rules of our culture, society needs to be informed about any significant transformations: such a "one-thousand-year autoevolution plan," say, would certainly count as a significant transformation. Yet offering information without providing any proper justification amounts to imposing future plans on people, instead of persuading them about the need to implement such plans. However, any such justification will only really be understandable to those who have extensive knowledge in the fields of medicine, evolutionary theory, anthropology, and population genetics.

Another unique aspect of this technology is that it enables us to obtain different results with regard to different traits of the system. It would be relatively easy, for example, to aim to distribute widely high intelligence as a natural characteristic of the species—although one that is not as frequent as one could wish. This would have tremendous significance in an era of mental competition between humans and machines. The most difficult thing to achieve by means of the previously described method would be a deep transformation of man's systemic organization. What kind of transformation are we talking about here? According to some researchers (e.g., Dart), we are "genetically burdened," or rather, we are characterized by the "asymmetry" of our desires for "good" and "evil" as a result of the fact that our ancestors practiced cannibalism during three quarters of a million years—not in exceptional circumstances, when facing death from hunger (which is what "regular" predators do), but rather on a regular basis. This has been known for quite a long time, yet these days, cannibalism is seen as a creative factor in anthropogenesis. It is claimed today that herbivorousness does not accumulate "intelligence" because bananas do not encourage one to develop any tactical maneuvers that would allow one to assess situations or strategies for approaching, fighting, and chasing one's victims. This is why anthropoids kind of stopped at a certain stage of their development, while the fastest development occurred in primitive man, since he hunted those who were his match on the level of acuity. This resulted in the most intense elimination of "the slow," because a mentally retarded herbivore at worst had to starve now and again, while not too bright a hunter, chasing others that were similar to him, inevitably died quite quickly. The "cannibalistic invention" was therefore supposed to have been an accelerator of mental progress, because interspecies struggle only guaranteed the survival of those who had the quickest wit—a trait that facilitated a universal transfer of life experience on to new situations. In any case, the Australopithecus—as it is this particular hominid form we are talking about here—was omnivorous. The Stone Age culture was preceded by the osteodontokeratic[8] culture, since a long bone became the first club—which happened by accident, after it had been chewed on. Skulls and bones were Australopithecus's first tools and cudgels, while blood vapor accompanied the emergence of first rituals. This does not mean that we have inherited from our ancestors an "archetype of criminality," because accumulated knowledge that would induce one toward specific actions beyond basic drives is not transmissible in this way. However, constant fighting must have shaped the human brain and the human body. The "asymmetry" of cultural history is also rather puzzling, with good intentions quite regularly turning into evil, while the reverse metamorphosis did not take place. Indeed, in one of the dominant religions, blood still plays a significant role—in the doctrine of transubstantiation. If there is some merit behind such hypotheses, and if the contents of our brains was formed as a result of phenomena occurring across those hundreds of years, then some kind of species improvement—with regard to this previously mentioned "asymmetry"—would be truly desirable.

Of course, we do not know today *whether* it should actually be embarked on or *how* it should be done. "Marriage machines" could take thousands of years before they arrive at the desired state of events because they are solely capable of maximizing evolution's natural speed—which is, as we know, very slow. In any case, the situation is as follows: the reservations we have with regard to the prospect of autoevolutionary transformations are caused not just by the latter's range and scope but also by the fluid and seamless way in which they occur. "Cutting up people's brains and bodies" evokes disgust, whereas "machinic marriage counseling" seems to be quite an innocent intervention—yet these are just two paths of different lengths that can both lead to analogous results.

EXTRASENSORY PHENOMENA

We have not yet touched on many important issues in this book, while we have dealt with many others in a more cursory way than they deserve. If now that we are coming to an end, we are going to mention telepathy and other extrasensory phenomena, we will do this just to avoid an accusation that, in devoting so much time to problems of the future, we have relentlessly mechanized matters of the spirit—and have thus developed a blindness of sorts. Given that there is considerable interest in telepathy today, also among certain scientists, is it not possible that learning more about it will make us radically change our physicalist worldview—or even that phenomena of this kind will become open to interventions from designers? If a human being can experience telepathy, and if an electronic brain can be a bona fide replacement for a human, then we have reached a simple conclusion that even such a brain will manifest a propensity for extrasensory cognition, provided it has been designed this way. This takes us straight to various new technologies of information transfer via "telepathic channels"; to nonhuman "telepatrons" and "telekinetors"; and also to cybernetic clairvoyance.

I am quite familiar with the literature on ESP (extrasensory perception). The arguments raised against it by scholars such as Rhine or Soal, and gathered in a dry yet intelligently written book by Spencer Brown,[9] seem convincing to me. As we know, the phenomena from the 1900s that took place in the presence of "spiritualist mediums," so thoroughly examined by the scientists of the day, came to a halt more or less at a time when infrared devices were introduced—which enabled the observation of everything that happened even in a fully darkened room. It looks like "spirits" are afraid not just of darkness but also of infrared binoculars.

The phenomena examined by Rhine and Soal have nothing to do with "spirits." As telepathy, they stand for a transfer of information from one mind to another without the support of any material (sensory) channels. As crypthestesia, they stand for the mind obtaining information from material objects that remain hidden, veiled, and remote—also without any help from the senses. As psychokinesis, they stand for a spatial manipulation of material objects by means of a purely mental effort—once again without any material effector. Last but not least, as clairvoyance, they stand for predicting future states of material phenomena without resorting to conclusions drawn from known premises ("a spirit's glimpse into the future"). Such research, conducted primarily in Rhine's laboratory, provided a great amount of statistical material.

The control conditions are strict, and the statistical results are rather significant. In case of telepathy, the so-called Zener cards are used most frequently. In case of psychokinesis, a dice-rolling machine is produced, with an experimenter attempting to increase or decrease the number of pips with each roll.

Spencer Brown attacks statistical methods by saying that in long random sequences, certain series of results of low probability may reoccur. The probability of this occurring is greater the longer the sequence. Phenomena such as "streaks of good luck" (and of bad luck) are known to all those involved in hazard. Brown claims that as a result of pure accident, there may occur, in the process of continuing long random sequences, an expansion of almost any kind of significant deviation from the average of all results. This thesis is actually confirmed by the fact known to anyone who has worked on constructing so-called random number tables: at times the machinery that is supposed to produce such numbers with the most chaotic distributions produces a series of ten or even a hundred zeroes in a row. Of course, this can happen with any number. Yet this is indeed a random result: statistical techniques used by scientists in experiments are never "empty" because they (i.e., their formulae) are filled with the material content of phenomena. But long-term observation of random sequences that are completely empty, that is, that have no connection whatsoever with any material phenomena, can indeed lead to the emergence of what look like very peculiar deviations. The latter's irrelevance, that is, accidentality, can be proven by the fact that these deviations are not repeatable but rather, after a period of time, become dispersed "by themselves" and then disappear, after which the results that follow once again slightly oscillate for a very long time around the expected statistical average. Thus, if we are expecting a phenomenon that does not take place, using a random sequence in the experiment, in reality we are just observing the behavior of that sequence, detached from any material significations, whereby the "significant statistical deviations" expand every now and then, after which they disappear without trace. Brown's argumentation is exhaustive, but I shall not outline it here in full because there is something else that convinces me about the nonexistence of the discussed phenomena.

If telepathic phenomena were real, if they formed a unique channel of information transfer, free from the noise disturbance to which any transfer of sensory information is subject, then biological evolution would no doubt have used such phenomena because this would have considerably increased the chances of different species in their struggle for survival. It would have been much easier for a pack of predators, for example, wolves, chasing a victim in a dark wood and dispersed by the trees in their chase, to be pulled back into formation by its leader if it remained in telepathic contact with its companions—a contact that supposedly does not depend on atmospheric conditions, on visibility, or on the presence of any material barriers. Last but not least, evolution would not have needed to have resorted to the troublesome and elaborate methods aimed at finding partners of opposite sexes. A regular "telepathic tendency" would have replaced smell, sight, hydrolocation, and so on and so forth.

The only puzzling case is that of a moth that is able to attract sexual partners from several kilometers away. At the same time, we know how sensitive the olfactory and tactile organs of insects' feelers are. The moth attracts partners while being placed in a little wire cage; this phenomenon is not known to be repeated if the moth is placed in a hermetic dish. We have shown earlier how sensitive various senses in animals can be. It would have been an entirely futile effort on the part of Evolution if telepathic phenomena were subject to the laws of natural selection. As long as such selection is at work, any characteristics of an organism would be able to escape it once they have manifested themselves for the first time. The fact that some moths, humans, or dogs show signs of telepathy in experiments leads to the conclusion that it is a trait of living organisms. This means that Mesozoic ancestors should have had it too.

If, in the course of two to three billion years, evolution was unable to accumulate the phenomenon beyond the level that is barely detectable in thousands of experiments, then we do not even need to analyze the statistical tool itself to conclude that the issue at hand does not open up any future prospects for it. Indeed, no matter what environment we look into, we can see the great potential usefulness of telepathic phenomena, coupled with their complete absence.

Deep-sea fish live in total darkness. Instead of using their primitive phosphorescing organs, with which they illuminate their surroundings in a small radius when looking for prospective partners and trying to avoid their enemies, should they not rather resort to their telepathic location? Should there not be some very strong telepathic links between parents and their offspring? Yet a female, if its offspring are hiding somewhere, will look for them using sight and smell, not any kind of "telepathic sense." Should night birds not have developed successful telepathic communication? How about bats? We could offer hundreds of such examples. We can therefore safely stop talking about any future prospective developments of telepathic technology because even if there is a seed of objective truth, that is, some kind of unknown phenomenon, in the networks of statistical prophets, it has nothing to do with extrasensory perception.[10]

When it comes to psychokinesis, it is probably enough to say that any statistical experiments are unnecessary if we can just produce Einthoven's string galvanometer, which will be sensitive enough, and then ask some kind of spiritual athlete to move—by, say, one thousandth of a millimeter—the bunch of light reflected from the galvanometer's mirror and falling on the scale. The strength required to do this will be many thousand times smaller than the strength required to roll some dice and make them fall out of a pot onto a table in a way that changes the score, increasing or decreasing the number of pips with regard to the statistically predicted number. The psychokinetic athlete should be grateful to us for this idea because dice can only be affected for a short while before they fall out of the pot and get dispersed on the table, whereas, when acting on the galvanometer's extremely sensitive quartz string, our athlete will be able to concentrate for long hours or even days.

CONCLUSION

A book's conclusion is to some extent its summary. It may thus be worth pondering once again the eagerness with which I have shifted responsibility for the future Gnosis of our species onto the dead shoulders of nonexistent machines. Someone could ask whether this was not caused by some kind of frustration of which the author himself was not fully aware, a frustration resulting from the fact that—owing to historical and his own limitations—he was unable to penetrate science and its prospects. Consequently, he seems to have invented, or rather slightly modernized, a version of the famous *Ars Magna,*[1] which clever Lullus presented quite a long time ago, that is, in the year 1300, and which was rightly mocked by Swift in *Gulliver's Travels.*

Leaving the question of my incompetence aside, I would reply as follows. This book differs from pure fantasy in that it tries to provide a relatively firm basis for its hypotheses, while also considering that which exists to be the most solid thing. This is why it constantly refers to Nature, since this is where the "autonomous apsychic predictors" and the "intelligent device" are located—in the form of chromosomal roots and a brainlike crown that form the large tree of evolution. The enquiry as to whether we can imitate those predictors and that device is worth the effort because it is based on reason. Yet, when it comes to the actual possibility of designing the preceding, it is a nonissue because all those "devices" already exist—and they have passed, rather well, an empirical test that is billions of years old.

The question that remains concerns my privileging of the "chromosomal" model of nonintelligent causality over the "brain" model of intelligent causality. This decision was based purely on design factors regarding matter and information because, when it comes to their capacity, throughput, degree of miniaturization, economy of material, independence, efficacy, stability, speed, and last but not least, universality, chromosomal systems manifest superiority over brain ones, beating them in all the areas listed previously. They also lack any kind of formal linguistic limitation, which means that no complex problems of semantic or psychological nature are likely to arise in the course of their material operation. Finally, we also know that a direct confrontation between genotypic aggregates on a molecular level, which is supposed to optimize the results of their material causality according to the state of the environment, is possible—as demonstrated by every act of fertilization. Fertilization is an act of "taking a molecular decision" in a confrontation between two partly alternative "hypotheses" about a future state of the organism, whereby the gametes of both sexes are "carriers" of those hypotheses. The possibility of recombining in a similar fashion the material elements of a prediction does not result from imposing on ontogenetic processes some other processes that would be external to the ontogenetic ones but is rather built into the very structure of the chromosomes. Besides, genotypes are dedicated to the problem of prediction—which is so precious to science—in an exclusive and absolute way. The brain lacks any of those design features. Brains as systems that are more "definitely closed" than genotypic systems are unable to confront their full information content directly (the way chromosomes can), while a significant part of their high complexity, which is permanently involved in tasks connected with system control, cannot commit itself to "prediction work." Brains seem to be "finished" or "tested" models, or prototypes. They "just" need to be repeated, perhaps by using some selective amplification, if they are to become, in their synthetic versions, inducers of theory formation. Engaging specialized systems such as chromosomal ones in this task would in turn not only be extremely difficult but might also prove impossible. Yet the success of "heredity devices," measured in the number of bits in a time unit per a carrier's atom, is so high that it may be worth trying it, and not just with one single generation. In any case, what technologist is going to resist such a temptation? From twenty letters of amino acids Nature constructed a "pure" language, which expresses—via a slight rearrangement of nucleotide syllables—phages, viruses, bacteria, T. rexes, termites, hummingbirds, forests, and nations, as long as it has enough time at its disposal. This language, so perfectly atheoretical, anticipates not only the conditions at the bottom of the oceans and at the tops of the mountains but also the quantum character of light, thermodynamics, electrochemistry, echolocation, hydrostatics—and many other things we still know nothing about. It does so only "practically," because, though it causes everything, it does not understand anything—yet its lack of intelligence is much more productive than our wisdom. The chromosomal language is also unreliable; it is an extravagant dispenser of synthetic pronouncements about the properties of the world because it understands the world's statistical nature and acts according to it. It does not pay any attention to singular statements. What matters to it is the totality of expression over billions of years. It truly makes sense to learn such a language—because it constructs philosophers, while ours constructs only philosophies.

Krakow, August 1966

NOTES

TRANSLATOR'S INTRODUCTION

1   Peter Butko, "*Summa Technologiae*—Looking Back and Ahead," in *The Art and Science of Stanisław Lem,* ed. Peter Swirski (Montreal: McGill-Queen's University Press, 2006), 84.

2   Stanisław Lem, "Thirty Years Later," in *A Stanisław Lem Reader,* ed. Peter Swirski (Evanston, Ill.: Northwestern University Press, 1997), 70.

3   Butko, "*Summa Technologiae,*" 102.

4   Bernard Stiegler, *Technics and Time, 1: The Fault of Epimetheus,* trans. Richard Beardsworth and George Collins (Stanford, Calif.: Stanford University Press, 1998), 113. For more on this issue, see Joanna Zylinska, "Playing God, Playing Adam: The Politics and Ethics of Enhancement," *Journal of Bioethical Inquiry* 7 (2010): 2.

5   See N. Katherine Hayles, *How We Became Posthuman* (Chicago: University of Chicago Press, 1999); Cary Wolfe, "In Search of Posthumanist Theory: The Second-Order Cybernetics of Maturana and Varela," in *The Politics of Systems and Environments,* Part I, special issue of *Cultural Critique* 30 (1995): 33–70; and Bruce Clarke, *Posthuman Metamorphosis: Narrative and Systems* (New York: Fordham University Press, 2008).

6   Peter Swirski, "Stanisław Lem: A Stranger in a Strange Land," in Swirski, *A Stanisław Lem Reader,* 6.

7   Ibid.

8   Ibid., 10.

9   As Lem himself puts it in the essay titled "Small Robots," "we cannot exclude the possibility that machines equipped with sovereign will may at some point begin to resist us. I am not thinking, of course,

of a robot rebellion against humankind, so beloved by all primitive purveyors of cognitive magic. My only point is that, with the rise in the degree of behavioral freedom, one can no longer preserve the 'good and only the good,' because this very freedom can also give rise to a touch of 'evil.' We can see this in natural evolution only too well, and this reflection may perhaps temper our intention to endow robots with free will." Lem, "Small Robots," in *Lemistry: A Celebration of the Work of Stanisław Lem,* ed. Ra Page and Magda Raczynska (Manchester, U.K.: Comma Press, 2011), 15–16. This line of thought was already visible in Lem's essay "The Ethics of Technology and the Technology of Ethics" (originally published in 1967), in which he positions morality as a form of "applied technosocial control." Cited in Peter Swirski, "Reflections on Philosophy, Literature, and Science (Personal Interview with Stanisław Lem, June 1992)," in Swirski, *A Stanisław Lem Reader,* 50.

10 N. Katherine Hayles, "(Un)masking the Agent: Stanisław Lem's 'The Mask,'" in *The Art and Science of Stanisław Lem,* ed. Peter Swirski (Montreal: McGill-Queen's University Press, 2006), 29.

11 Peter Swirski, "Lem in a Nutshell (Written Interview with Stanisław Lem, July 1994)," in Swirski, *A Stanisław Lem Reader,* 115.

12 Ibid., 114.

13 See Stiegler, *Technics and Time, 1,* 180–95.

14 Jerzy Jarzębski, "Models of Evolution in the Writings of Stanisław Lem," in Swirski, *The Art and Science of Stanisław Lem,* 115.

15 Ibid., 111.

16 See Paisley Livingston, "Skepticism, Realism, Fallibilism: On Lem's Epistemological Themes," in Swirski, *A Stanisław Lem Reader.*

17 See Swirski, "Reflections on Philosophy, Literature, and Science," 31.

18 See Hayles, "(Un)masking the Agent," 43.

19 Swirski, "Lem in a Nutshell," 93.

20 Swirski, "Reflections on Philosophy, Literature, and Science," 56.

21 Jarzębski, "Models of Evolution," 105.

22 Ibid., 115.

23 Andy Sawyer, "Stanisław Lem—Who's He?" in Page and Raczynska, *Lemistry,* 258.

24 Swirski, "Reflections on Philosophy, Literature, and Science," 27.

25 Swirski, "Lem in a Nutshell," 116.

26 Sarah Kember and Joanna Zylinska, *Life after New Media: Meditation as a Vital Process* (Cambridge, Mass.: MIT Press, 2012).

27 Butko, "*Summa Technologiae*," 83–84.

28 Cited, in Polish, on Lem's official website (http://www.lem.pl).

29 Swirski, "Reflections on Philosophy, Literature, and Science," 54.

30 Lem, "Thirty Years Later," 68.

31 Stanisław Lem, "Dwadzieścia lat później" [Twenty Years Later], in Lem, *Summa Technologiae*, 4th exp. ed. (Lublin, Poland: Wydawnictwo Lubelskie, 1997), 327.

32 Swirski, "Reflections on Philosophy, Literature, and Science," 55–56.

1. DILEMMAS

1 [Although the prophet Isaiah–inspired version of the phrase that mentions beating swords into plowshares as an act of replacing warfare with peaceful coexistence is more frequently evoked (Isaiah 2:4), the Bible also mentions the opposite activity (Joel 3:10), which is the one Lem is referencing here.—Trans.]

2 [Ormuzd (the good spirit) and Ahriman (the evil spirit) are the two main deities of the Zoroastrian religion.—Trans.]

3 [This term, now largely obsolete, refers to an autonomous nuclear power station, with energy produced via hydrogen fusion. It reflects Lem's concern with trying to find some alternative sources of energy for our civilization—an issue that returns throughout the volume.—Trans.]

4 [Early 1960s.—Trans.]

5 [Lem is referring to a 1930 sci-fi novel, *Last and First Men: A Story of the Near and Far Future,* by British writer Olaf Stapledon.—Trans.]

2. TWO EVOLUTIONS

1 [As Polish nouns are marked for gender, with *technology* being designated feminine, this passage entails an interesting implicit personalization of technology not just on the level of the argument but also on the level of language. This ambiguity consequently allows Lem to pose the "Who causes whom?" question. Unfortunately, this additional layer of meaning is lost in the English translation, when technology is rendered as an *it* rather than a *she* in the subsequent sentence.—Trans.]

2 [Lem is referring here to Fred Hoyle (1915–2001), a British astronomer known for coining the term *big bang* and for his repudiation of

the big bang hypothesis. Hoyle was a proponent of the "steady state" cosmos, a theory according to which a continuous creation of matter propelled the expansion of the Universe—even though the Universe was seen as remaining in an infinitely steady state. For the average density of the Universe to remain constant, matter for Hoyle had to be created in those new areas where space was expanding.—Trans.]

3 [A videorama is a multiscreen video installation surrounding the viewer. Circarama is the name of a movie theater using a circle-vision 360 degree film technique, which was refined by the Walt Disney Company in the 1950s. It allows for what we would today describe as an "immersive" cinematic experience.—Trans.]

4 See Davitashvili, *Teoria Polowogo Otbora*.

5 See Smith, *Theory of Evolution*.

6 Interesting results could be provided by an attempt to draw a tree model of technical evolution. When it comes to its overall shape, it would most likely resemble a tree we would draw to represent bio-evolution (i.e., it would have a primary uniform trunk and an ever denser crown). The difficulty is that the current increase in knowledge is caused by interspecies hybridization in technology but not in biology. Any randomly chosen areas of human activity are capable of cross-fertilization (we have the "crossbreeds" of cybernetics and medicine, of mathematics and biology, etc.), yet biological species, having become fixed, cannot reproduce through crossbreeding. As a result, the *speed* of technoevolution is characterized by constant acceleration, which is much higher than that of biological evolution. What is more, long-term predictions in the field of technoevolution are difficult owing to the occurrence of sudden turns—which are totally unexpected and unpredictable (one could not predict the emergence of cybernetics before it happened). The number of "technological species" emerging at any given time is a function of all the existing species, which is something we cannot say about bioevolution. Sudden turns in technoevolution are not equivalent to biological mutations because the former have much more significance. For example, physics is currently pinning great hopes on the investigation into neutrinos, which have actually been known to us for a long time, but it is only recently that the ubiquity of their influence on numerous processes in the Universe (e.g., on the formation of stars) as well as their often decisive role in those processes has been understood. Certain types of stars at the point of departure from the state of equilibrium (e.g.,

before a supernova explosion) are capable of neutrino emission, which significantly exceeds their total luminosity emission. This does not apply to stationary stars such as the Sun (whose neutrino emission, conditioned by beta decay phenomena, is much smaller than the energy emitted in luminous radiation). But astronomy is now particularly hopeful about its research into supernovae, whose role in the overall development of the Universe, in the emergence of the elements—especially heavy ones—and also in the genesis of life seems unique. This is why there exists a certain possibility that neutrino astronomy, which does not use any conventional apparatuses such as the telescope or the reflector, will be able to replace optical astronomy, which currently prevails. (Radio astronomy is yet another potential competitor.)

The question of the neutrino probably contains many other secrets. Research into this area may lead to the discovery of some yet unknown energy sources. (This would involve changes of a very high energy level, which, among other things, characterize the transformation of the electron–positron pair into the neutrino–antineutrino pair or so-called neutrino braking radiation [*Bremsstrahlung*].) The picture of the Universe as a whole may undergo a radical transformation: if the number of neutrino particles is indeed as high as some claim, the evolution of the Universe will be conditioned not so much by the unequally distributed islands of galaxies in a given space as by neutrino gas, which will uniformly fill that space.

All these issues are both promising and debatable. They illustrate how unpredictable the development of knowledge is and how it would be wrong to assume that we already know with great certainty many fundamental laws that determine the nature of the Universe and also that any further discoveries would only fill in with detail this already more or less complete picture. Yet the current situation looks as follows: we possess solid and rather certain knowledge about a number of branches of technology, but this only applies to widely used technologies, which constitute the material foundation of the Earth's civilization. However, we seemingly know even less than we did only a few dozen years ago about the nature of the micro- and macrocosm, about the prospects of the emergence of new technologies, or about cosmogony and planetogony. The reason for this is that a number of diametrically opposed hypotheses and theories are competing against one another in these fields (e.g., the hypothesis about the enlargement of the Earth, about the role of supernovae in the creation of planets and elements, about supernova types).

The preceding consequence of scientific progress is only apparently a paradox because ignorance can mean two different things. First, it refers to all this we do not know, while we also have no idea about our lack of knowledge (the Neanderthal man did not know anything about the nature of electrons, nor did he have any inkling about the possibility of their existence). This is, so to speak, "total" ignorance. Second, ignorance can also mean the awareness of the existence of a problem, coupled with the lack of knowledge with regard to its solution. Progress undoubtedly diminishes the ignorance of the first type, i.e., the "total" kind, while also increasing the ignorance of the second type, or the set of questions without answers. This last statement does not solely refer to the realm of human activity; i.e., it is not an assessment of man's theoretical and cognitive practices. It also, to a certain extent, applies to the whole of the Universe (because the multiplication of questions as a result of the increase in knowledge implies a particular structure of this Universe).

At the current stage of development, we are inclined to assume that such an "infinitesimal–labyrinthine" character is an immanent trait of Everything That Exists. But accepting such an assumption as a heuristic ontological thesis would be quite risky. The historical development of man has been too brief to allow for the formulation of similar theses as "certain truths." The knowledge of a large number of facts, and of the links between them, may lead to a kind of "cognitive peak," after which the number of questions without answers will start to decrease (rather than increase, as has been continuously occurring so far). For someone who can count to a hundred, there is actually no practical difference between a quintillion and infinity. Man as a researcher of the Universe is more like someone who has just learned arithmetic than like a mathematician who is freely juggling infinities. Besides, we could find out the "finite" formula for the structure of the Universe (if it actually exists) on having reached the "gnoseological culmination" as discussed earlier. But a constant and continuous increase in the number of questions does not yet determine the issue because maybe it is only civilizations, which have had more than, say, a hundred billion years of constant development, that reach such a "cognitive peak," so all the presuppositions made on this matter earlier are groundless...

7　[Lem is referring here to a sixteen-hundred-year-old iron pillar in a mosque complex in Delhi, which has never suffered from corrosion. Apparently a thin film composed of iron, oxygen, and hydrogen has protected the cast iron pillar against rust.—Trans.]

8　See Koestler, *Lotus and the Robot*.

9　A random statistical approach to the problem of technogenesis complies with the currently popular application of game theory (the foundations of which were laid by John von Neumann) to various social issues. I have in fact drawn on such models several times in this book myself. And yet the real complexity of the problem cannot be contained within probabilistic schemas. In systems with a high level of organization, even very small structural changes can have significant consequences. Additionally, the problem of "amplification" arises here. We can speak about "spatial amplification," for which the lever serves as a model, whereby "small movement" turns into "big movement" with its help. There is also "temporal amplification," an example of which can be found in embryonic development. We are still lacking a field such as topological sociology, whose task would be to study relationships between actions of individuals and topologically conceived social structures. Such structures can probably have "amplification" effects, i.e., an individual's thoughts or actions can find propitious conditions for their proliferation in society. This phenomenon can even snowball. (Cybernetics is only beginning to show interest in phenomena of this kind, which occur in very complex systems such as society or the brain—where they take the form of, e.g., epilepsy in the latter.) This also works the other way round—certain structures can "extinguish" singular actions. I have touched on this issue in my *Dialogues* [Lem, *Dialogi*].

Of course, freedom of action depends within a *given* social structure on the place an individual occupies in it (a king has more degrees of freedom than a slave). Such differentiation is rather trivial because it does not bring anything new to the analysis of system dynamics: in turn, *different* structures favor or impede individual actions (e.g., scholarly thinking) in various ways. This problem is actually situated at the crossroads of sociology, psychosociology, information theory, and cybernetics. We still have to wait a while for some significant developments in this field. The probabilistic model proposed by Lévi-Strauss is erroneous if we take it literally. Its value lies in the fact that it postulates the introduction of an objective method into the history of science and technology, a field that used to be dominated by a more "humanist" mode of problem solving, one that proclaimed: "The human spirit, having experienced failures and successes throughout the course of history, has eventually learned to read from the Book of Nature," etc.

Lévi-Strauss is undoubtedly correct in emphasizing the importance of "information hybridization," an international exchange of spiritual goods. A single culture is a lone player, adopting a particular strategy. It is only when a coalition of different cultures emerges that strategic enhancement (or the exchange of experiences) takes place—which considerably increases the chance of "technological victory." Let us cite Lévi-Strauss: "A culture's chance of uniting the complex body of inventions of all sorts which we describe as a civilization depends on the number and diversity of the other cultures with which it is working out, generally involuntarily, a common strategy. Number and diversity..." [Lévi-Strauss, *Race and History,* 41].

But this kind of collaboration is not always possible. And it is not always that a culture will be "closed," i.e., isolated, as a result of its geographical location (on an island, as is the case with Japan, or behind the wall of the Himalayas, as is the case with India). A culture can "close itself" on a structural level, thus successfully yet unknowingly blocking its path to technological progress. A geographical location is no doubt of great significance. This was particularly the case with Europe, where multinational cultures, which had emerged in close proximity, exerted considerable influence over one another (as the history of wars tells us, for example...). But such an element of chance is not a satisfactory explanation. A universally valid methodological rule should state that statistical regularities need to be reduced to deterministic ones as far as possible. Restarting trials in the face of initial failures (let us recall, e.g., futile attempts on the part of Einstein and his collaborators to "determine" quantum mechanics) is not a waste of time because statistics *can* (although it does not have to) be just a foggy picture, a blurry close-up, and not an exact reflection of the regularity of a phenomenon. Statistics allow us to *predict* the number of accidents depending on the weather, the day of the week, etc. But a *singular* approach allows us to prevent accidents better (because each separate accident is caused by *determined* principles such as poor visibility, faulty brakes, or excessive speed).

A creature from Mars, observing the circulation of the "vehicle liquid," with its "motorcar-bodies" on terrestrial motorways, could easily consider this to be a purely statistical phenomenon. This creature would then consider the fact that Mr. Smith—who goes to work by car every day—suddenly turns back while halfway there, an "indeterministic" phenomenon. But he is turning back because he has left his briefcase at home. This was a "hidden parameter" of the phenomenon. Another person does not reach his destination because he has remembered that he has an important meeting or because he has noticed that the engine is overheating. Thus various purely deterministic factors can offer a total picture of a certain median behavior of a large mass of elementary phenomena—a mass that is only apparently homogeneous. The creature from Mars could suggest to the terrestrial engineers that they should widen the roads, which would facilitate the circulation of the "vehicle liquid" and decrease the number of accidents.

As we can see, a statistical overview also allows us to outline pragmatically useful theories. Yet taking into account the "hidden parameters" would allow us to improve on them radically: Mr. Smith should be thus advised always to leave his briefcase in his car; the other driver to make a note of any important meetings in his diary; the third one to have his car undergo a safety check well in advance. The mystery of the fixed percentage of cars that never reach their destination disappears once we have identified the hidden parameters. The mystery of variable culture-forming strategies among humans similarly disappears once we have examined the topology and the information aspect of their functioning. As the Russian cyberneticist Gelfand rightly observed, also in extremely complex phenomena, we can detect their relevant and irrelevant parameters. We often continue with an investigation by desperately trying to determine some new irrelevant parameters! This is the case, e.g., with the research into correlations between cycles of solar activity and cycles of economic prosperity (e.g., as conducted by Huntington). [Lem is referring here to the work of American geographer Ellsworth Huntington, author of books such Civilization and Climate (1915) and Mainsprings of Civilization (1945).—Trans.] It is not that such a correlation does not exist—it has actually been identified; it is just that there are too many such correlations. Huntington is quoting such a high number of them in his book that the problem of the drivers of progress drowns in the ocean of correlations.

Not taking correlations, or irrelevant variables, into account is at least as significant as investigating the relevant ones. Of course, we do not know in advance which variables are relevant and which ones are not. But a dynamic and topological approach allows us to give up the analytical method, which proves unsuitable in this case.

10 [Sebastian von Hoerner (1919–2003) was a German astrophysicist and radio astronomer who conducted extensive research into extraterrestrial intelligence and the formation of stars. The term *psychozoic,* which Lem uses frequently through the course of his argument, refers to the "era of man," i.e., the "psychozoic era."—Trans.]

11 [A medieval land of milk and honey. Lem uses the German term, *Schlaraffenland,* in the Polish original.—Trans.]

3. CIVILIZATIONS IN THE UNIVERSE

1 See Baumsztejn, "Wozniknowienije obitajemoj planiety."

2 Ashby, *Introduction to Cybernetics.*

3 Drawing conclusions about the probable development of a civilization in the Universe on the basis of the *indiscernibility* of any signals or astroengineering phenomena can, of course, be compared to "drawing conclusions" about the existence of the wireless telegraph in ancient Babylon (because the archeologists did not find any wires among the ruins, this must mean that the local population used radio waves ...).

The following response can be given to the charge presented previously. As is discussed in note 1 of chapter 4, exponential growth of a civilization is not possible for a prolonged period of time. The hypothesis that claims that, after a short period of technological development lasting several thousand years, civilizations are annihilated assumes an absurd kind of determinism. (It postulates that *every* civilization must become extinct promptly because if only 99.999 percent of them were to become extinct, the remaining 0.1 per mil would lead to the formation of large, expansive radiations, which would extend to whole galaxies in a short period of time, i.e., time measured in thousands of years.)

We are therefore left with a third hypothesis—one about the exceptionally rare occurrence of psychozoics (between one and three per galaxy). It contradicts the fundamental postulate of cosmogony: that of the homogeneity of conditions in the Universe as a whole and, consequently, that of the highly probable conclusion that the Earth, the Sun, and also we ourselves are regular, and thus rather frequent, phenomena.

This is why the hypothesis about the "isolation" of civilizations from the Universe so that their activity cannot be easily observed on an astronomical scale seems to be the most probable one. It has therefore been selected as the dominant hypothesis in the course of writing this book.

4   All the hypotheses discussed here take as their starting point the model of the Universe proposed by Shklovsky, i.e., that of the "pulsating" Universe in which, after the phase of "red" escape of galaxies, comes their "blue" concentration. A single stroke of such a "cosmic engine" takes around twenty billion years. There also exist some other cosmogonic models, e.g., the one outlined by Lyttleton. It fulfils the conditions of the "perfect cosmogonic principle," a principle that assumes that the observable state of the Universe is *always* the same, i.e., that an observer would always see the same picture of galaxy escape we are seeing today. This model runs against a number of difficulties on an astrophysical level, not to mention the fact that it assumes the creation of matter from nothing (in the cubic capacity that equals that of a room, one atom of hydrogen would have to come into being once every hundred million years). One usually does not draw on biological arguments when discussing cosmogonic models, yet we should recognize that such an infinitely old and constant Universe introduces another paradox. If the Universe had been in existence in a similar state to the one in which it finds itself now for an infinite amount of time, an infinite number of civilizations should have emerged in it. No matter how great and terrifying the limitations to the existence of such civilizations, we can assume that a randomly chosen small fraction of them go beyond their astroengineering stage, thus making the existence of intelligent beings on them independent from the limited life span of their home stars. This would be enough to conclude that an infinite number of civilizations should currently exist in the Universe (since a randomly chosen fraction of infinity is itself an infinity). This paradox, too, leads indirectly to the acceptance of the hypothesis about the changeability of the states of the Universe over time.

Incidentally, biogenesis does not have to take place solely within a planetary system, with a central star as its energy source. As Harlow Shapley has observed in the *American Scholar* [Shapley, "Crusted

Stars and Self-warming Planets"], there exist fluid transitions between stars and planets; there also exist some very small stars and some very large planets. Besides, the existence in the Universe of a great number of "in-between" bodies—i.e., some old and very small stars, which have a constant surface (crust) and are heated by their interior, which is cooling down more slowly—is highly probable. Such bodies, as Shapley postulates, can also experience the emergence of various forms of homeostasis, i.e., life. This kind of life would not resemble the planetary forms of life owing to a number of significant differences between their respective physical conditions: the mass of such "star planets" is quite significant when compared to the mass of the Earth (otherwise they would cool down too quickly). Besides, such planets do not have their own Sun: they are lone bodies, trapped in permanent darkness, which means that any forms of life emerging on them would probably not develop the sense of sight.

We have not thoroughly covered this rather convincing hypothesis because our focus has been not so much on the discovery of all possible forms of the emergence of life and civilization but only on the discovery of those forms whose evolutions are likely to resemble the evolution on Earth. We have been referring to the Universe as an arbiter that is supposed to predict the possible future directions of our own civilization.

5 [And also in 1974. This remark was added by Lem in the fourth edition of *Summa*.—Trans.]

6 [The term "Metagalaxy," which in *Summa* stands for the entire physical Universe, is no longer used by contemporary astronomy.—Trans.]

7 The octet theory, which introduces order into the existing chaos among elementary particles, has led, among other things, to the postulation of the existence of special particles, which Gell-Mann called "quarks." (The word *quark* does not mean anything; it was invented by Joyce and used in his book *Finnegans Wake*.) According to octet theory, all elementary particles consist of quarks—particles that are much bigger than protons and that together show an enormous mass deficit. Despite intense research, so far, it has not been possible to detect the existence of those hypothetical quarks in a free state. Some researchers are inclined to think that we are only dealing with a very useful mathematical fiction here. [The existence of quarks has since been proved in accelerator experiments.—Trans.]

4. INTELECTRONICS

1 E.g., de Solla Price, *Science since Babylon.*

2 The problems of exponential growth are going to determine the future
of civilization to a more significant extent than is widely believed
today. The exponential growth of the number of intelligent beings,
or of (technological and scientific) information, is possible. The exponential growth of information and energy can take place as long
as the number of all living beings remains relatively stable. Every
civilization is most likely attempting to increase to the maximum
the pace of growth of scientific and technological information and
probably also the pace of growth of accessible energy sources. We
can think of no reason that would remove the motivation behind
this process. A civilization that is entering an astronautic phase is
becoming exceptionally "greedy" for energy. This is because galaxy
flights (outside its own solar system) require an amount of energy
that is comparable to a fraction of solar power, if the relativistic
effects caused by nearing the speed of light—which enable return
flights (planet–star–planet) within one generation's lifetime (i.e., that
of the ship's crew)—are to materialize themselves. Thus, even if we
were to limit the number of beings living on the planet, civilization's
demand for energy would grow rapidly.

When it comes to the amount of information gain, even crossing the barrier of information growth does not open up the kind of
freedom of population growth that could be desired. Many experts
today actually notice detrimental future consequences of an excessive demographic expansion (i.e., an increase in the number of living
beings). Yet they see first of all difficulties connected with the need
to feed and provide on a material level (clothing, accommodation,
transportation, etc.) for the planet's population—which is growing
exponentially. At the same time, problems concerning the sociocultural
development of a civilization that is growing exponentially have not
yet been thoroughly examined by anyone, as far as I know. But in
the long run, they may turn out to be a determining factor as far as
the need to limit the birthrate is concerned, even if it was possible
to provide accommodation and food for billions of humans thanks
to technological innovation.

Dyson's example, astrophysics—which postulates constructing
a "Dyson sphere," an empty ball made out of material taken from

large planets and positioned at a distance of one astronomical unit from the Sun—seems typical here. He claims that already after a few thousand years of its existence, *every* civilization is forced by some objective factors (primarily by population growth) to surround its sun with such a thin-walled spherical shell. This would both enable it to absorb all the energy coming from solar radiation and to create an enormous space in which the inhabitants of that civilization would be able to settle. Since the inner surface of that shell, which would be turned toward the Sun, would be about a billion times bigger than the surface of the Earth, it would be capable of accommodating a billion times as many people as the Earth is capable of housing. Thus about three to eight trillion people could live inside a "Dyson sphere" at any one time.

Dyson is so convinced about the inevitability of having to construct "circumsolar spheres" that he suggests we look for them in the Universe because such a sphere should be seen as a place that emits constant radiation at a temperature of around 300 K (assuming that it transforms the radiation energy of its sun into various kinds of energy needed for industrial purposes, with the latter types of energy ultimately leaving the sphere in the form of heat radiation).

This is one of the most astounding cases of "orthoevolutionary" reasoning I know of. Having calculated nothing less than the amount of matter contained in all the planets of our system, solar radiation, etc., Dyson arrived at a result that proved that such an astroengineering project would be possible to accomplish (since the amount of matter would be enough to construct the sphere, as all the power of solar emission could be used in this way). Yes, this would no doubt be possible. Yet this argument silently assumes that first, population growth up to trillions of beings is desirable, and second, that it is possible in a sociocultural sense (even if we assume that it can be achieved on a technical level).

All living beings, including intelligent ones, have been granted a reproductive tendency by bioevolution, whereby gains from the growing birthrate should exceed losses caused by mortality. Yet the fact that people *could* increase their number exponentially does not mean that they *should* do so.

We should point out that even a Dyson sphere does not enable exponential growth for an infinite period of time. When the number of beings living on it exceeds a dozen or so trillion, there will be a need either to halt any further growth or to look for some other territories where space colonization could take place (e.g., in nearby star systems). We can thus conclude first of all that a Dyson sphere only defers the problem of the regulation of birthrate but does not solve it. Then we have to realize that every society is a self-organizing system. Even though we do not yet know anything about the maximum possible size of such systems, it is certain that they cannot expand for an infinite period of time. The most populous system we know is the human brain: it consists of around twelve billion elements (neurons). Systems consisting of around a trillion elements are probably a possibility, but it is highly doubtful whether there can exist homogenous systems that would consist of trillions of elements. After a certain threshold is crossed, degradation, divisions, and, consequently, sociocultural disintegration become inevitable. This is not to try to find an answer naïvely to the question of what those trillions living on the inner surface of a Dyson sphere would be expected to be doing, or to attempt to find "tasks and jobs" for them (although the future of those creatures seems rather deplorable: the surface of the sphere—as demonstrated by calculations that estimate the amount of material needed per each unit of the surface—must be relatively thin and uniform, and thus we can forget about any kind of "landscape," about mountains, forests, rivers, etc.). The point is that trillions of coexisting creatures cannot have one shared culture, one sociocultural tradition that would at least slightly resemble something we are familiar with from the history of humankind. A Dyson sphere cuts off the starry sky; it also involves the liquidation of planets and hence entails giving up on the conditions present on them. It is an artificial creation, something like a city that has been enlarged a billion times and that surrounds with itself the center of its system: its star. Several simple calculations easily demonstrate that achieving a relative order within its limits, as well as providing its inhabitants with supplies that are necessary for their existence, is only possible provided the inhabitants of that sphere basically stay close to their birthplace throughout their lives.

Those creatures would not be able to travel owing to purely physical reasons (let us imagine that a Dyson sphere contains some "attractive places"; these would be visited not by millions of tourists, the way it happens today, but by hundreds of millions). And because a technical civilization means an increase in the number of mechanical and technical devices per living unit, the surface of a Dyson sphere would not be so much a city as a factory hall or a machine park one billion times bigger than the Earth's surface. We could keep listing various such "inconveniences," to use a euphemism, of a trillion-sized existence as long as we wish. When we understand it as an increase in individual freedoms rather a decrease in them, we thus reduce the whole idea of progress to absurdity. Thus the "freedom of unlimited reproduction" we would gain (and which would only be superficial anyway, as demonstrated earlier) would indeed be a peculiar kind of freedom if we had to sacrifice a whole lot of other freedoms on its altar.

A civilization does not stand for an increase in *all* possible kinds of freedom. Culinary freedom for cannibals, freedom to self-harm, and lots of other freedoms have already been crossed out from the *magna charta liberatum* of a society that is undergoing technological development. It is actually hard to understand why reproductive freedom would have to remain intact even if it was to lead to a total immobilization of individuals, to smashing the cultural tradition, to giving up, literally, on the Earth's and sky's beauty. The vision of trillions of "Dyson spheres" as the main developmental direction of all intelligent beings in the Universe seems to me to be not any less monstrous than von Hoerner's vision of self-liquidating psychozoics. Besides, a civilization that is undergoing exponential growth is actually not possible because it would populate the whole of the visible Universe up to the most remote metagalactic clusters within a few hundred thousand years. And thus, if a Dyson sphere is only to defer the introduction of the regulation of the birthrate for a few thousand years, then we have to say that this is a truly terrible price to pay for an unwillingness to act, dictated by common sense at the right time.

I have summarized Dyson's idea mainly to show its ridiculousness rather than because of any factual interest it may evoke. "A Dyson sphere" is impossible to construct, as demonstrated by the astronomer W. D. Davidov (*Priroda*, no. 11, 1963). It is not possible either as a spherical thin-walled sphere, or as a system of ring straps, or in the shape of domes, because in none of these cases can it become a dynamically durable construction, even for an extremely short period of time.

3 Ashby, *An Introduction to Cybernetics*.

4 See Pask, "A Proposed Evolutionary Model."

5   [Lem is referring here to the popular children's rhyme "Centipede," by the Polish writer Jan Brzechwa.—Trans.]

6   [The Tarpeian morality refers to the Tarpeian rock in ancient Rome. It functioned as an execution place, from which criminals and those who suffered from mental or physical disabilities were hurled down.—Trans.]

7   [Beer was actually British.—Trans.]

8   Wiener, *Cybernetics*.

9   [Vinoba Bhave (1895–1982) was a religious leader and social reformer in India who was recognized as a spiritual follower to Mahatma Ghandi. He was an advocate of nonviolence and human rights.—Trans.]

10 [Arthur Eddington and James Jeans were two British astronomers who studied stellar models and who were involved in a dispute over what it meant to do science "properly."—Trans.]

11 [British philosopher Gilbert Ryle (1900–76) was an author of, among other works, *The Concept of the Mind* (Chicago: University of Chicago Press, 1949), which offered a critique of the Cartesian theory of the mind–body dualism (dubbed by Ryle "the dogma of the ghost in the machine").—Trans.]

12 [The Vistula is Poland's principal river, which traverses the country from the south (the Tatra Mountains) to the north (the Baltic Sea).—Trans.]

13 See Charkiewicz, "O cennosti informacii."

14 Some very interesting remarks about the current "geocentrism" of chemistry have been made by Professor J. Chodakov (*Priroda*, no. 6, 1963, published by the Russian Academy of Sciences). He points out that properties of elements are relative because they can only convey the relation of one element to others. Thus, e.g., *flammability* is a relative term: we consider hydrogen to be flammable because it burns in oxygen's atmosphere. If the Earth's atmosphere consisted of methane the way the atmospheres of large planets do, then we would consider hydrogen to be an inflammable gas and oxygen a flammable one. It is a similar case with acids and alkalis: if we swap water for another solvent, the substances that appear in the environment as acids will become alkalis, weak acids will become strong, etc. Even the "metalicity" of an element, i.e., the degree to which it shows properties of metals, is measured by the relation between this element and oxygen. Oxygen, as once observed by Berzelius, is an axis around which all

our chemistry revolves. Yet oxygen occupies an exceptional place on Earth, and not just in relation to all the existing elements. The fact that there is so much of it on Earth has determined the development of the "geocentric" chemistry we currently have. If the Earth's crust consisted of different elements, and if its depressions were filled with liquids other than water, we would have a different classification of elements, and we would evaluate their chemical properties differently. On planets such as Jupiter, nitrogen replaces oxygen in its role of an element with a negative electric charge. Oxygen cannot play a significant role on planets like this because of its scarcity. On such bodies, water is replaced by ammonia, which emerges out of the amalgamation of hydrogen and nitrogen; limestone is replaced with amido-cyanide of lime, quartz with nitric compounds of silicon and aluminum, etc. The meteorology of a "nitrogen" planet must also be different, while all those relations must no doubt determine the processes of self-organization (or bioevolution) in such environments—which can lead to the emergence of hypothetical (for now) nonprotein living organisms.

15 Taube, *Computers and Common Sense*, 59–60.

16 Ibid., 69.

17 See *Woprosy Filozofji*, no. 4, 1966.

18 [In English in the original.—Trans.]

5. PROLEGOMENA TO OMNIPOTENCE

1 In fact, there are no "simple systems." All systems are complex. Yet in practice, this complexity can be ignored because it has no consequence for what matters to us. In a regular clock—which consists of a dial, mainspring, hairspring, and ratchets—processes such as crystallization, material fatigue, corrosion, transmission of electric charges, shrinkage and expansion of parts, etc., take place. These processes are practically of no consequence for the working of the clock as a simple machine, one we use to measure time. Similarly, we can ignore thousands of parameters that can be isolated in every machine and every object—but only up to a point, of course, because these parameters (whose existence is real, even though it is not being taken into account by us) at a certain point will have changed so much that the machine will stop working.

Science consists in identifying meaningful variables, while ignoring insignificant ones. A complex machine is one in which many parameters *cannot* be ignored because they all contribute in a substantial way to its working. The brain is one such machine. It does not mean that a machine of this kind, if it is a regulator—such as the brain—would have to consider *all* the parameters. One could list practically an infinite number of those. If the brain were to take all of them into account, it would not be able to perform its actions. The brain "does not have to" consider the parameters of individual atoms, protons, and electrons from which it is constructed. The complexity of the brain, and also of any regulator, or, more broadly, any machine, is not an advantage but rather a necessary evil. It is a response on the part of its constructor, evolution, to the complexity of the environment in which they exist, because only the great variability of the regulator can cope with the great complexity of the environment. Cybernetics is a science that deals with regulating the states and dynamics of actual systems *despite* their complexity.

2 [In Greek mythology, Scylla and Charybdis were two monsters who lived on opposite ends of the Strait of Messina between Italy and Sicily. Scylla was believed to be a six-headed monster, whereas Charybdis was imagined as a whirlpool. They were seen as a danger to sailors traversing the Strait. The myth gave birth to the phrase "between Scylla and Charybdis," which describes a situation in which one has to choose between two equally unattractive options.—Trans.]

3 See Shapiro, "O kwantowanii prostranstwa."

4 [A line from a Polish verse has been replaced by one from Lewis Carroll's "Jabberwocky" to convey the sense of Lem's argument better.—Trans.]

5 See Bohm, *Quantum Theory,* 1951.

6 See Parin and Bajewski, *Kibernetika w miedicinie i fizjologii.*

6. PHANTOMOLOGY

1 [In his 1991 essay, Lem, "Thirty Years Later," 81, added that "in 1991 it will be a helmet."—Trans.]

2 Olds and Milner, "Positive Reinforcement."

3 Magoun, *Waking Brain,* 63.

4 [Podbipięta is a character in the 1884 historical novel *With Fire and Sword,* by the Polish writer Henryk Sienkiewicz. It is set in seventeenth-century Poland, during the period of the Tatar raids.—Trans.]

5　["Nothing is in the understanding that was not earlier in the senses." This is a central doctrine of empiricism. When talking about phantomatics, Lem then replaces "senses" with "nerves."—Trans.]

6　It cannot be verified empirically whether, during a given procedure, a complete "killing of personality" happened at T1 time—to be replaced with a new personality from T2 time. The reason for this is that what is involved in the process is a continuous cerebromatic "change of pathway." Because it is impossible to detect the killing of an earlier personality, we need a ban on such procedures. Small "corrections" are not likely to annihilate personality, but, like with the paradox faced by a bald man, it is hard to tell when a situation changes from an innocent bit of retouching to a crime.

7　It is possible to "transfer" all the experiences and memories from an earlier personality to a future, artificially created one. On the surface, it guarantees the continuity of existence, yet those experiences and memories will be just reminiscences that will not "fit" the new personality. I am aware that the categorical position I take on this issue is subject to dispute.

8　See Shields, *Monozygotic Twins*.

9　Ibid.

7. THE CREATION OF WORLDS

1　See Amarel, "An Approach to Automatic Theory Formation."

2　Even though it may seem strange, there are many contradictory viewpoints with regard to what a scientific theory actually is. This happens even within a single ideological circle. Views held by creators of science are by no means more credible than those held by a famous artist about his own creative method. Psychological motivation can be a source of the secondary rationalization of the thought process—which the author himself is unable to re-create in detail. And thus, e.g., Einstein was entirely convinced about the objective existence of the external world that is independent from man and also about the fact that man can get to know its design structure. This could be understood in many ways. Every new theory is no doubt a step forward with regard to the previous one (as Einstein's theory of gravitation was with regard to Newton's theory), but this does not necessarily mean that there exists, i.e., that there can exist, an "ultimate theory" that will represent the end of knowledge. The postulate of the unification of phenomena in a unified theory (e.g., field theory) seems to have worked in the course of the evolution of classical physics—which has moved from theories concerning individual aspects of phenomena toward more and more unified overviews. Yet it does not have to be like this in the future: even creating a unified field theory that would cover both quantum and gravitational phenomena would not serve as proof (for the fact that the postulate of uniformity is fulfilled by Nature) because it is impossible to get to know *all* phenomena and thus to find out whether such a new (nonexistent) theory also applies to them. In any case, a scientist cannot work on the assumption that he is only creating a certain transient and impermanent link in the cognitive process, even if he was to hold such a philosophical position. A theory is "valid for a certain period of time": the whole of the history of science demonstrates this. Then it gives way to another theory. It is entirely possible that there exists a threshold of theoretical constructions that the human mind cannot cross *by itself*—but that it will be able to cross with the help of, say, an "intelligence amplifier." A road toward future progress thus opens up before us, yet we still do not know whether some objective laws that cannot be overcome (such as, e.g., the speed of light) will not impede the construction process of such "amplifiers."

3   Among systems studied by technical cybernetics, there is a class of systems that are so similar to the brain on the level of their general design principles that they are called "biological." These are systems that could have emerged though natural evolution. In turn, none of the machines we construct could have emerged through evolution because they are not capable either of autonomous existence or of autoreproduction. Only a biological system, i.e., a system that is adapted to the environment during every moment of its existence, can emerge via the "evolutionary method." Such a system not only conveys through its construction the present goals it serves but also shows the evolutionary path that has been followed to enable its emergence. Gears, wires, rubber, etc., cannot by themselves become conjoined to form a dynamo. A metazoon develops from a single cell not because the current tasks of its life demand this but because protozoa existed before tissue organisms and were thus able to become conjoined to form groups (colonies). As a result, biological organisms are *homogeneous,* unlike regular machines. Thanks to this,

a biological regulator can work even without any kind of functional localization.

Let us take a look at an example from Ivakhnenko's *Technical Cybernetics*. We place a digital machine on a cybernetic "tortoise." The machine does not have any "receptors"; it only has a device that measures the *quality* of its actions. In moving around the room, such a "tortoise" will look for a place in which warmth, light, vibrations, and similar disturbances affect the quality of the device's functioning to the *smallest* degree. This system has no "senses"; it does not "perceive" light, temperature, etc. It feels the presence of stimuli "through its whole being," which is why we classify it as a biological system. When the change in temperature affects adversely a given part of the machine, the device that measures the quality of its functioning, in seeing the temperature drop, will start the tortoise—which will begin to wander around, looking for a "better" spot. In another place, a vibration will disturb the functioning of *another* part of the machine, yet it will produce the same reaction: the tortoise will walk away, looking for optimal conditions. A system does not need to be programmed by taking all possible disturbances into account: a designer may not predict, for example, the electromagnetic influences, yet if they affect negatively the machine's functioning, the tortoise will start searching for some propitious "living" conditions. Such a system works by trial and error—a method that fails when a problem is too complicated or when a disturbance has a delayed negative effect (e.g., radioactivity). Since *adaptation* is not always equivalent to *cognition*, a biological regulator does not have to represent "the perfect model of a gnostic machine." It is entirely possible that such an ideal machine will need to be found not in a class of biological regulators but in one of the other classes of the complex systems that cybernetics deals with.

4 A probabilistic and statistical approach to the method of information transfer gets us closer, in mathematical terms, to the problem of bisexuality and the damaging effects of inbreeding, i.e., the crossbreeding of closely related individuals. The probability that a certain number of individuals will have the same kind of genetic damage (recessive mutations) is higher the closer these individuals are related, and if they come from the same parents, then such a probability is the highest. In the latter case, the possibility of the emergence of phenotypic mutants is the highest—of course, as long as the genetic

information of the individuals concerned is *already* damaged. The inbreeding of individuals whose genotypes are free from damage will not have any negative consequences.

If we have a series of various linotypes involved in printing mathematical texts, whereby every typesetting error introduces confusion into the content, then it is clear that on comparing the same texts that have been printed by different linotypes, we shall obtain material that will enable a full reconstruction of the original information, because it is highly unlikely for different machines to cause errors in exactly the same places in the text. If, however, we have a series of absolutely identical linotypes that, as a result of their inherent design fault, always cause the same errors, then the comparison (or "reading") of the texts that have been typeset on them will not allow us to reconstruct the information because it will always be damaged in the same places. Of course, if linotypes make no errors whatsoever, then the problem disappears. The same applies to the transfer of biological information.

5 Schmalhausen, "Osnovy evolucyonnogo processa v svietje kibernetiki," 1960.

6 [Lem is, of course, referring here to the twentieth century.—Trans.]

7 "To demonstrate the consistency of a given system is to prove that no sentence *A* exists in it that would permit us to deduce both *A* and *non-A* from within this system.

To demonstrate the completeness of a given system is to prove that any sentence within this system is deducible, either as itself or as its negation." Marković, *Formalizm w logice współczesnej*, 52n.

8 No matter how fascinating and promising the prospect of starting an "information farm" may seem to an optimist, this phenomenon, actualized within a civilization, certainly does not offer a panacea for all problems. First of all, a "farm" can actually exacerbate, rather than diminish, a crisis resulting from the *excess* of information. So far, humanity has not experienced any crises of excess (except for an excess of disasters, plagues, etc.), which is why it is impossible for us to envisage effective behavior in a situation when we are faced not just with one but with hundreds or even thousands of possible ways of proceeding. We are talking about a situation in which we could, e.g. (as we shall be notified by the current "information set" produced by the "farm"), act in way *A, B, C, D, E,* etc., with each of these ways of proceeding promising a lot but simultaneously excluding

all the other ways of proceeding (let us imagine that a human being can be reconstructed on a biological level so that he becomes almost indestructible, yet this will cause a very radical slowing down of the birthrate because, since no one or almost no one will be dying, the world will gradually become too small). The criteria that currently determine practical ways of proceeding can go out of date (e.g., the criterion of economic profitability or energy saving does not apply anymore when a material process that is practically inexhaustible is the source of energy). Besides, when basic needs are being fully satisfied, the problem "what to do next," i.e., whether we should create some new needs and, if so, what kind, comes to the fore. No "information farm" can, of course, provide a solution to such dilemmas because such a "farm" only offers alternative ways of proceeding; it reveals what can be achieved but not whether we *should* proceed at all. Such a decision-making process must not be mechanized; this would only be possible if social mentality were to undergo a transformation that would cut it off from what we understand as "human" today. Increasing "information freedom," i.e., the number of possible ways of proceeding, implies increasing responsibility for the decisions and choices made. Giving up on such choices, while physically no doubt possible (via an electrobrain ruler who decides what should happen to humanity), seems unacceptable for nonphysical reasons.

Second, an "information farm" cannot really provide "knowledge about everything," "total knowledge," or "all possible knowledge." We could, of course, envisage the whole hierarchy or complex structure of information-carrying and information-gathering cells, among which some would act as elementary "scoops" of factual knowledge, and of the knowledge about the relations between world facts; others would examine the relations between relations, looking for regularities of a higher level; still others would occupy themselves with collating results delivered by the previous ones so that, at the "output" of this whole gigantic feedback pyramid, we would only have the kind of information that can, within any broad range offered, prove useful for the civilization that started this evolutionary machine. Yet, ultimately, the very activity of this "pyramid farm" cannot depart too far from what constitutes the material and spiritual content of life (broadly conceived) in this particular civilization *at a given stage of its development*. Otherwise, having become detached from its human matrix, such a farm would produce information that would not only be useless but also incomprehensible, i.e., information that could not be translated into the language used by that civilization. In any case, another reason why this phenomenon of "becoming detached," "leaping into the future," and experiencing "information eruption" would be a disaster rather than an actual developmental leap is that, as soon as the "information farm" has progressed too far ahead of the current state of knowledge about that civilization, the criteria for *eliminating* irrelevant information would disappear. In this way, the "farm" itself would be immediately transformed into a "bomb"—if not a megabyte then at least a gigabyte one; it would become a giant, with oceans of information produced by it leading to a most unusual deluge.

To understand this better, let us imagine that in the Neolithic period, or maybe even in the early Middle Ages, an "information farm" opens and starts to supply knowledge about twentieth-century technologies: about atomic technology, cybernetics, radio astronomy, etc. Without a doubt the civilization of that day would not be able to accept, grasp, digest, or actualize even a small fraction of such an avalanche of information. It would be even less able to make appropriate, i.e., sensible, tactical, and strategic, decisions (as to whether to produce nuclear weapons, or whether to develop a wide range of new technologies, or whether to limit itself to a few selected ones or even a single one, etc.).

From the most optimistic point of view, an "information farm"— should it be possible to actualize at all—is a device that remains "connected to the world" and that, in examining this world in a particular way, is trying to find out what is materially possible (i.e., what can be actualized). And thus such a "farm" can establish that it is possible to construct a laser or a neutrino electric transformer; that it is possible to somehow change, say, the speed at which time passes, the gravitational field; that seemingly irreversible processes (such as certain biological ones) can actually be reversed, etc. Such a farm will be particularly useful if it is presented with specific tasks, although outlined in a general manner. However, left to its own devices, it will quickly produce the excess of information in which it will drown, together with its creators. The crux of the matter is that it is a "mindless" farm and that it equally produces information about relations that are extremely important to a civilization (e.g., about the possibility of traveling across galaxies) and about completely irrelevant relations (e.g., about the fact that Jupiter's clouds can be dyed golden). As long as the selectors at the information farm are under the active influence of intelligent beings, they are capable of selecting information effectively. Should they, however, depart (i.e., become detached) from such modes of rational selection through entering the domain of "omniscience," or knowledge about all facts, information deluge would become inevitable. We have to be aware that information of all kinds, including that which is useful, grows very rapidly. Let us imagine the "farm" comes up with the possibility of actualizing brain transplants from the body of one creature to that of another. If the farm occupies itself with this problem, this will involve discovering a whole range of new facts and phenomena, the whole "brain transplant technology," etc. Yet of what use will this be if such a problem does not actually interest that particular civilization? As a result, the "farm" may easily find itself "smothered" by masses of entirely useless information. Just think about the vast areas of technology, physics, electronics, and various areas of artistic creativity that are currently covered by television all over the world. If the "farm" could come up with another device that would be ready to serve a similar role within that civilization, the decision as to whether it would be worth spending the time working on such a technology and whether it would be worth actualizing it would have to be made already during the initial stage of information gathering; otherwise, the "farm" will become a producer of a billion "possible inventions" that no one is going to find useful.

We should mention another problem here—a particularly burning one today, although seemingly trivial—which is related to the production of scientific information: the harmonization of the technologies that interested parties can use to find the information that has already been "extracted" from Nature and recorded in print. This problem results, among other things, from the exponential growth of specialized libraries, while no preventative measures—publishing abbreviated information summaries (so-called abstracts), synopses, temporary announcements, etc.—are able to ensure the effective supply of significant information to adequately trained persons who need it. This is because if repeating the experiments that have already been conducted somewhere else turns out to be cheaper and faster than finding an appropriate publication, if a scientist can predict that the information he needs is not hidden "in Nature" but is rather stored on the shelves of some unknown libraries, the research process itself is put in doubt because its results, covered by piles of printed paper, cannot reach those who need them the most. The interested parties themselves sometimes do not fully realize how damaging this phenomenon is because they usually manage to keep abreast of the publications in their discipline. Yet it is well known that *crisscrossing* information from various disciplines is a most productive thing in research. This may mean that in the collections of learned books on different continents, there already exists a great amount of information, bits of which, simply when placed alongside one another and approached all at once by a well-trained expert, would lead to some new and valuable generalizations. An increase in specialization, its internal-disciplinary and gradual differentiation, counteracts this process. The professional librarian is no more capable of replacing the truly outstanding expertise with regard to a specialist area because no librarian is able to decide which results from entirely separate disciplines should first be sent to particular researchers. Even more so, an expert librarian cannot be replaced by an automated catalog or by any other automatic technologies available today because algorithmic methods still turn out to be worthless when it comes to making selections from among the "avalanche of information." The saying that discoveries are made twice today—once when being published and the second time when the (perhaps already long-) published announcement is discovered by a group of experts—has become a cliché. If today's technologies of recording, protecting, and addressing information are not radically streamlined within the next fifty years, we shall be threatened with a vision that will be both grotesque and frightening: that of a world covered up with piles of books and of a humanity turned into busy librarians.

On this "library front," methodology understood as a set of directions for finding knowledge in the world should give way to "ariadnology"—which would stand for a guide to the labyrinth of the already assembled knowledge. A machine librarian, sending the right kinds of information to the right kinds of people, cannot be a device that lacks "comprehension" and that is based on frequency analysis. (Such attempts—to detect the "value" of works through counting how often they get mentioned by experts in the bibliographies of their publications, or to address works in which certain terms appear frequently enough, in a mechanical way, by means of selection—are already under way.) Research has shown that even an expert from a similar area is unable to sort a given range of works in a rigorous manner, as argued, e.g., by J. Kemeny. Yet, owing to its wide-ranging insight, a machine librarian that would be capable of "comprehending" things would inevitably have to be a better specialist than all the researchers put together... these are the kinds of paradoxes that encapsulate this (ever-worsening) situation. It seems that the crisis of information distribution will lead to the strengthening of publication criteria in the future so that their original selection prevents the flooding of the professional market with insignificant works, works that are only produced to obtain an academic degree or to satisfy someone's ambition. We can even think that presenting banal works will at last be declared a menace and thus will be seen as a violation of scientists' professional ethics because such works create nothing but "noise" that prevents us from receiving valuable information—which is vital if knowledge is to grow further. An information farm that would lack an effective "addressing sieve" would most certainly lead to a paper deluge. Such a catastrophe of excess would make any further research impossible. This is why the automatization of cognitive processes, at least within the field of librarianship and publishing, is a more urgent task.

9 [This is a central doctrine of empiricism, which means, "Nothing is in the understanding that was not earlier in the senses."—Trans.]

10 [This is an Old Roman saying, which means that one should first eat and then philosophize.—Trans.]

11 [Lem is engaging here again in his favorite activity of playing with words. This comparison sounds better in Polish, where the suffixes for cosmogony and electronics are the same (*kosmogonika* and *elektronika*).—Trans.]

12 See Brown, *Probability and Scientific Interference.*

8. A LAMPOON OF EVOLUTION

1 In the *New York Times*, May 20, 1963.

2 A fragmentary critique of evolutionary design solutions may at times look like a "lampoon of ignorance" because even today, we do not fully understand organ mechanics (e.g., the immensely complicated working of the heart). Only just first steps have been taken toward creating precise mathematical models of biological structures; e.g.,

Wiener and Rosenbluth have created a mathematical theory of the fibrillation of the heart muscle. Yet it seems groundless and premature to criticize the designs that we do not understand that well. At the same time, our very fragmentary knowledge about the complexity of various evolutionary solutions cannot hide the fact that in many cases, biological complexity is a consequence of the insistent application of the already worked out organic formula from one kind of organism into another—which developed later. A designer who would wish to make the whole of future cosmonautic technology dependent only on rocket engines using chemical fuels would ultimately have to construct ships and propeller devices of immense size and of equally astounding complexity. He could no doubt achieve some degree of success, but this would be a mere display of technological acrobatics. It would certainly not be the most sensible solution, especially given that a great number of difficulties and complications could be avoided by radically *eliminating* the idea of chemical propulsion and by undertaking a revolutionary switch to engines of a different type (nuclear, annihilation, magnetohydrodynamic, ion, etc.).

Complexity—which is a consequence of a certain conservatism entailed in the idea that underpins creative activity, and of "conceptual inertia," i.e., an unwillingness (or inability) to undertake any sudden and radical changes—can be seen as redundant from a designer's point of view. This is because a designer will have his eye on the best results possible, without wanting to trouble himself with any premises he does not have to take into account. Like evolution, a contemporary rocket designer must overcome his problems by working through a number of complications that seem redundant from the point of view of future (e.g., nuclear) technology. Yet he will abandon all those complications as soon as some kind of technological development enables him to actualize nuclear, photon, or another kind of nonchemical propulsion. Evolution, in turn, for the obvious reasons discussed here, cannot really "abandon" any solutions in a similarly drastic way. We can say that after several billion years of its existence and activity, we really cannot expect it to come up with any new solutions that would match in their perfection the solutions it created in its early stages. This very situation allows us to mount a critique of its design solutions, even if we do not fully understand their complexity, because we perceive this complexity to be a consequence of the creative method actualized by evolution—a method with which some other, simpler and more effective methods are able to compete. It is a shame that evolution itself is unable to actualize them, yet evolution's loss may be man's gain—in his guise as a designer of the future.

Apart from a strictly design-focused aspect, this problem also has another dimension, which I am not going to discuss here. Man (and I am just expanding on the preceding remark) does not know himself well—either on a biological or on a psychological level. No doubt the saying "man, the unknown" (which became the title of Alexis Carrel's book) is to some extent true. It is not just his body as a "biological machine" but also his mind that contains mysterious and unexplained contradictions. We could ask whether we should seriously consider the possibility of transforming "the natural *Homo sapiens* model" before fully getting to grips with its actual design and value. Will interventions into the human genotype (and this is just a random example) not eliminate some potentially valuable genotypic traits that we know nothing about, together with harmful ones?

This would be a return to the problem of "biological design" on our part, a problem that (somewhat more traditionally) is represented by the debate between eugenicists and their opponents. Will the elimination of, say, epilepsy not get rid of both epileptics and Dostoyevskys?

It is astounding how abstract similar discussions can be. Any action whatsoever, as discussed in this book (and we are certainly not claiming "authorship" over this idea), is based on incomplete knowledge, because this is the nature of our world. If we thus wanted to delay "species reconstruction" until we are in possession of "total" knowledge about this world, we would have to wait all eternity. Partial unpredictability of the consequences of any action, and thus its potential weakness—while discrediting action completely for certain philosophical approaches—became a foundation within philosophy for the thesis about "the superiority of nonaction over action." (This is a really old idea, which can be traced back, across the continents and centuries, all the way to Zhuangzi.) Yet such critiques, and the apotheosis of "nonaction," are possible, among other things, thanks to some specific actions that took place across the period of thousands of years. It is actually as a result of those actions that civilization developed—and, along with it, speech and writing (which enable the formulation of any judgment and thought). A philosopher who is a supporter of radical conservatism (e.g., nonaction in a biological or technical sphere) is like a millionaire's son who, not having to worry about providing for his living thanks to his father's fortune, criticizes any possessions. If he is consistent, he should give up on all of those riches. An opponent of "bioconstructionism" cannot limit himself to opposing any "plans for man's redesign" but should rather give up on all of civilization's achievements, such as medicine and technology, and retreat to the forest on all four. All the solutions and methods he does *not* criticize or oppose (e.g., the method of medical intervention) actually used to be opposed from standpoints similar to the one he has adopted. It is only the passage of time, and thus the acceptance of their effectiveness, that led to them being included in our civilization's treasure chest, which is why they do not evoke opposition from anyone anymore.

It is not our intention by any means to offer, either here or anywhere else, an apology for "revolutionary reconstructionism." We just think that any arguments against history are pointless. If man had been able to control and regulate consciously the development of his civilization, the latter may have ended up being more perfect, less paradoxical, and more capable than it is now. But this was not possible because, in creating and developing it, he was simultaneously creating and developing himself as a socially oriented being.

An opponent of bioconstructionism may reply that unique singular existence is priceless, which is why we should not, in our ignorance, manipulate genotypes by trying to free them from traits we consider harmful and by introducing others. He may also want to say that his argument can only be proved in a world that does not exist and that is different from ours. In our world, as a result of the global political situation, the Earth's atmosphere has been poisoned over decades by radioactive waste. The majority of respected geneticists and biologists would insist that this would lead to the emergence of numerous mutations in subsequent generations and that every nuclear test explosion meant a certain number of deformations, illnesses, and premature deaths caused by cancer, leukemia, etc. Also, those explosions were not to serve anything except increasing the nuclear potential of the interested parties. This kind of politics, continued up to this day by some states that deem themselves civilized, has produced at least thousands (or, more probably, tens of thousands) of victims. *This* is the world in which we live—and *this* is the world in which we are discussing the problems of bioconstructionism. We should not think that all this that results from the global regulatory failure does not bear on our conscience or on our "civilization balance" and that, disregarding this state of events, we should proceed with great caution (which leads to complete nonaction) in the areas under our control.

Man is a "mysterious" creature only when we assign some kind of "author," i.e., a personified creator, to him. In this case, the numerous biological and psychological contradictions of human nature imply the secretive and mysterious motivation of our "creator." But if we decide that we emerged as a result of evolutionary trials and efforts across millions of years, this "mysteriousness" is simply reduced to a catalog of solutions that it was possible to implement under given evolutionary and historical conditions. We can then begin to consider how we should work on redirecting the processes of self-organization to eliminate everything that brings suffering to our species.

All of this does not, of course, amount to suggesting an equivalence between man and any material object to be constructed or any technical product to be improved. The aura of moral responsibility must envelop the field of bioconstructionism—which is an area of great risk (but also perhaps of equally great hope). Yet, given that man caused himself so much pain and suffering in previous centuries (and not only then) as a result of his uncontrolled activity on the scale of society and civilization, it is high time he took such a conscious and responsible risk as soon as the state of his knowledge allows for it, even if this knowledge remains incomplete.

3   See Ivachnenko, *Cybernetyka Techniczna*.

4   The "antistatistical" standpoint of the designer outlined in the main body of the text is already out of date, in fact. Device reliability cannot be discussed separately from statistical techniques. Technological progress has led to a situation in which serial (i.e., mass) production is accompanied by an increase in the complexity of manufactured devices. If each of the elements in the system (which consists of five hundred elements) is 99 percent reliable, then the system as a whole is just 1 percent reliable (as long as all those elements are necessary for the organism to function). Maximum reliability to be achieved is proportional to the number of elements squared, as a result of which it is impossible to obtain a reliable product, especially when it takes the form of a highly complex system. Systems that are "connected" to

a human being as a regulator (an airplane, a car) are less susceptible to damage because the human's plastic behavior often allows him to compensate for the lost function. But in a "humanless" system, such as an intercontinental bullet or, more generally, an automatic system (a digital machine), there is absolutely no room for such plasticity—which is why the lower reliability of such systems is caused not just by the higher number of elements in them and by the novelty of technologies being actualized but also by the absence of the human—who would have been able to "buffer" the vasomotor symptoms of the damage. As a result of the rapid progress in the area of design, reliability theory has become quite advanced today (see, e.g., Chorafas, *Statistical Processes and Reliability Engineering*). The methods it currently uses are usually "external" with regard to the final product (calculations, repeated trials, measurement of an average time span between errors, examination of the time scale of the aging of different elements, quality control, etc.). Evolution also uses "external control" in the form of natural selection as well as some "internal methods": the doubling of devices, embedding a local self-repairing feature into those devices, and also a self-repairing feature that is subordinated to the support and control functions of the superior centers, and—most important, it seems—the use of devices with maximum plasticity as regulators. The fact that, despite the overall reliability of these methods, organisms so often fail is to a large extent caused by Evolution's "reluctance" to use a significant excess of design-focused information transfer (as expressed by Dancoff's principle).

As a matter of fact, 99 percent of all the sufferings and ailments connected with old age result from the manifestation of unreliability in an ever greater number of the organism's systems (the loss of teeth, of muscular tension, of sight and hearing, local tissue atrophy, degenerative processes, etc.). In the future, the main technical direction in which work on the prevention of the unreliability of devices will develop will most likely match the evolutionary direction, the main difference being that Evolution tends to "embed" "counterreliability" devices into its creations, while a human designer is more inclined to use methods that are external to the final product so as not to overcomplicate it with the number of elements. The working criteria are in fact very different in both cases—"material costs," e.g., are irrelevant for Evolution, which is why the amount of wasted hereditary material (sperm, eggs) does not matter to it, as long as there is enough of such material to ensure the continuation of the species. Studying the evolution of various technical tools reveals that achieving successful functioning (i.e., obtaining high reliability) is a process that occurs long after the optimal overall solution has been found (in principle, i.e., in its overall design, the airplanes from the 1930s or even from the 1920s were very similar to the present ones—as machines that are heavier than air, that fly thanks to the carrying force produced on the wings, that are put in motion by a jet engine with electric ignition, that have the same control system as today's planes, etc.). The success involved in flying across the oceans did not result from increasing, e.g., the size of the planes (because large planes were already being constructed then—sometimes even larger than today) but only from managing to achieve their high reliability—which was still impossible to achieve then.

The number of elements, which grows exponentially, radically decreases the reliability of very complex devices—hence the great difficulties involved in building complex devices such as a multi-stage rocket or a mathematical machine. Besides, there are limits to increasing reliability through doubling devices and information transfer. A device equipped with the best protection possible does not have to represent an optimal solution. It is similar to the problem of the strength of steel cable: if it is too long, no increase in its thickness will help, because the cable is going to break under its own weight. Apart from some possible hidden factors, the problems with functioning caused by an exponential growth in unreliability pose a limit to building systems of immense complexity (say, electric digital machines consisting of hundreds of billions or trillions of elements).

An important question arises as to whether it will ever be possible to manufacture devices that will be capable of crossing this "threshold of reliability," i.e., that will be able to trump the evolutionary solutions. This does not seem likely. We come across similar kinds of limits at almost every level of material phenomena, and thus also in solid body physics, in molecular engineering, etc. Aging at the level of tissues and cells is considered by many biophysicists to be a cumulative effect of "elementary molecular errors," of the "atomic lapses" that a living cells allows itself in the course of its existence—so that those "errors" eventually lead the system as a whole beyond the limit point of reversibility. If this is indeed the case, then we could in turn ask whether the statistical nature of the laws of microphysics, the singular uncertainty of results that every material fact, even the simplest one, is burdened with—e.g., the disintegration of a radio-active atom, the conjoining of atomic particles, the absorption of these particles by atomic nuclei—does not result from the fact that everything that happens is *"unreliable,"* that even atoms and their "building" parts—protons, neutrons, mezons—which are treated as "machines," or systems manifesting the *regularity* of behavior, do not in themselves constitute "reliable" elements of the construct we call the Universe or "reliable devices" in the form of chemical molecules, solid bodies, liquids, gases. In other words, does the statistical unre-liability of functioning not lie at the core of all the laws of Nature that are discovered by Science? And does it result from the fact that the Universe is designed in a rather similar way to the Tree of Evolu-tion—according to the principle whereby a "Certain System" (i.e., a relatively reliable one) is made from "Uncertain Elements"? And that the kind of "polarity" of the cosmic structure (matter, antimatter, positive participles, negative particles, etc.) is to some extent necessary, because *no other universe would be possible*—owing to the looming "unreliability of function," which would prevent it from undergoing any kind of evolution, thus permanently keeping it at the stage of "original chaos"? Such a (partly fictitious, we admit) positing of the problem may seem to involve some anthropomorphizing, or at least it may seem to open the door to the debate about the "Designer of the Universe," i.e., the "Creator of Everything." Yet this is not the case because, having determined that Evolution had no personified originator, we can still discuss its design work, i.e., the principle of building relatively reliable systems from highly unreliable parts, as discussed earlier.

5   [The reconnection of two previously separate blood vessels.—Trans.]
6   Simpson, *Major Features of Evolution.*
7   Goldschmidt, *Material Basis of Evolution.*
8   [The term *osteodontokeratic,* describing the tool kit of our hominid ancestor, *Australopithecus africanus,* was put forward by the Aus-tralian archaeologist Raymond Dart.—Trans.]
9   Brown, *Probability and Scientific Interference.*
10 Owing to the considerable interest in extrasensory phenomena, it may be worth adding the following. People narrate with great gusto stories of "prophetic dreams" as well as incidents that happened to them or their relatives that are supposed to prove the existence of telepathy, cryptesthesia, etc. We should therefore clarify that such

accounts, even if collected by eyewitnesses, do not have any value from a scientific point of view. Contrary to popular belief, their rejection by science does not in any way result from the scientist's contempt toward "a man in the street." It simply results from the demands of the scientific method. Let us first look at a simple example borrowed from Brown: imagine that five hundred psychologists in some country begin to examine the presence of telepathy by statistical methods. According to statistics, half of them will obtain results that are below the average or just average, whereas the other half will get results that show positive deviations from the statistically expected results. Now imagine that a hundred out of those five hundred psychologists obtain exceptionally significant results; this will confirm their inkling that "something is really going on." Out of those one hundred, a half will get insignificant results from further tests, which will lead them to abandon the study, whereas the other half will strengthen their belief that they have detected some telepathic phenomena. Ultimately, there will be five or six people left on the battlefield—those who have obtained positive results *several* times in a row. They are already "lost": it is impossible to explain to them that they themselves have become victims of statistics, which they used as their weapon.

Generally speaking, individual cases cannot have any significance for science because a simple calculation demonstrates that when several billion people dream at night, it is highly probable that the content of their dreams will be "fulfilled" at least in *several hundred* cases out of those several billion. If we add the natural vagueness and haziness of dreams, as well as their fleeting character and the tastes of the public that loves "puzzling" phenomena, the further propagation of similar stories becomes understandable. When it comes to entirely incomprehensible phenomena, such as observing "phantoms," etc., or suspending the laws of nature (i.e., "miracles"), science considers them to be illusions, hallucinations, delusions, etc. This should not offend the concerned, because scientists do not care about some "academic victory" but only about benefiting science. Science is too solid an edifice, constructed by means of too great a collective and comprehensive effort, for scientists to be willing to give up on accepted truths for the sake of some unverifiable phenomena (unverifiable mainly due to their *unrepeatability*), i.e., for the sake of the first, second, or tenth version of phenomena that contradicts the basic laws of nature discovered over centuries. Science only occupies itself with *repeatable* phenomena, and only thanks to this fact is it able to *predict* phenomena similar to the one under examination—which is what cannot be said about ESP.

Myself, I consider the "evolutionary" argument to be a conclusive one. No matter how many people would have seen, heard, or experienced "telepathic phenomena," this number is close to zero when compared to the number of "experiments" "conducted" by natural evolution during the existence of the species over the period of a billion years. The fact that evolution did not manage to "accumulate" any telepathic traits means that there was nothing to accumulate, eliminate, or coagulate. We also hear that such phenomena are not supposed to be just a characteristic of higher organisms, such as humans and dogs, but also of organisms such as insects. But the evolution of insects lasted several million years; it was a rather adequate time to fill the whole class of hexapods just with telepaths because it is hard to imagine a more useful trait in the struggle for survival than the possibility of obtaining information about one's environment and the organisms existing in it without the help of sensory organs and just "via a telepathic information channel." If Rhine's or Soal's statistics examine anything, this "anything" probably stands for certain dynamic structures of the human mind, which has been subjected to the test of "guessing" long random series. The results obtained can testify that, in a way that is sometimes incomprehensible to us, a system such as the brain may now and again "accidentally" come across the most efficient strategy for guessing sequences of this kind so that it gets results that are slightly above the average. By saying this, I have already said too much, because it is equally possible that we have a coincidence of two pseudorandom series (drawing Zener cards and "drawing" their equivalents imagined by the person tested) as a result of a "streak of good luck"—and nothing else.

CONCLUSION

I [*Ars Magna* was a treatise by a thirteenth-century Majorcan philosopher and writer Ramon Llull (in Latin, Lullus), outlining his method for combining religious and philosophical properties of things and for classifying various sciences. The work is seen as an early example of a logical method of knowledge production.—Trans.]

BIBLIOGRAPHY

Amarel, Samuel. "An Approach to Automatic Theory Formation." In *Principles of Self-Organization,* edited by Heinz von Foerster and George W. Zopf. New York: Pergamon Press, 1962.

Ashby, W. Ross. *An Introduction to Cybernetics.* London: Chapman and Hall, 1956.

———. *Design for a Brain.* London: Chapman and Hall, 1952.

Baumsztejn, A. I. "Wozniknowienije obitajemoj planiety." *Priroda* 12 (1961).

Bellamy, Edward. *Looking Backward: 2000–1887.* New York: New American Library, 1960.

Blackett, Patrick M. S. *Military and Political Consequences of Atomic Energy.* London: Turnstile Press, 1948.

Bohm, David. *Quantum Theory.* New York: Prentice Hall, 1951.

Brown, G. Spencer. *Probability and Scientific Interference.* London: Longman, 1958.

Carrel, Alexis. *Man, the Unknown.* New York: Harper, 1935.

Charkiewicz, A. A. "O cennosti informacii." *Probliemy Kibernetiki* 4 (1960).

Chorafas, Dimitris N. *Statistical Processes and Reliability Engineering.* Princeton, N.J.: Van Nostrand, 1960.

Davitashvili, L. S. *Teoria Polowogo Otbora.* Moscow: Izdatel'stvo Akademii Nauk SSSR, 1961.

De Latil, Pierre. *Thinking by Machine: A Study of Cybernetics.* Translated by Y. M. Golla. London: Sidgwick and Jackson, 1956.

De Solla Price, Derek J. *Science since Babylon.* New Haven, Conn.: Yale University Press, 1961.

Goldschmitdt, Richard. *The Material Basis of Evolution.* New Haven, Conn.: Yale University Press, 1940.

Hoyle, Fred. *Black Cloud.* New York: New American Library, 1957.

Huntington, Ellsworth. *Civilization and Climate.* New Haven, Conn.: Yale University Press, 1915.

———. *Mainsprings of Civilization.* New York: John Wiley, 1945.

Ivachnenko, Alexey Grigorevich. *Cybernetyka Techniczna.* Warsaw: Wydawnictwo Naukowe, 1962. Also published as *Technische kybernetik* by VEB, Berlin, 1962.

Jennings, Herbert Spence. *Behavior of the Lower Organisms.* New York: Columbia University Press, 1906.

Koestler, Arthur. *The Lotus and the Robot.* London: Hutchinson, 1960.

Leiber, Fritz. *The Silver Egghead.* New York: Ballantine Books, 1961.

Lem, Stanisław. *Dialogi.* Cracow, Poland: Wydawnictwo Literackie, 1957.

Lévi-Strauss, Claude. *Race and History.* Race Question in Modern Science. Paris: UNESCO, 1952.

Magoun, Horace W. *The Waking Brain.* Springfield, Ill.: Thomas, 1958.

Marković, Mihailo. *Formalizm w logice współczesnej.* Warsaw: PWN, 1962.

Olds, James, and Peter Milner. "Positive Reinforcement Produced by Electrical Stimulation of Septal Area and Other Regions of Rat Brain." *Journal of Comparative and Physiological Psychology* 47, no. 6 (1954): 419–27.

Parin, W. W., and R. M. Bajewski. *Kibernetika w miedicinie i fizjologii.* Moscow: Medgiz, 1963.

Pask, Gordon. "A Proposed Evolutionary Model." In *Principles of Self Organization,* edited by Heinz von Foerster and George W. Zopf. New York: Pergamon Press, 1962.

Schmalhausen, I. I. "Osnovy evolucyonnogo processa v svietje kibernetiki." *Problemy Kibernetiki* 4 (1960).

Shapiro, J. S. "O kwantowanii prostranstwa i wriemieni w tieorii 'elemientarnych' czastic." *Woprosy Filosofii* 5 (1962).

Shapley, Harlow. "Crusted Stars and Self-warming Planets." *American Scholar* 31 (1962): 512–15.

Shields, James. *Monozygotic Twins: Brought Up Apart and Brought Up Together.* Oxford: Oxford University Press, 1962.

Shklovsky, Iosif S. *Wsieljennaja, zizn, razum* [Universe, Life, Intelligence]. Moscow: Izdatel'stvo Akademii Nauk SSSR, 1962.

Simpson, G. G. *The Major Features of Evolution.* New York: Columbia University Press, 1953.

Smith, John Maynard. *The Theory of Evolution.* London: Penguin Books, 1962.

Stapledon, Olaf. *Last and First Men: A Story of the Near and Far Future.* London: Methuen, 1930.

Taube, Mortimer. *Computers and Common Sense: The Myth of Thinking Machines.* New York: Columbia University Press, 1961.

Thirring, Hans. *Die Geschichte der Atombombe.* Vienna: Neues Oster-reich Zeitungs-und-Verlags- gesellschaft, 1946.

Turing, Alan. "Computing Machinery and Intelligence." *Mind* LIX, no. 236 (1950): 433–60.

von Bertalanffy, Ludwig. *Problems of Life: An Evaluation of Modern Biological and Scientific Thought.* New York: Harper, 1952.

Wiener, Norbert. *Cybernetics: Or Control and Communication in the Animal and the Machine.* 2nd rev. ed. Cambridge, Mass.: MIT Press, 1961. First published 1948.

INDEX

52, 62–63, 101, 228, 277, 286, 295, 351, 356, 370; of survival, 62, 254–55, 327, 349–50

chaos, 31, 105, 155, 159–60, 164, 276, 289, 308, 311, 374n7, 397n4

civilization(s), ix–x, 3–4, 9, 12, 29, 36–38; 41–42, 44–46, 48, 56–57, 61–67, 71–77, 80–83, 85–89, 100–1, 109, 114, 119–22, 135, 149–51, 155, 176, 178, 188, 195, 202–3, 206, 211, 217, 221, 227, 235–36, 241, 262, 283–84, 286–87, 295–96, 298, 304, 334, 352, 368n6, 370n9, 372–73n3, 373–74n4, 375–78n2, 392–94n2; alien/extraterrestrial, 38, 44, 47, 49–53, 68–69, 77, 167; density, 44, 46–47, 71, 77; Earth's/human, x, xiii, 3, 9, 12, 30, 33, 41, 43, 46, 64, 66, 71, 88, 96, 104, 121, 235, 287, 367n6, 393n3; Neolithic, 36; technical, 12, 18, 34

coal, 9, 34, 70, 82–83, 87, 91, 101–2, 299

cognition, xvi–xvii, 81, 113–14, 242, 266, 284, 355, 384n3

Cold War, xiv

complexity, 19, 22, 30, 33, 91–93, 95, 98, 123, 147–48, 151–52, 156, 164, 166, 173, 196–97, 230, 238, 247–48, 257, 261, 274, 295, 299, 301, 306, 315, 326, 328–29, 336–37, 346, 360, 369n9, 380–81n1, 391n2, 394–96n4,

communication, 44, 73, 85, 109, 134, 172, 218, 221, 272, 298, 349, 357

competition, 15–16, 39, 114, 297, 353

conflict, xiv, 32, 147

consciousness, ix–x, 54, 113, 129–32, 140, 198–99, 203, 211–12, 219, 224, 228, 237, 246, 248, 270, 279, 293, 322–27

creation, 54, 127, 155–56, 183, 185, 188, 191, 235, 237, 279, 285, 289, 293–95, 299, 315, 343, 366n2, 368n6, 373n4

cybernetics, xiii–xiv, 8, 37, 60, 85, 87, 89–90, 92, 94, 96, 99, 101, 103, 107, 113, 132, 137, 167, 218, 220,

230, 235, 261, 284, 366n6, 369n9, 381n1, 383–84n3; biocybernetics, 331–32

decline, 14, 17, 28–29, 52, 80, 89

De Latil, Pierre, 12–13

design, xii, 8, 15, 17, 23, 25, 28, 40, 54, 59, 92–96, 108–9, 127, 130, 135, 138, 155, 164, 168, 178, 184–85, 189, 235, 238, 258, 263, 277, 281, 284, 286–87, 289–90, 294, 300–1, 303–4, 307–9, 314, 316–17, 319, 320, 323, 327–28, 330, 335–37, 339–41, 343, 345, 347, 350, 359–60, 383n3, 385n4, 390–92n2, 395n4, 396–97n

designer/Designer, xx, 17–19, 22–23, 29, 47, 97–98, 103, 127, 131, 141, 156, 159, 168–71, 189, 274, 276, 279, 282, 288–89, 291, 299, 307, 313, 315, 319–20, 324, 328–29, 333, 335, 339–40, 342–43, 345, 391–92n2, 394n4, 395–97n4

destruction, 4, 52, 129, 232, 320

determinism, 12, 42, 49, 73, 123, 127, 186, 372n3

diversity, 54, 71, 110, 161, 241, 260–62, 266, 289, 330, 370n9

Dyson sphere, 45, 75, 375–78n2

East, the, 34, 121, 123

electricity, 30, 33, 70, 150, 235, 299, 307, 318

emergence of life, xi, 19–21, 60, 72, 374n4

empiricism, 27, 33, 89, 114, 121, 135, 280, 287–88, 382n5, 390n9

energy, 5, 7, 14, 56, 72–74, 82–83, 87, 113, 133, 147, 151, 156–58, 162, 166, 173–74, 184, 188, 236, 268, 291–95, 298–99, 307, 309, 312, 338, 365n3, 367n6, 375–76n2; atomic/nuclear, 4, 9, 28, 37, 45–46, 51, 67, 70, 74, 82–83, 90, 188, 240, 298, 318, 344; solar 45, 58, 70, 188

entropy, xiii, 5, 134, 136, 187, 188, 295, 310, 318; antientropy, 254, 281

(continued from page ii)

Stanisław Lem (1921–2006) is best known to English-speaking readers as the author of *Solaris*. The author of many other science fiction novels translated into over forty languages, he was also a severe critic of the sci-fi genre, which he perceived as unimaginative, predictable, and focused on a narrow idea of the future. Drawing on scientific research, Lem's own works are deeply philosophical speculations about technology, time, evolution, and the nature—and culture—of humankind. In his later life, Lem abandoned writing fiction to focus on various discursive pieces, from short magazine columns to full-length treatises.

Joanna Zylinska is professor of new media and communications at Goldsmiths, University of London. She is the author and editor of many books on technology, culture, and ethics, including *Bioethics in the Age of New Media* (2009). Most recently, she coauthored *Life after New Media: Meditation as a Vital Process* (2012).